Advanced Insights into Acute Pancreatitis

Advanced Insights into Acute Pancreatitis

Integrating Research, Diagnosis, and Pathophysiology for Enhanced Clinical Strategies

Edited by

Hiroyuki Tomita

Department of Tumor Pathology, Gifu University Graduate School of Medicine, Center for One Medicine Innovative Translational Research, Gifu University Institute for Advanced Study, Gifu, Japan

Hideshi Okada

Department of Emergency and Disaster Medicine, Gifu University Graduate School of Medicine, Center for One Medicine Innovative Translational Research, Gifu University Institute for Advanced Study, Gifu, Japan

ELSEVIER

ACADEMIC PRESS

An imprint of Elsevier

Academic Press is an imprint of Elsevier
125 London Wall, London EC2Y 5AS, United Kingdom
50 Hampshire Street, 5th Floor, Cambridge, MA 02139, United States

ISBN: 978-0-443-44736-5

For information on all Academic Press publications visit our website at https://www.elsevier.com/books-and-journals

Publisher: Megan Ball
Editorial Project Manager: Vaishali Thapa
Production Project Manager: Aishwarya R
Cover Designer: Greg Harris

Typeset by TNQ Tech

Contents

Contributors

Tetsuya Fukuta
Graduate School of Medicine, Gifu University, Department of Emergency and Disaster Medicine, Gifu, Japan

Keisuke Iwata
Department of Gastroenterology, Gifu Municipal Hospital, Gifu, Japan

Shota Iwata
First Department of Internal Medicine, Gifu University Hospital, Gifu, Japan

Yoshinori Kakino
Gifu University Graduate School of Midicine, Emergency and Disaster Medicine, Gifu, Japan

Ryo Kamidani
Department of Emergency and Disaster Medicine, Gifu University Graduate School of Medicine, Gifu, Japan; Saitama Children's Medical Center, Department of Pediatric Critical Care, Saitama, Japan

Fumiya Kataoka
Department of Gastroenterology, Matsunami General Hospital, Gifu, Japan

Masaya Kawaguchi
Gifu University, Department of Radiology, Gifu, Japan

Yuichiro Kitagawa
Gifu University Hospital, Advanced Critical Care Center, Gifu, Japan

Takuya Koizumi
First Department of Internal Medicine, Gifu University Hospital, Gifu, Japan

Masashi Kuno
Gifu University, Graduate School of Medicine, Department of Gastroenterological Surgery, Gifu, Japan

Akinori Maruta
First Department of Internal Medicine, Gifu University Hospital, Gifu, Japan

Tomotaka Miura
Gifu University Hospital, Advanced Critical Care Center, Gifu, Japan

Takahito Miyake
Advanced Critical Care Center, Gifu University Hospital, Gifu, Japan

Hisataka Moriwaki
Department of Gastroenterology, Matsunami General Hospital, Gifu, Japan

Tsuyoshi Mukai
Department of Gastroenterology, Matsunami General Hospital, Gifu, Japan

Yosuke Ohashi
First Department of Internal Medicine, Gifu University Hospital, Gifu, Japan

Takafumi Okubo
Department of Tumor Pathology, Gifu University Graduate School of Medicine, Gifu, Japan

Mitsuru Okuno
Department of Gastroenterology, Matsunami General Hospital, Gifu, Japan

Masahito Shimizu
First Department of Internal Medicine, Gifu University Hospital, Gifu, Japan

Kunihiro Shirai
Kagawa Prefectural Central Hospital, Department of Emergency Medicine, Kagawa, Japan

Akio Suzuki
Gifu University Hospital, Department of Pharmacy, Gifu, Japan; Gifu Pharmaceutical University, Laboratory of Advanced Medical Pharmacy, Gifu, Japan

Kodai Suzuki
Gifu University, Graduate School of Medicine, Department of Infection Control, and the Advanced Critical Care Center at Gifu University Hospital, Gifu, Japan

Keiko Suzuki
Gifu University Graduate School of Medicine, Department of Emergency and Disaster Medicine, Gifu, Japan; Gifu Pharmaceutical University, Practical Pharmaceutical Research Center, Gifu, Japan

Nobuyuki Tetsuka
Gifu University, Graduate School of Medicine, Department of Infection Control, and the Advanced Critical Care Center at Gifu University Hospital, Gifu, Japan

Hiroyuki Tomita
Department of Tumor Pathology, Gifu University Graduate School of Medicine, Gifu, Japan

Eiichi Tomita
Department of Gastroenterology, Matsunami General Hospital, Gifu, Japan

Shinya Uemura
First Department of Internal Medicine, Gifu University Hospital, Gifu, Japan

Fuminori Yamaji
Graduate School of Medical and Dental Sciences, Science Tokyo, Department of Emergency and Disaster Medicine, Tokyo, Japan

CHAPTER

Basic research in acute pancreatitis

1

Takafumi Okubo

Department of Tumor Pathology, Gifu University Graduate School of Medicine, Gifu, Japan

1.1 Pathophysiology of acute pancreatitis

1.1.1 Anatomy of the pancreas

The pancreas is located in the upper abdomen behind the stomach and touches the posterior abdominal wall. It is anatomically divided into three parts: the head, body, and tail. The lower, protruding portion of the pancreatic head is known as the uncinate process. The pancreatic head extends to the left margin of the superior mesenteric vein. The remaining portion is divided equally between the body and tail. The pancreatic head is surrounded by the duodenum. The pancreatic tail is adjacent to the spleen and left kidney. The pancreatic duct extends throughout the entire pancreas. It originates in the tail, travels toward the right side of the body, and branches into the main pancreatic duct and the accessory pancreatic duct in the head. The main pancreatic duct then joins the common bile duct within the duodenal wall to form the papilla of Vater, which opens into the descending duodenum. The papilla of Vater contains the sphincter of Oddi, which surrounds the pancreatic and bile ducts. This sphincter regulates the discharge of bile and pancreatic juice and prevents reflux. The accessory pancreatic duct opens into the duodenum at the minor duodenal papilla, located above the papilla of Vater (Drake et al., 2014).

A capsule of loose connective tissue surrounds the pancreas. Septa extend from this capsule into the pancreas, forming indistinct lobules. Within these lobules, a loose connective tissue stroma surrounds units of pancreatic parenchyma. The pancreas is surrounded by adipose tissue (Ross, 2010) (Fig. 1.1).

1.1.2 Histology of the pancreas

The pancreatic lobule primarily consists of exocrine acini and ducts. Most lobules contain endocrine islets of Langerhans. In acinar tissue, cuboidal acinar cells are arranged in clusters. These cells contain zymogen granules and rough endoplasmic reticulum (ER), giving the cytoplasm a granular appearance under a light microscope. The ductal system originates from center cells within the acini and connects to interstitial ducts. The ductal system gradually converges and thickens to form intra- and interlobular ducts, ultimately becoming the main pancreatic duct or the accessory pancreatic duct (Fukushima, 2017; Ross, 2010).

Advanced Insights into Acute Pancreatitis. https://doi.org/10.1016/B978-0-443-44736-5.00003-6

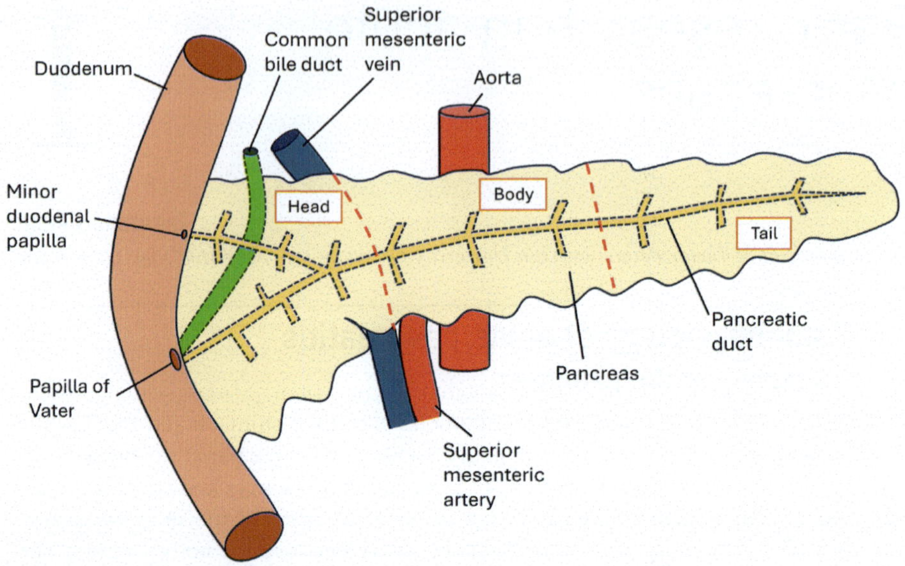

FIGURE 1.1 Macroscopic anatomy of the pancreas.

The pancreas is divided into three parts: The head, body, and tail. The pancreatic duct branches into two ducts within the pancreatic head, opening into the duodenum at the minor duodenal papilla and the papilla of Vater.

Islets of Langerhans appear as groups of pale blue cells with clear borders within intensely stained acini in hematoxylin and eosin (H&E)-stained specimens. The islets primarily consist of α cells, β cells, δ cells, and PP cells. These cells are accompanied by a rich network of fenestrated capillaries and are arranged in short, irregular cords (Fukushima, 2017).

Depending on the level of aging, adipose tissue is present between lobules to varying degrees.

1.1.3 Pancreatic function

The pancreas has two functions: exocrine and endocrine. The exocrine function involves the acinar cells and the ductal system. These cells contain eosinophilic enzyme granules that hold various digestive enzymes in an inactive state. These enzymes include:

- proteolytic endopeptidases (trypsinogen and chymotrypsinogen);
- proteolytic exopeptidases (procarboxypeptidase and proaminopeptidase);
- amylase, which digests starch;
- lipase, a lipid-digesting enzyme;
- deoxyribonuclease and ribonuclease, which digest nucleic acids.

These enzymes are secreted into the duodenum through the pancreatic duct and papilla. They are then activated within the small intestinal lumen, where they digest the three major nutrients: carbohydrates, proteins, and lipids (Ross, 2010). Additionally, bicarbonate ions are secreted by ductal epithelial cells. These ions are secreted into the duodenum as pancreatic juice, which neutralizes gastric acid and maintains an alkaline environment in the small intestine that facilitates the action of digestive enzymes. Hormones that regulate pancreatic exocrine function include secretin and cholecystokinin. These hormones are secreted in the duodenum and jejunum. When acidic digestive contents enter the duodenum, these two hormones are released into the bloodstream. They reach the pancreas and stimulate the secretion of pancreatic enzymes. Furthermore, pancreatic function is controlled by the autonomic nervous system. The sympathetic nervous system regulates pancreatic blood flow, and the parasympathetic nervous system stimulates acinar and intercalated cell activity. Endocrine function involves the cells found in the islets of Langerhans. The islets primarily contain four types of cells: β cells, α cells, δ cells, and PP cells. Beta cells secrete insulin, alpha cells secrete glucagon, delta cells secrete somatostatin, and PP cells secrete pancreatic polypeptide. Insulin acts on the liver, skeletal muscle, and adipose tissue. Insulin lowers blood glucose levels by facilitating glucose uptake, promoting glycogen synthesis, and enhancing glycolysis within cells. Glucagon, on the other hand, increases blood glucose levels by promoting the release of glucose into the bloodstream, stimulating gluconeogenesis in the liver, and accelerating glycogen breakdown. Somatostatin inhibits the secretion of insulin and glucagon. Pancreatic peptide hormones are believed to play a role in regulating appetite, as well as pancreatic exocrine and endocrine functions. The pancreatic islets are innervated by both sympathetic and parasympathetic nerves. Parasympathetic stimulation promotes the secretion of both insulin and glucagon, while sympathetic stimulation promotes glucagon release and inhibits insulin release. Furthermore, incretins (GLP-1 and GIP), which are secreted from the small intestine in response to nutrient intake, promote insulin secretion and inhibit glucagon secretion in accordance with rising blood glucose levels (Betts et al., 2022).

1.1.4 Overview of the pathogenesis of acute pancreatitis

Acute pancreatitis begins with the abnormal activation of trypsin, an exocrine pancreatic enzyme. Trypsinogen, the inactive form of trypsin, is normally secreted from the pancreas and converted into trypsin in the duodenum. However, during acute pancreatitis, trypsinogen converts into trypsin within the pancreas. This activates other pancreatic enzymes, such as phospholipase and elastase, leading to pancreatic autodigestion (Wu, 2000). Cytokine and chemokine release from damaged acinar cells triggers an inflammatory response. This mobilizes immune cells to the damaged site, where they release additional inflammatory mediators and reactive oxygen species. This worsens tissue damage and inflammation (Mansfield, 2012). This inflammation can spread systemically, leading to microcirculatory abnormalities that cause vasodilation, capillary leakage, and edema, as well as coagulation abnormalities that cause tissue hypoxia and arterial hypotension. Ultimately, this can result in multiple organ failure due to inflammation and mitochondrial dysfunction, which lead to cell injury and tissue damage. Furthermore, infection complicating pancreatic necrosis can cause sepsis, which can lead to organ failure similar to that seen in other septic conditions.

1.1.5 Causes of acute pancreatitis

Most cases of acute pancreatitis are caused by alcoholic pancreatitis, cholelithiasis, or idiopathic pancreatitis. Other causes include postendoscopic retrograde cholangiopancreatography (ERCP) and related procedures, pancreaticobiliary maljunction, dyslipidemia, drug-induced pancreatitis, neoplastic causes, abdominal trauma, and hereditary factors. The link between alcohol abuse and pancreatic disease has been recognized for over a century. It is well known that increased alcohol consumption increases the risk of developing pancreatitis. Ethanol and its metabolites are known to have numerous harmful effects on acinar cells. However, they alone do not cause acute pancreatitis. Rather, they are thought to lower the threshold for damage caused by other factors. Acinar cell functions affected by ethanol include calcium signaling, zymogen secretion, autophagy, cellular regeneration, the unfolded protein response, and mitochondrial membrane integrity (Clemens et al., 2014). Factors that contribute to alcoholic pancreatitis include smoking, high-fat diets, obesity, genetics, and infectious pathogens. Cholelithic pancreatitis may occur when gallstones compress the septum between the distal bile duct and the pancreatic duct. This obstructs the pancreatic duct. Alternatively, gallstones may deposit on the papilla of Vater. This leads to bile reflux into the pancreatic duct. In either case, this can lead to increased pressure within the pancreatic duct. Pancreatic duct injury is believed to result from the reflux of pancreatic juice and bile secretions associated with duct obstruction and increased intraductal pressure, as well as abnormal secretion by acinar cells. This results in the release of pancreatic enzymes into the interstitium, which causes the pancreas to digest itself, triggering acute pancreatitis (Lightner, 2001). Furthermore, it has recently become clear that some cases of acute pancreatitis previously considered idiopathic actually involve hereditary pancreatitis. Hereditary pancreatitis causes acute or chronic pancreatitis in young individuals with no history of alcohol consumption. Causative mutations have been identified in the PRSS1, SPINK1, CFTR, CTRC, and CPA1 genes, as well as in the calcium-sensing receptor (CASR) and claudin-2 (CLDN2) genes. These mutations are thought to cause pancreatitis by leading to impaired trypsin self-digestion, increased trypsin autoactivation, impaired proenzyme secretion, and reduced trypsin inhibition (Panchoo, 2022). The incidence of post-ERCP pancreatitis is reported to be 10.2%. Risk factors include endoscopic sphincterotomy (EST), difficult cannula insertion, and pancreatic ductography (Fogel and Sherman, 2014). Pancreaticobiliary maljunction is a congenital anomaly in which the pancreatic and bile ducts join outside the duodenal wall. Since sphincter function remains unaffected, bile refluxes into the pancreatic duct. This is believed to cause acute pancreatitis (Kamisawa et al., 2003). Regarding dyslipidemia, elevated blood triglyceride levels, particularly those exceeding 500 mg/dL, have been reported to increase the risk of acute pancreatitis. Excessive triglycerides are hydrolyzed by pancreatic lipase, resulting in the buildup of large quantities of free fatty acids (FFA) in the pancreas. These free fatty acids aggregate into micelles. This process induces ischemia within the pancreas, triggers acidosis, and activates trypsinogen through cathepsin B activation in lysosomes to form active trypsin. This leads to pancreatic autodigestion and injury (Qiu et al., 2023). Additionally, free fatty acids are believed to contribute to the development of acute pancreatitis by directly damaging pancreatic acinar and vascular endothelial cells. According to the Badalov classification, drugs that cause drug-induced pancreatitis are classified into Classes I to IV based on reporting frequency and reproducibility (Badalov et al., 2007). Class Ia drugs are specifically associated with pancreatitis recurrence upon readministration where other causes have been excluded. Examples include mesalamine, dipentum (azodisalicylic acid), bezafibrate, cannabis, carbimazole, codeine, enalapril, isoniazid, metronidazole, pravastatin, Premarin, procainamide, pyritinol, simvastatin, stibogluconate, sulfamethoxazole, sulindac, tetracycline, and valproic acid. Tumors are relatively rare, but intraductal papillary mucinous neoplasm (IPMN) is frequently reported as a tumor that causes acute pancreatitis. This is believed to occur because the tumor secretes large quantities of mucus that obstruct the main pancreatic duct and its branches. Pancreatic ductal adenocarcinoma is also relatively common. Conversely, serous cystadenoma and mucinous cystadenoma are less common causes of acute pancreatitis. However, cases have been reported in which these tumors compress the pancreatic duct, leading to acute pancreatitis (Gou et al., 2021).

1.1.6 Pathological classification of acute pancreatitis

According to the revised Atlanta classification, acute pancreatitis is subclassified as interstitial edematous pancreatitis or necrotizing pancreatitis (Banks et al., 2013; Nakanuma, 2020).

1.1.6.1 Interstitial edematous pancreatitis

In 90%–95% of acute pancreatitis cases, the pancreas enlarges due to edema associated with inflammation, either diffusely or locally. Histologically, interlobular interstitial edema, acinar-ductal metaplasia associated with acinar cell damage, and fat necrosis are observed (Marstrand-Daucé et al., 2023). Intralobular interstitial edema is thought to result from increased capillary permeability due to inflammatory cytokines, such as IL-1, IL-6, and TNF-α; microthrombus formation; and increased hydrostatic pressure associated with arterial constriction and venous congestion. Additionally, a mechanism involving the activation of the kallikrein system by pancreatic enzymes, which promotes vasodilation and increased permeability through bradykinin, is thought to contribute to interstitial edema. Acinar-ductal metaplasia refers to the transformation of acinar cells into duct-like cells. This metaplasia occurs when acinar cells are damaged. It is characterized by decreased expression of acinar-specific transcription factors (e.g., Ptf1a, Mist1, and Nr5a2) and digestive enzymes (e.g., carboxypeptidase and amylase) and increased expression of ductal markers (e.g., Sox9, Hnf1b, Hnf6, Pdx1, CA19-9, CAII, CD133, and osteopontin). Acinar-ductal metaplasia can occur in cases of acute or chronic pancreatitis, as well as in cases of autoimmune or obstructive pancreatitis caused by tumors. It is believed to be a protective mechanism that temporarily minimizes tissue damage caused by the overproduction of pancreatic digestive enzymes. Once the stimulus causing metaplasia is removed, the damaged tissue may return to normal. However, if the stimulus promoting metaplasia persists, acinar-ductal metaplasia may progress to dysplasia (Pan-IN) or adenocarcinoma. Steatonecrosis is the necrosis of peripancreatic and interlobular fat tissue caused by activated lipase from acinar cells. Steatonecrosis consists of degenerated fat cells, called ghost cells, that exhibit increased eosinophilia and indistinct nuclei. Steatonecrosis is also characterized by the breakdown of triglycerides due to lipase activation and calcification resulting from the binding of fatty acids and calcium (saponification). Interstitial edematous pancreatitis is a mild form in which the lobular outline is preserved. Within the lobules, acinar cell loss is observed, but necrotic cells or cellular debris are absent. Protein plugs often form within the lobules and in the interlobular pancreatic ducts.

1.1.6.2 Necrotizing pancreatitis

Approximately 5%–10% of acute pancreatitis cases present with the severe form, necrotizing pancreatitis (Nakanuma, 2020). Large, map-like areas of necrosis form within the pancreatic parenchyma and peripancreatic adipose tissue, obscuring the lobules. Necrosis and degeneration frequently occur in arterial and venous walls and are sometimes accompanied by hemorrhage due to vascular wall injury. Conversely, ischemic necrosis may occur due to thrombus formation in arteries. Protein plugs form within the pancreatic ducts, and neutrophilic infiltration may be present.

Furthermore, Klöppel's pathological classification of pancreatic damage in acute pancreatitis is divided into types 1 through 3. Type 1 involves necrosis between and around the acini and is described as "autodigestive fat necrosis due to enzyme effusion and activation." This type is the most common and is seen in conditions such as alcoholic pancreatitis and choledocholithiasis-induced pancreatitis. Type 2 is primarily caused by circulatory disorders, such as shock, and involves ductal and periductal necrosis. Type 3 is a rare form caused by infection, characterized by disseminated acinar necrosis (Klöppel, 2004).

1.1.7 Local complications of acute pancreatitis

Fluid collections and cystic structures may form in association with acute pancreatitis. The revised Atlanta classification system categorizes these complications into four types based on time since disease onset, morphology, and relationship to disease type: acute peripancreatic fluid collection (APFC), acute necrotic collection (ANC), pancreatic pseudocyst (PPC), and walled-off necrosis (WON) (Banks et al., 2013; Nakanuma, 2020). An APFC is a peripancreatic fluid collection observed in interstitial edematous pancreatitis within 4 weeks of symptom onset. APFC is not associated with peripancreatic necrosis or pseudocyst formation. It is usually sterile. It usually resolves spontaneously; however, if it persists beyond 4 weeks, it may progress to a PPC. ANC refers to necrosis and fluid accumulation in the pancreatic parenchyma or peripancreatic tissue and is observed in necrotizing pancreatitis within 4 weeks of symptom onset. ANC may be associated with destruction of the main pancreatic duct within the necrotic parenchyma and may involve infection. PPC refers to fluid accumulation with the formation of a well-defined capsule around the pancreas. This occurs in interstitial edematous pancreatitis after 4 weeks from the onset of symptoms. There are no solid components present, and necrosis is absent or minimal. One theory suggests that leakage of pancreatic fluid due to destruction of the pancreatic duct or its intrapancreatic branches causes persistent localized fluid accumulation. WON refers to necrotic tissue surrounded by reactive tissue walls, as seen in necrotizing pancreatitis that occurs 4 weeks or more after the onset of symptoms. It is encapsulated necrosis of pancreatic parenchyma or peripancreatic tissue with a well-defined inflammatory wall and may be associated with infection.

1.1.8 Transition from acute to chronic pancreatitis

In some cases, acute pancreatitis progresses to chronic pancreatitis. Repeated inflammation is believed to cause irreversible damage, resulting in the loss of both endocrine and exocrine functions. Histopathological findings include reduced acinar cells, lobular atrophy, and fibrosis within and between lobules, as well as in the peripancreatic adipose tissue. Other findings include the formation of protein plugs within the pancreatic ducts and sloughing of the ductal epithelium. It should be noted that even when fibrosis occurs, the pancreatic ducts and islets of Langerhans often remain within the lobules. Studies indicate that 17% of initial acute pancreatitis cases lead to recurrent pancreatitis, with approximately 8% of patients progressing to chronic pancreatitis within 5 years (Yamamiya et al., 2022). Risk factors for transitioning from acute to chronic pancreatitis include alcohol consumption and smoking. The precise mechanism by which acute pancreatitis progresses to chronic pancreatitis remains unclear. Why some patients develop chronic pancreatitis and suffer irreversible damage while others do not remains unclear.

1.2 Experimental models of acute pancreatitis

1.2.1 Overview of animal models for acute pancreatitis

Acute pancreatitis can become severe and involve complications or organ failure. Currently, treatment remains primarily supportive with no fundamentally specific therapeutic approach. This is thought to be due to the many unclear aspects of the pathogenesis of acute pancreatitis. Additionally, the rapid progression of acute pancreatitis makes capturing the early pathophysiology in human clinical cases difficult. Furthermore, obtaining pancreatic tissue at different stages of the inflammatory process in humans is challenging. This necessitates the use of animal models to elucidate the pathogenesis.

A crucial aspect of animal models of acute pancreatitis is using species that are physiologically and pathophysiologically similar to humans. Many animal species have been used as models, including rodents, dogs, cats, guinea pigs, and zebrafish; however, mice and rats are the most commonly employed (Rosales-Muñoz et al., 2025). Regarding differences between mice/rats and humans: The pancreas has exocrine and endocrine functions, a similar cellular composition, and secretes pancreatic juice into the duodenum. However, mice and rats differ from humans in that they have large interlobular ducts in each of the three lobules, as well as a long common duct formed by the bile and pancreatic ducts. The bile duct also functions as a pancreatic duct (Dolenšek et al., 2015). Additionally, mice have a gallbladder, while rats do not, demonstrating species-specific differences (Hyun and Lee, 2014). It is essential to understand these differences before using them in experiments.

Here, animal models are categorized as either noninvasive or invasive based on whether invasive procedures such as surgery are performed (Table 1.1).

1.2.2 Noninvasive animal models

There are three broad categories of noninvasive animal models of acute pancreatitis: pharmacological, dietary, and genetic.

1.2.2.1 Pharmacological

1.2.2.1.1 Caerulein

Caerulein, a peptide similar to cholecystokinin (a gastrointestinal hormone that promotes pancreatic secretion), was first isolated from the skin extract of the Australian green tree frog. Methods of administration include intravenous, subcutaneous, and intraperitoneal injection. Repeated administration of doses 10 to 100 times higher than the physiological equivalent stimulates the CCK1 receptor (CCK1R), which is a G-protein-coupled receptor found on pancreatic acinar cells. This activates trypsinogen and inflammatory signaling pathways (Gorelick and Lerch, 2017). This leads to excessive secretion of amylase and lipase, inducing pancreatitis. Caerulein-induced pancreatitis is characterized by cytoplasmic vacuolation and necrosis of acinar cells, edema, and infiltration of inflammatory cells into the pancreas. This resembles human edematous pancreatitis and is similar to acute pancreatitis (AP) induced by scorpion venom or organophosphorus pesticides (Rosales-Muñoz et al., 2025; Kim, 2008). Furthermore, inflammation peaks relatively quickly, and recovery is rapid. Advantages of this model include low cost and high reproducibility. However, the caerulein-induced model produces relatively mild acute pancreatitis and does not lead to multiorgan failure or fatal outcomes, which differs from

Table 1.1 Animal models for acute pancreatitis.

Category	Model/Induction method	Mechanism/Features	Advantages	Limitations/Disadvantages	References
Noninvasive models	Pharmacological				
	Caerulein	Overstimulation of the CCK1 receptor induces trypsinogen activation and inflammatory signaling pathways.	Low cost and high reproducibility.	The model produces relatively mild acute pancreatitis.	Li et al. (2009), Rosales-Muñoz et al. (2025), Gorelick and Lerch (2017), Kim (2008)
	Ethanol	Ethanol alone generally does not cause acute pancreatitis; it is often combined with other factors such as caerulein or LPS.	Can reproduce the risk factor of alcohol consumption in human pancreatitis. Allows for noninvasive administration methods.	Does not easily induce acute pancreatitis on its own.	Rosales-Muñoz et al. (2025), Vonlaufen et al. (2007), Criddle et al. (2006), Patel et al. (2016)
	Basic amino acids (e.g., L-arginine)	Intraperitoneal injection of large amounts of basic amino acids causes acinar cell necrosis.	A single dose is often sufficient to induce acute necrotizing pancreatitis.	Cases of acute pancreatitis due to excessive amino acid intake in humans are rare.	Majdoub et al. (2025), Rosales-Muñoz et al. (2025), Vrolyk and Singh (2020)
	Lipopolysaccharide (LPS)	LPS, is administered to excessively stimulate the immune response and cause organ dysfunction.	Can be used as a model of pancreatitis with systemic inflammatory response syndrome (SIRS).	LPS-induced inflammation can extend beyond the pancreas to other organs, necessitating consideration of systemic effects.	Kellum and Ronco (2023), Vaccaro et al. (2000), Fu et al. (2018), Fortunato and Gates (2000)

Continued

Table 1.1 Animal models for acute pancreatitis.—*cont'd*

Category	Model/Induction method	Mechanism/Features	Advantages	Limitations/Disadvantages	References
	Dietary				
	High-fat diet	Used to investigate the aggravation of acute pancreatitis induced by other methods (e.g., caerulein).	Allows for studying the link between risk factors such as obesity and metabolic syndrome and acute pancreatitis.	It is difficult to create an ideal model for hypertriglyceridemia in mice and rats due to their very high lipoprotein lipase activity.	Qiu et al. (2023), Hong et al. (2020), Lin et al. (2023), Kimura et al. (2019), Pan et al. (2017)
	Ethionine-supplemented, choline-deficient (CDE) diet	Feeding a diet containing the abnormal amino acid ethionine and lacking choline induces fatal acute hemorrhagic pancreatic necrosis.	Can induce severe pancreatitis through a noninvasive method.	The diet is costly, and food intake requires monitoring.	Rosales-Muñoz et al. (2025), Matsumoto et al. (2013), Lombardi (1976), Gilliland and Steer (1980)
	Genetic				
	Genetically modified mice	Uses mice with genetic modifications, such as knockout, overexpression, or conditional/tissue-specific expression of genes.	Allows for detailed analysis of the role of specific molecular mechanisms in pancreatitis.	Models targeting specific genes may not fully reproduce the pathophysiology of nonhereditary pancreatitis in humans.	Merry and Petrov (2018)
Invasive models	Pancreatic duct ligation	Induces acute pancreatitis by ligating the bile or pancreatic duct.	Reproduces pancreatitis associated with cholelithiasis and biliary-pancreatic duct reflux.	The procedure is complex, there are limitations in reproducibility, and results vary greatly by species.	Rosales-Muñoz et al. (2025), Samuel et al. (1994, 2010), Lerch et al. (1993)

Bile salt-induced	Mimics acute pancreatitis caused by bile acid reflux through retrograde injection of bile salts into the pancreatic duct.	The ability to induce acute necrotizing pancreatitis is a key advantage. Can be used in genetically modified mice.	Drawbacks include the technical difficulty of the procedure and a variable mortality rate depending on the concentration of the bile salt.	Rosales-Muñoz et al. (2025), Wan et al. (2012), Liu et al. (2009)
Closed Duodenal Loop	Increases duodenal lumen pressure by ligating the duodenum and causing reflux of bile and pancreatic juice.	Used in research investigating the relationship between acute pancreatitis and intestinal bacteria.	This model is rarely used due to the necessity of surgical intervention and ongoing debate about bacterial involvement.	Silva-Vaz et al. (2019)
Ischemia-reperfusion	Triggers acute necrotizing pancreatitis by temporarily blocking blood flow to the pancreas and then allowing reperfusion.	Can be used to study pancreatitis where ischemia is a cause.	The severity of pancreatitis depends on the duration of ischemia and reperfusion.	Rosales-Muñoz et al. (2025), Schanaider et al. (2015)
Trauma	Induces pancreatitis by applying direct compression or contusion to the pancreas, resulting in congestion, edema, and necrosis.	Serves as a model for traumatic pancreatitis. The intensity and area of impact are reportedly adjustable.	The primary mechanism is "direct injury," which differs from the mechanisms in most human cases of acute pancreatitis.	Hailin et al. (2023)
ERCP-induced	Mimics post-ERCP pancreatitis by injecting a contrast medium into the pancreatic duct of rats, which demonstrates edema and hemorrhaging.	Useful for research on the pathogenesis and prevention of post-ERCP pancreatitis.	Injury may result from multiple factors, including the chemical toxicity of the contrast agent and physical damage from the injection.	He et al. (2003)

human cases (Rosales-Muñoz et al., 2025; Chen et al., 2018; Wan et al., 2021). To overcome this limitation, researchers sometimes combine caerulein with other techniques capable of inducing acute pancreatitis, such as lipopolysaccharide (LPS) injection, enterokinase injection, or pancreatic duct ligation (Li et al., 2009).

1.2.2.1.2 Ethanol

Because alcohol consumption increases the risk of acute pancreatitis in humans, animal models that use ethanol are widely employed. Ethanol-induced models use relatively noninvasive administration methods, such as the intravenous, intraperitoneal, and oral routes, as well as more invasive methods, such as intragastric or direct bile duct administration (Rosales-Muñoz et al., 2025). While ethanol increases the risk of acute pancreatitis, it generally does not cause it alone. Therefore, animal models often combine ethanol administration with caerulein or lipopolysaccharide (LPS) administration, or involve administration of ethanol to genetically modified mice. However, one study found that administering 48% ethanol into the bile duct led to acute pancreatitis in a model (Unal et al., 2015). One theory of alcohol-induced pancreatic injury suggests that reactive oxygen species (ROS), which are generated during the oxidation of acetaldehyde (a metabolic product), damage acinar cells (Vonlaufen et al., 2007). Other alcohol metabolites include acetaldehyde and fatty acid ethyl esters. These esters are thought to contribute to the onset of acute pancreatitis through inositol trisphosphate receptor-dependent calcium release from the ER and impaired removal of cytoplasmic calcium by a dysfunctional calcium ATPase pump, which is caused by impaired mitochondrial ATP production (Criddle et al., 2006). However, some reports indicate that its toxicity is milder than that of fatty acids produced by lipolysis (Patel et al., 2016).

1.2.2.1.3 Basic amino acids

The intraperitoneal injection of large amounts of basic amino acids, such as L-arginine, L-ornithine, and L-lysine, activates pancreatic enzymes and causes acinar cell necrosis. L-arginine is the most frequently used of these amino acids. Following the administration of basic amino acids, histological findings report that enzyme precursor granules disappear from some pancreatic acinar cell areas 6 hours postadministration. Inflammatory cells and acinar cell necrosis appear 72 h later. By 96–120 h postadministration, only a very small portion of intact acinar cells remain. This pattern is considered similar to that seen in acute necrotizing pancreatitis (Vrolyk and Singh, 2020). While a single dose is often sufficient to induce acute necrotizing pancreatitis, multiple doses may sometimes be administered. The mechanism by which basic amino acids cause acute pancreatitis is not well understood, but it potentially involves protein synthesis inhibition, excessive nitric oxide production, and lipid peroxidation. Specifically regarding protein synthesis inhibition, it is thought that a decrease in polyamine concentration caused by the inhibition of ornithine decarboxylase (an enzyme involved in polyamine synthesis) by basic amino acids leads to delayed nucleic acid synthesis and, consequently, protein synthesis inhibition (Rosales-Muñoz et al., 2025). Human cases of acute pancreatitis suspected to be associated with L-arginine-containing protein product use have been reported. However, cases of acute pancreatitis due to excessive amino acid intake in humans are rare. The low clinical relevance is a limitation of this animal model.

1.2.2.1.4 Lipopolysaccharide

Lipopolysaccharide is a component of the outer membrane of Gram-negative bacteria. When Gram-negative bacteria invade the body, their death or destruction releases lipopolysaccharides. These free lipopolysaccharides excessively stimulate the immune response, causing sequential or simultaneous multiple organ dysfunction. For this reason, they are called endotoxins and are also causative agents of septic shock (Kellum and Ronco, 2023). Animal models of acute pancreatitis induced by intraperitoneal administration of lipopolysaccharide have been reported (Vaccaro et al., 2000). It should be noted that LPS-induced inflammation can extend beyond the pancreas to other organs, necessitating consideration of the effects of systemic inflammatory response syndrome (SIRS) and multiple organ failure (Fu et al., 2018). Cases combining lipopolysaccharide administration with ethanol-induced acute pancreatitis models have also been reported (Fortunato and Gates, 2000).

1.2.2.2 Dietary

1.2.2.2.1 High-fat diet

Obesity and metabolic syndrome are considered risk factors for acute pancreatitis. Studies have been conducted where mice fed a high-fat diet were then induced to develop acute pancreatitis through caerulein administration, intraductal sodium taurocholate administration, L-arginine administration, etc., followed by investigation of the severity of the pancreatitis (Qiu et al., 2023). Multiple reports indicate that pancreatitis becomes more severe in mice fed a high-fat diet compared with those not fed such a diet (Hong et al., 2020; Lin et al., 2023). It should be noted that mice and rats possess very high lipoprotein lipase activity. This makes it difficult to create an ideal animal model for hypertriglyceridemia (triglyceride levels $>$1000 mg/dL) simply by feeding a high-fat diet. This difference in lipoprotein lipase activity is considered a key distinction between animal models and humans (Kimura et al., 2019; Pan et al., 2017).

1.2.2.2.2 Ethionine-supplemented, choline-deficient diet

This model induces acute pancreatitis in mice and rats by feeding them a diet containing ethionine, an abnormal amino acid similar to methionine, which occurs naturally, and lacking in choline, a phospholipid and a major component of cell membranes. This diet is also used to induce nonalcoholic steatohepatitis (Matsumoto et al., 2013). As a model of pancreatitis, reports indicate that fatal acute hemorrhagic pancreatic necrosis (with fatty necrosis) develops within 5 days of starting the CDE diet (Lombardi, 1976). Furthermore, increased digestive enzyme content within pancreatic acinar cells and reduced amylase secretion have been confirmed (Gilliland and Steer, 1980). The CDE diet model exhibits findings that are common to other animal models, such as inhibited digestive enzyme secretion and the formation of cytoplasmic vesicles containing lysosomal hydrolases that colocalize with digestive enzymes. It should be noted that the severity of pancreatic injury in the CDE diet model depends heavily on the mouse's sex, age and body weight (Rosales-Muñoz et al., 2025). Although noninvasive, the CDE diet is costly and requires monitoring of food intake.

1.2.2.3 Genetic

Numerous acute pancreatitis models using genetically modified mice are employed. Types of genetic modification include: Global knockout models: Complete deletion of a gene in all tissues. Transgenic/overexpression models: The introduction or overexpression of specific mutant genes (e.g., mutant human PRSS1) or regulatory genes. Conditional knockout/tissue-specific expression models: Gene modification is limited to acinar cells or specific developmental stages, using systems such as Cre/loxP. Genes targeted for modification include:

- Trypsinogen-related genes (PRSS1 and SPINK1/SPINK3)
- Disease-causing genes associated with pancreatitis-like phenotypes (e.g., cystic fibrosis, polycystic kidney disease and liver dysfunction linked to mutations in keratin genes such as CFTR, KIF3A and KRT8)
- Genes related to inflammation and cytokines (TGFB, COX2, IL1B, IKKB, CREB2 (ATF4), AIRE and LT) These include genes related to inflammation and cytokines (TGFB, COX2, IL1B, IKKB, CREB2 (ATF4), AIRE, LT), proteins acting as sensors when ER stress occurs (PERK) and autophagy-related genes (ATG5, ATG7, LAMP2). Although some genes have been identified as being causative of hereditary pancreatitis, models that target genes involved in regulating ER stress and autophagy are used as they effectively reproduce nonhereditary pancreatitis (Merry and Petrov, 2018).

1.2.3 Invasive animal models

1.2.3.1 Pancreatic duct ligation

This model induces acute pancreatitis by ligating either the bile duct or the pancreatic duct. However, experimental results may vary between species due to differences in body size and the anatomical structure of the bile and pancreatic ducts. In rats, ligation of the distal bile duct rapidly induces acute pancreatitis, obstructive jaundice and cholangitis. However, ligation of the pancreatic duct alone does not produce the characteristic features of acute pancreatitis, such as necrosis and inflammation. Instead, it has been reported to induce changes resembling chronic pancreatitis, including atrophy and apoptosis of acinar and ductal tissues (Samuel et al., 1994). Additionally, models combining caerulein administration to induce acute pancreatitis have been reported. In mice, however, pancreatic duct ligation induces acute pancreatitis with a 100% mortality rate (Samuel et al., 2010). Opossums are commonly used in pancreatic duct ligation models. Their pancreatic duct structure involves the convergence of pancreatic ducts into the main pancreatic duct, which then flows into the long common bile duct and ultimately enters the duodenum at the papilla. This structure is relatively similar to human pancreatic and bile ducts (Lerch et al., 1993). Consequently, it is considered easier to reproduce the mechanisms involved in cholelithiasis-induced pancreatitis, such as bile reflux or sphincter of Oddi obstruction. Advantages of the pancreatic duct ligation model include avoiding drugs that affect the whole body and reproducing pancreatitis associated with cholelithiasis and biliary pancreatic duct reflux. However, the procedure is complex and there are limitations in reproducibility (Rosales-Muñoz et al., 2025).

1.2.3.2 Bile salt-induced

Retrograde injection of bile salts into the pancreatic duct can mimic acute pancreatitis caused by bile acid reflux. Various types of bile salt have been used for this purpose, including sodium chenodeoxycholate (Na-CDC), sodium taurocholate (Na-TC), sodium glycodeoxycholate (Na-GDC), sodium taurodeoxycholate (Na-TDC) and taurolithocholic acid 3-sulfate (TLC-S). Although sodium taurocholate is the most frequently used agent, taurolithocholic acid 3-sulfate is considered the most effective for inducing acute pancreatitis (Wan et al., 2012). In recent years, retrograde bile salt injection has become applicable to mouse models, enabling its use in genetically modified mice (Rosales-Muñoz et al., 2025). The ability to induce acute necrotizing pancreatitis is a key advantage of this model. However, drawbacks include the technical difficulty of the procedure and the variable mortality rate among experimental animals, depending on the concentration of the injected bile salt. A simpler method involves injecting bile salts uniformly immediately beneath the pancreatic capsule. This model has also been reported to produce hemorrhagic necrotizing pancreatitis (Liu et al., 2009).

1.2.3.3 Closed duodenal loop

This model increases duodenal lumen pressure and causes reflux of bile and pancreatic juice by ligating the proximal and distal portions of the duodenum, just beyond the confluence of the common bile duct and pancreatic duct. This model has been reported to induce interlobular edema and hemorrhagic pancreatic necrosis. Additionally, some studies use models that insert a bile duct cannula to avoid involving the bile in pancreatitis, thereby inducing necrotizing hemorrhagic pancreatitis. This model is also used in research investigating the relationship between acute pancreatitis and intestinal bacteria because duodenal transection causes changes in mucosal barrier function, leading to small intestinal bacterial overgrowth. However, due to the necessity of surgical intervention, the resulting increase in pressure within the closed duodenum and pancreatic duct, and the ongoing debate about bacterial infection involvement, this model is rarely used (Silva-Vaz et al., 2019).

1.2.3.4 Ischemia-reperfusion

Ischemia is one cause of acute pancreatitis. Models that induce ischemia-reperfusion by temporarily blocking blood flow to the pancreas can trigger acute necrotizing pancreatitis. Examples of these models include clamping the splenic artery, retrograde injection of 20 μm polystyrene microspheres into the distal splenic artery, and inducing either complete or partial pancreatic ischemia (Rosales-Muñoz et al., 2025; Schanaider et al., 2015). The severity of pancreatitis is thought to depend on the duration of ischemia and reperfusion.

1.2.3.5 Trauma

This model induces pancreatitis by applying compression or contusion to the pancreas (Hailin et al., 2023). After laparotomy and removal of the pancreas in rats, compression is applied using an impact device. Histopathological examination after compression revealed congestion, edema, and necrosis in pancreatic parenchymal and interstitial cells. Additionally, rupture and leakage of the pancreatic duct resulted in the influx of pancreatic fluid containing large amounts of amylase and lipase into the abdominal cavity. This was accompanied by secondary peritoneal infection. The intensity and area of impact are reportedly adjustable. It should be noted that physical injury models, such as contusion or compression, differ somewhat from the mechanisms typically seen in human cases of acute pancreatitis. The primary mechanism is "direct injury," and this must be considered.

1.2.3.6 Endoscopic retrograde cholangiopancreatography (ERCP)

Post-ERCP pancreatitis is one of the most common ERCP complications. One model that mimics this condition involves injecting a contrast medium into the pancreatic duct of rats. This model has been reported to demonstrate edema and hemorrhaging in the rat pancreas 24 h after injection (He et al., 2003). Factors that may contribute to injury include the chemical toxicity of the contrast agent, destruction of the pancreatic duct, and rupture of acini due to the forced injection of the contrast agent.

1.3 Molecular mechanisms and signaling pathways

1.3.1 Overview

Since Chiari first proposed the concept of pancreatic autodigestion by activated enzymes in 1896, the trypsin-centered hypothesis has been considered the primary mechanism of acute pancreatitis pathogenesis. However, in recent years, various alternative mechanisms have been proposed, including abnormal calcium signaling, mitochondrial dysfunction, endoplasmic reticulum (ER) stress, an impaired ER stress response, autophagy defects, and NF-κB activation (Singh and Garg, 2016). However, ectopic activation of trypsin is considered only one aspect of the pathogenesis, and a multicentric theory has been proposed, suggesting that many phenomena progress in parallel (Sah et al., 2012). The fundamental mechanisms of pathogenesis are not fully understood, and few of the proposed mechanisms have been applied clinically. However, animal models, such as genetically modified mice, have illuminated the molecular mechanisms and related signaling pathways of acute pancreatitis (Fig. 1.2).

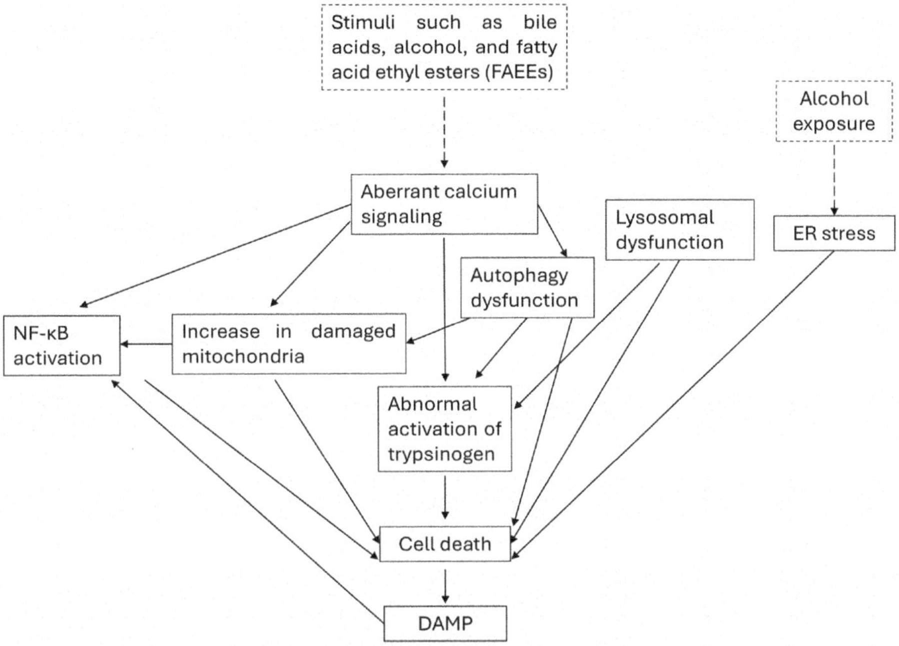

FIGURE 1.2 Flowchart illustrating the molecular mechanisms and signaling pathways involved in the onset of acute pancreatitis.

The molecular mechanisms and signaling pathways associated with the onset of acute pancreatitis are thought to be intricately interrelated. However, many aspects of the pathogenesis of acute pancreatitis remain unclear.

1.3.2 Premature trypsinogen activation in acinar cells and macrophages

Abnormal trypsinogen activation in acute pancreatitis is considered a central mechanism in its pathogenesis. Two mechanisms for converting trypsinogen to trypsin have been reported: trypsinogen autoactivation and conversion by cathepsin B.

However, in experimental pancreatitis, trypsinogen is activated solely by cathepsin B, suggesting that autoactivation contributes little. The colocalization hypothesis has been proposed as a mechanism for abnormal trypsinogen activation. This theory posits that the colocalization of trypsinogen and cathepsin B leads to the conversion of trypsinogen into trypsin. However, the detailed mechanism remains unidentified. Currently, mechanisms implicated in abnormal trypsinogen activation include lysosomal and autophagy dysfunction, which will be discussed later. Trypsinogen activation primarily occurs within acinar cells in the pancreas. However, it is also possible that large amounts of trypsinogen leak into the stroma and are activated by cathepsin B secreted from acinar cells (Hartwig et al., 1999). Furthermore, trypsinogen activation in pancreatitis has been shown to occur in activated macrophages within the pancreas. One proposed mechanism is that macrophages in the necrotic pancreatic area either phagocytose vesicles containing enzymes from damaged acinar cells or endocytose trypsinogen, converting it to trypsin in a cathepsin B-dependent manner (Gukovskaya et al., 2024).

1.3.3 Aberrant calcium signaling

Calcium signaling in pancreatic acinar cells plays a role in both the onset of acute pancreatitis and the physiological secretion of enzymes (Gryshchenko et al., 2018). Acetylcholine, which is released from nerve terminals near acinar cells through vagus nerve stimulation, and cholecystokinin, which is secreted from the duodenum and jejunum, promote calcium release from the endoplasmic reticulum (ER) into the cytoplasm. This leads to a localized increase in calcium concentration, or calcium spike, primarily at the apical pole. This spike causes zymogen granules to fuse with the cell membrane and release enzymes from the vesicles into the acinar lumen. Cytoplasmic calcium is expelled by two pumps: the sarcoplasmic reticulum Ca^{2+}-ATPase (SERCA) and the plasma membrane Ca^{2+}-ATPase (PMCA) (Brini and Carafoli, 2009). SERCA transports calcium from the cytoplasm into the ER, and PMCA moves calcium from the cytoplasm to the extracellular space, thereby maintaining the cytoplasmic calcium concentration. In acute pancreatitis, however, the increase in calcium concentration, normally confined to the apical pole, is thought to spread throughout the entire cytoplasm. Stimuli such as bile acids, alcohol, and fatty acid ethyl esters (FAEEs)—which are nonoxidative metabolites of alcohol—activate the IP_3 receptor and ryanodine receptor on the ER, promoting calcium release from the ER. Furthermore, these stimuli inhibit the SERCA and PMCA pumps, which transport calcium out of the cytoplasm. When calcium within the ER is depleted, STIM, which is present on the ER membrane, senses the depletion and moves to the plasma membrane. There, it opens the store-operated Ca^{2+} channels TRPC3 and ORAI, which promotes calcium influx from the extracellular space. These mechanisms increase the overall intracellular calcium concentration, which leads to the activation of proteolytic enzymes, NF-κB activation, autophagy failure, mitochondrial dysfunction, cytokine release, and cell necrosis (Singh and Garg, 2016).

1.3.4 Mitochondrial dysfunction

Mitochondria play a role in producing ATP through oxidative phosphorylation, as well as in regulating cytoplasmic calcium concentration, redox reactions, reactive oxygen species production, apoptosis, and other processes. Regarding calcium concentration regulation in particular, they exert cytoplasmic calcium buffering by taking calcium into the mitochondria through the mitochondrial Ca^{2+} uniporter. However, excessive cytoplasmic calcium accumulation stimulates the opening of the mitochondrial permeability transition pore (MPTP). Opening of the MPTP allows molecules smaller than 1.5 kDa (kDa), including protons, to freely pass into the mitochondria. This causes the proton gradient to collapse, resulting in reduced ATP production and eventual ATP depletion (Chen et al., 2024; Chen et al., 2025). Under conditions of ATP depletion, the SERCA and PMCA pumps function less effectively, reducing calcium efflux from the cytoplasm and creating conditions in which calcium accumulates more readily. Furthermore, MPTP opening causes an influx of water and solutes into the mitochondria, increased production of reactive oxygen species (ROS) within the mitochondria, and the release of apoptosis-related proteins. The influx of water and solutes leads to mitochondrial swelling, ultimately resulting in apoptosis or necrosis. ROS are generated as byproducts in the mitochondrial electron transport chain. However, excessively produced ROS within the mitochondria trigger oxidative stress responses, causing cellular damage and increased inflammatory reactions. This can activate various inflammatory signaling pathways, such as NF-κB and the NLRP3 inflammasome. ROS also contribute to pancreatic fibrosis in pancreatitis and affect mtDNA stability and genetic information transmission; mtDNA is susceptible to direct ROS attack, which leads to oxidative damage and mutations that impair mitochondrial function. This creates a vicious cycle in which mitochondrial dysfunction and ROS production escalate further. Apoptosis-related proteins, such as cytochrome *c*, are released from mitochondria into the cytoplasm through the opening of the mitochondrial permeability transition pore (MPTP). This activates the caspase family, ultimately inducing apoptosis (Mignotte and Vayssiere, 1998).

1.3.5 Endoplasmic reticulum (ER) stress

The ER is a cell organelle that performs various functions, such as synthesizing lipids and steroids, controlling the quality of proteins (folding and maturation), storing calcium, and detoxifying. Protein quality control occurs after translation when structural and secretory proteins are transported into the lumen of the vesicles. Post-translational modifications and appropriate folding by multiple chaperones and folding enzymes then occur within the vesicles, resulting in the formation of mature proteins. The ER plays an important role in ensuring quality control before proteins reach their sites of action. Misfolded proteins are repaired or degraded by a process called the ubiquitin-proteasome pathway (Zhang et al., 2025). ER stress occurs when the protein folding capacity of the ER becomes saturated and abnormal proteins accumulate within it. ER stress induces a compensatory mechanism called the unfolded protein response (UPR). The UPR can resolve ER stress if it is reversible or induce cell death if it is irreversible. Intracellular signaling for the UPR is initiated by three transmembrane sensor proteins: inositol-requiring enzyme 1 (IRE1, both the α and β isoforms), protein kinase RNA-like ER kinase (PERK), and activating transcription factor. The activated UPR signaling pathway maintains ER homeostasis by repairing protein folding defects and promoting proteolysis. However, if ER stress persists, the UPR induces cell death. Pancreatic acinar cells synthesize and secrete large amounts of digestive enzymes and constantly require high ER activity. Consequently, they are highly susceptible to the effects of ER stress. It is thought that alcohol exposure induces ER stress and the UPR in the pancreas. ER stress is now recognized as a key event in alcohol-induced pancreatitis (Singh and Garg, 2016; Li et al., 2022).

1.3.6 Impaired autophagy

Autophagy is one of the primary intracellular degradation mechanisms. It is a process whereby cytoplasmic components are transported to lysosomes for degradation. There are three types of autophagy: macroautophagy, microautophagy, and chaperone-mediated autophagy. Macroautophagy is the most representative type. Macroautophagy uses autophagosomes, which are intermediate organelles. An isolation membrane encloses a portion of the cytoplasm containing soluble substances and organelles to form autophagosomes. The autophagosome then fuses with a lysosome to form an autolysosome, which degrades the enclosed substances. Research has shown that autophagy selectively degrades intracellular organelles, such as mitochondria, lysosomes, and the ER, as well as intracellular bacteria and the p62 protein (Mizushima and Komatsu, 2011). Autophagy that occurs in damaged mitochondria is called mitophagy. Mitophagy suppresses the harmful effects of damaged mitochondria, such as excessive ROS production. Furthermore, acinar cells are prone to accumulating misfolded proteins, which can induce ER stress. However, ER stress is reduced by the degradation of misfolded protein aggregates through autophagy (Antonucci et al., 2015). It is also thought that autophagy removes zymogen granules, which contain digestive enzyme precursors such as amylase, trypsin, and lipase, to prevent their excessive accumulation in acinar cells (Iwahashi et al., 2018). In contrast, autophagy is impaired in acute pancreatitis. Although autophagosome and autolysosome formation is maintained, lysosomal function is compromised due to decreased proteolytic activity and pathological changes in lysosomal membrane proteins, such as lysosome-associated membrane protein (LAMP). This leads to delayed autophagy flux. In autolysosomes that have internalized trypsinogen under impaired lysosomal function, trypsinogen and activating factors, such as cathepsin B, colocalize internally. This promotes inappropriate trypsin activation within pancreatic cells, ultimately leading to autodigestion. Additionally, abnormally large vacuoles accumulate in pancreatitis; these vacuoles are believed to be autolysosomes containing substances that are difficult to degrade (Iwahashi et al., 2018). Additionally, impaired mitophagy leads to an increase in damaged mitochondria. These mechanisms collectively exacerbate acute pancreatitis.

1.3.7 Activation of NF-κB

NF-κB is a term that refers to a group of transcription factors, including RelA (p65), RelB, c-Rel, p105/p50 (NF-κB1), and p100/p52 (NF-κB2). NF-κB is involved in inflammation, innate and adaptive immune responses, development, cell survival, and proliferation (O'Dea & Hoffmann, 2009). NF-κB exists in the cytoplasm in an inactive form, bound to inhibitory molecules. Upon stimulation by cytokines such as IL-1 or TNFα, however, these inhibitory molecules degrade, allowing NF-κB to rapidly translocate to the nucleus and induce the expression of target genes (Ghosh and Karin, 2002). NF-κB activation has been suggested to play a key role in the inflammatory pathway of acute pancreatitis pathogenesis. NF-κB activation occurs when calcineurin, phospholipase C, an altered mitochondrial membrane potential, and MyD88 activate the IκB kinase (IKK) complex (IKKα and IKKβ), which polyubiquitinates the NF-κB inhibitor IκB. This releases NF-κB, allowing it to translocate to the nucleus (Jakkampudi et al., 2016; Luo et al., 2025). Calcineurin and phospholipase C are activated in response to increased cytoplasmic calcium concentration. Diacylglycerol, produced by phospholipase C activation, contributes to PKC activation. Meanwhile, damage-associated molecular patterns (DAMPs) are involved in MyD88 activation. NF-κB activation is also mediated by tumor necrosis factor (TNF)-α, interleukin (IL)-1β, IL-2, IL-6, IL-18, various chemokines (e.g., IL-8, MIP-1, MCP-1), platelet-activating factor, and adhesion molecules. NF-κB activation can lead to increased expression of intercellular adhesion molecule 1 (ICAM1). TNF-α release may lead to pancreatic necrosis, and increased TNF-α is thought to be involved in the progression to systemic inflammatory response syndrome (SIRS). Additionally, reports indicate that TNF-α and IL-1β can activate NF-κB (Wang et al., 2009). Increased ICAM1 expression promotes neutrophil infiltration into pancreatic tissue. Although reactive oxygen species are known to activate NF-κB, many aspects of the mechanism remain unclear (Jakkampudi et al., 2016).

1.3.8 DAMP

In acute pancreatitis, acinar cells may undergo necrosis. In association with this process, DAMPs can be released from the cells. Triggers for DAMP release include necrosis, apoptosis, necroptosis, pyroptosis, and autophagy associated with cell damage. Many endogenous substances originating from multiple intracellular compartments have been identified as potential DAMPs. Key examples include high mobility group box 1 (HMGB1), histones, nuclear DNA (nDNA), heat shock protein 70 (HSP70), mitochondrial DNA (mtDNA), N-formyl peptides, and transcription factor A (TFAM, a mitochondrial HMGB1 homolog) (Kang et al., 2014). HMGB1 is primarily localized within the nucleus and functions as a protein that stabilizes chromatin structure and gene transcription (Yanai et al., 2013). Its effects vary depending on its redox state: reduced HMGB1 promotes cell migration, while disulfide-linked HMGB1 induces cytokine production. In contrast, oxidized HMGB1 loses the aforementioned immune activities. Furthermore, reduced HMGB1 induces autophagy, while oxidized HMGB1 induces caspase-dependent apoptosis. Given the variability of redox states within tissues, HMGB1's effects may change depending on the state. Recent reports also indicate that HMGB1 induces neutrophil extracellular trap (NET) formation, leading to NLRP3 inflammasome activation and enhanced IL-1β production in macrophages (Wu et al., 2021). Most DAMPs are recognized by pattern recognition receptors (PRRs), which are membrane-bound or cytoplasmic and are expressed by both immune and nonimmune cells. These PRRs include Toll-like receptors (TLRs), Nod-like receptors (NLRs), RIG-I-like receptors (RLRs), the receptor for advanced glycation end products (RAGE), and P2X7, among others. These PRRs trigger downstream signaling pathways, inducing sterile inflammation. An early inflammatory response to a DAMP involves the release of TNF-α and IL-1β upon stimulation of immune cells. DAMPs induce TNF-α and IL-1β release through TLR and NF-κB activation. TNF-α and IL-1β then enhance the inflammatory response by activating macrophages and regulating the release of other inflammatory mediators, such as IL-6, IL-8, and MIF (Singh and Garg, 2016).

AI disclosure

During the preparation of this work, the author(s) used DeepL translate the Japanese manuscript into English. After using this tool/service, the author(s) reviewed and edited the content as needed and take(s) full responsibility for the content of the publication.

References

Antonucci, L., Fagman, J.B., Kim, J.Y., Todoric, J., Gukovsky, I., Mackey, M., Ellisman, M. H., Karin, M., 2015. Basal autophagy maintains pancreatic acinar cell homeostasis and protein synthesis and prevents ER stress. Proceedings of the National Academy of Sciences of the United States of America 112 (45), E6166. https://doi.org/10.1073/pnas.1519384112.

Badalov, N., Baradarian, R., Iswara, K., Li, J., Steinberg, W., Tenner, S., 2007. Drug-induced acute pancreatitis: an evidence-based review. Clinical Gastroenterology and Hepatology 5 (6). https://doi.org/10.1016/j.cgh.2006.11.023.

Banks, P.A., Bollen, T.L., Dervenis, C., Gooszen, H.G., Johnson, C.D., Sarr, M.G., et al., 2013. Classification of acute pancreatitis—2012: revision of the Atlanta classification and definitions by international consensus. Gut 62 (1). https://doi.org/10.1136/gutjnl-2012-302779.

Betts, J.G., Desaix, P., Johnson, E., Johnson, J.E., Korol, O., Kruse, D., Poe, B., Wise, J., Womble, M.D., Young, K.A., 2022. Anatomy and physiology. OpenStax, Rice University.

Brini, M., Carafoli, E., 2009. Calcium pumps in health and disease. Physiological Reviews 89 (4), 1341–1378. https://doi.org/10.1152/physrev.00032.2008Italy.

Chen, F., Xu, K., Han, Y., Ding, J., Ren, J., Wang, Y., Ma, Z., Cao, F., 2024. Mitochondrial dysfunction in pancreatic acinar cells: mechanisms and therapeutic strategies in acute pancreatitis. Frontiers in Immunology 15. https://doi.org/10.3389/fimmu.2024.1503087.

Chen, K.L., Chen, J.L., Chi, J.L., Zhou, Z.G., Li, Y., 2018. An improved acute pancreatitis apoptosis/necrosis in vitro model of acinar cells. Journal of Sichuan University (Medical Science Edition) 49 (5), 765–770.

Chen, X., Zhong, R., Hu, B., 2025. Mitochondrial dysfunction in the pathogenesis of acute pancreatitis. Hepatobiliary and Pancreatic Diseases International 24 (1), 76–83. https://doi.org/10.1016/j.hbpd.2023.12.008.

Clemens, D.L., Wells, M.A., Schneider, K.J., Singh, S., 2014. Molecular mechanisms of alcohol associated pancreatitis. World Journal of Gastrointestinal Pathophysiology 5 (3), 147.

Criddle, D.N., Murphy, J., Fistetto, G., Barrow, S., Tepikin, A.V., Neoptolemos, J.P., Sutton, R., Petersen, O.H., 2006. Fatty acid ethyl esters cause pancreatic calcium toxicity via inositol trisphosphate receptors and loss of ATP synthesis. Gastroenterology 130 (3), 781–793. https://doi.org/10.1053/j.gastro.2005.12.031.

Dolenšek, J., Rupnik, M.S., Stožer, A., 2015. Structural similarities and differences between the human and the mouse pancreas. Islets 7 (1), e1024405. https://doi.org/10.1080/19382014.2015.1024405.

Drake, R.L., Vogl, A.W., Mitchell, A.W., et al., 2014. GRAY'S Anatomy for Students, Second. Elsevier.

Fogel, E.L., Sherman, S., 2014. ERCP for gallstone pancreatitis. New England Journal of Medicine 370 (20). https://doi.org/10.1056/NEJMc1403445.

Fortunato, F., Gates, L.K., 2000. Alcohol feeding and lipopolysaccharide injection modulate apoptotic effectors in the rat pancreas in vivo. Pancreas 21 (2), 174–180. https://doi.org/10.1097/00006676-200008000-00011.

Fu, Q., Zhai, Z., Wang, Y., Xu, L., Jia, P., Peng, X., Liu, C., Zhang, X., Qin, T., Zhang, H., 2018. NLRP3 deficiency alleviates severe acute pancreatitis and pancreatitis-associated lung injury in a mouse model. BioMed Research International 2018, 1–10. https://doi.org/10.1155/2018/1294951.

Fukushima, 2017. Histology directly linked to pathological diagnosis. In: Oda (Ed.), Pathology and Clinical Medicine (in Japanese), 35. Bunkodo.

Ghosh, Karin, 2002. Missing pieces in the NF-kappaB puzzle. Cell 109 (Suppl. l).

Gilliland, L., Steer, M.L., 1980. Effects of ethionine on digestive enzyme synthesis and discharge by mouse pancreas. American Journal of Physiology—Gastrointestinal and Liver Physiology 239 (5), G418. https://doi.org/10.1152/ajpgi.1980.239.5.g418.

Gorelick, F.S., Lerch, M.M., 2017. Do animal models of acute pancreatitis reproduce human disease? Cellular and Molecular Gastroenterology and Hepatology 4 (2), 251–262. https://doi.org/10.1016/j.jcmgh.2017.05.007.

Gou, Liu, Xiao, Li, Xu, Song, et al., 2021. A narrative review of a type of pancreatitis worthy of attention: acute pancreatitis associated with pancreatic tumors-current problems and future thinking. Gland Surgery 10 (7). https://doi.org/10.21037/gs-21-400.

Gryshchenko, O., Gerasimenko, J.V., Peng, S., Gerasimenko, O.V., Petersen, O.H., 2018. Calcium signalling in the acinar environment of the exocrine pancreas: physiology and pathophysiology. The Journal of Physiology 596 (14), 2663–2678. https://doi.org/10.1113/jp275395.

Gukovskaya, A.S., Lerch, M.M., Mayerle, J., Sendler, M., Ji, B., Saluja, A.K., Gorelick, F.S., Gukovsky, I., 2024. Trypsin in pancreatitis: the culprit, a mediator, or epiphenomenon? World Journal of Gastroenterology 30 (41), 4417–4438. https://doi.org/10.3748/wjg.v30.i41.4417.

Hailin, W., Li, H., Zhirong, Z., Wang, Q., Jingdong, L., Ruiwu, D., 2023. Establishment of a multifunctional impact system and a study of a pancreatic trauma model in rats based on controlling the injury area. Heliyon 9 (6), e17010. https://doi.org/10.1016/j.heliyon.2023.e17010.

Hartwig, W., Jimenez, R.E., Werner, J., Lewandrowski, K.B., Warshaw, A.L., Fernández–del Castillo, Carlos, 1999. Interstitial trypsinogen release and its relevance to the transformation of mild into necrotizing pancreatitis in rats. Gastroenterology 117 (3), 717–725. https://doi.org/10.1016/s0016-5085(99)70466-x.

He, Z.-J., Winston, J.H., Yusuf, T.E., Micci, M.-A., Elfert, A., Xiao, S.-Y., Pasricha, P.J., 2003. Intraductal administration of an NK1 receptor antagonist attenuates the inflammatory response to retrograde infusion of radiological contrast in rats: implications for the pathogenesis and prevention of ERCP-induced pancreatitis. Pancreas 27 (1), e13. https://doi.org/10.1097/00006676-200307000-00018.

Hong, Y.P., Yu, J., Su, Y.R., Mei, F.C., Li, M., Zhao, K.L., Zhao, L., Deng, W.H., Chen, C., Wang, W.X., 2020. High-fat diet aggravates acute pancreatitis via TLR4-mediated necroptosis and inflammation in rats. Oxidative Medicine and Cellular Longevity 2020. https://doi.org/10.1155/2020/8172714. Available from:

Hyun, Lee, 2014. Experimental models of pancreatitis. Clinical Endoscopy 47 (3). https://doi.org/10.5946/ce.2014.47.3.212.

Iwahashi, K., Hikita, H., Makino, Y., Shigekawa, M., Ikezawa, K., Yoshioka, T., Kodama, T., Sakamori, R., Tatsumi, T., Takehara, T., 2018. Autophagy impairment in pancreatic acinar cells causes zymogen granule accumulation and pancreatitis. Biochemical and Biophysical Research Communications 503 (4), 2576–2582. https://doi.org/10.1016/j.bbrc.2018.07.018.

Jakkampudi, A., Jangala, R., Ratnakar Reddy, B., Mitnala, S., Nageshwar Reddy, D., 2016. Rupjyoti Talukdar, NF-κB in acute pancreatitis: mechanisms and therapeutic potential. Pancreatology 16 (4), 477–488. https://doi.org/10.1016/j.pan.2016.05.001.

Kamisawa, Matsukawa, Amemiya, Tu, Egawa, Okamoto, et al., 2003. Pancreatitis associated with pancreaticobiliary maljunction. Hepatogastroenterology 50 (53).

Kang, R., Lotze, M.T., Zeh, H.J., Billiar, T.R., Tang, D., 2014. Cell death and DAMPs in acute pancreatitis. Molecular Medicine 20 (1), 466–477. https://doi.org/10.2119/molmed.2014.00117.

Kellum, J.A., Ronco, C., 2023. The role of endotoxin in septic shock. Critical Care 27 (1). https://doi.org/10.1186/s13054-023-04690-5. Available from:

Kim, H., 2008. Cerulein pancreatitis: oxidative stress, inflammation, and apoptosis. Gut and Liver 2 (2), 74–80. https://doi.org/10.5009/gnl.2008.2.2.74.

Kimura, N., Kikumori, A., Kawase, D., Okano, M., Fukamachi, K., Ishida, T., Nakajima, K., Shiomi, M., 2019. Species differences in lipoprotein lipase and hepatic lipase activities: comparative studies of animal models of lifestyle-related diseases. Experimental Animals 68 (3), 267–275. https://doi.org/10.1538/expanim.18-0176.

Klöppel, 2004. Acute pancreatitis. Seminars in Diagnostic Pathology 21 (4). https://doi.org/10.1053/j.semdp.2005.07.001.

Lerch, M.M., Saluja, A.K., Rünzi, M., Dawra, R., Saluja, M., Steer, M.L., 1993. Pancreatic duct obstruction triggers acute necrotizing pancreatitis in the opossum. Gastroenterology 104 (3), 853–861. https://doi.org/10.1016/0016-5085(93)91022-A.

Li, H., Wen, W., Luo, J., 2022. Targeting endoplasmic reticulum stress as an effective treatment for alcoholic pancreatitis. Biomedicines 10 (1). https://doi.org/10.3390/biomedicines10010108. Available from:

Li, Y.Y., Ochs, S., Gao, Z.R., Malo, A., Chen, C.J., Lv, S., Gallmeier, E., Göke, B., Schäfer, C., 2009. Regulation of HSP60 and the role of MK2 in a new model of severe experimental pancreatitis. American Journal of Physiology—Gastrointestinal and Liver Physiology 297 (5), G981. https://doi.org/10.1152/ajpgi.00225.2009China.

Lightner, 2001. Pathophysiology of gallstone pancreatitis. Frontiers in Bioscience.

Lin, T.Y., Zhang, Y.F., Wang, Y., Liu, Y., Xu, J., Liu, Y.L., 2023. Nonalcoholic fatty liver disease aggravates acute pancreatitis through bacterial translocation and cholesterol metabolic dysregulation in the liver and pancreas in mice. Hepatobiliary and Pancreatic Diseases International 22 (5), 504–511. https://doi.org/10.1016/j.hbpd.2022.07.004.

Liu, Z.H., Peng, J.S., Li, C.J., Yang, Z.L., Xiang, J., Song, H., Wu, X.B., Chen, J.R., Diao, D. C., 2009. A simple taurocholate-induced model of severe acute pancreatitis in rats. World Journal of Gastroenterology 15 (45), 5732–5739. https://doi.org/10.3748/wjg.15.5732.

Lombardi, B., 1976. Influence of dietary factors on the pancreatotoxicity of ethionine. American Journal Of Pathology 84 (3), 633–648.

Luo, M., Jin, T., Fang, Y., Chen, F., Zhu, L., Bai, J., Ding, J., 2025. Signaling pathways involved in acute pancreatitis. Journal of Inflammation Research 18, 2287–2303. https://doi.org/10.2147/JIR.S485804.

Majdoub, Baig, Pettet, Sullivan, 2025. Unique Case of Acute Pancreatitis Linked to L-Arginine Supplement Use In A Bodybuilder. Advances in Clinical Medical Research & Healthcare Delivery.

Mansfield, 2012. Pathophysiology of acute pancreatitis: potential application from experimental models and human medicine to dogs. Journal of Veterinary Internal Medicine 26 (4). https://doi.org/10.1111/j.1939-1676.2012.00949.x.

Marstrand-Daucé, Lorenzo, Chassac, Nicole, Couvelard, Haumaitre, 2023. Acinar-to-Ductal Metaplasia (ADM): On the Road to Pancreatic Intraepithelial Neoplasia (PanIN) and Pancreatic Cancer. International Journal of Molecular Sciences 24 (12). https://doi.org/10.3390/ijms24129946.

Matsumoto, M., Hada, N., Sakamaki, Y., Uno, A., Shiga, T., Tanaka, C., Ito, T., Katsume, A., Sudoh, M., 2013. An improved mouse model that rapidly develops fibrosis in non-alcoholic steatohepatitis. International Journal of Experimental Pathology 94 (2), 93–103. https://doi.org/10.1111/iep.12008.

Merry, T.L., Petrov, M.S., 2018. The rise of genetically engineered mouse models of pancreatitis: a review of literature. Biomolecular Concepts 9 (1), 103–114. https://doi.org/10.1515/bmc-2018-0011.

Mignotte, B., Vayssiere, J.L., 1998. Mitochondria and apoptosis. European Journal of Biochemistry 252 (1), 1–15. https://doi.org/10.1046/j.1432-1327.1998.2520001.x.

Mizushima, N., Komatsu, M., 2011. Autophagy: renovation of cells and tissues. Cell 147 (4), 728–741. https://doi.org/10.1016/j.cell.2011.10.026.

Nakanuma, 2020. Pancreatic Pathology Textbook for Physicians Examining the Pancreas (in Japanese). Nankodo.

O'Dea, E., Hoffmann, A., 2009. NF-κB signaling. Wiley Interdisciplinary Reviews: Systems Biology and Medicine 1 (1), 107–115.

Pan, Y., Li, Y., Gao, L., Tong, Z., Ye, B., Liu, S., Li, B., Chen, Y., Qi, Y., Meng, L., Wang, Y., Liu, G., Lu, G., Li, W., Li, J., 2017. Development of a novel model of hypertriglyceridemic acute pancreatitis in mice. Scientific Reports 7 (1). https://doi.org/10.1038/srep40799.

Panchoo, 2022. Hereditary pancreatitis: An updated review in pediatrics. World Journal of Clinical Pediatrics 11 (1). https://doi.org/10.5409/wjcp.v11.i1.27.

Patel, K., Chandra, D., Noel, P., Trivedi, R.N., Oliveira, C.de, Singh, V.P., 2016. Fatty acid ethyl esters are less toxic than their parent fatty acids generated during acute pancreatitis. American Journal Of Pathology 186 (4), 874–884. https://doi.org/10.1016/j.ajpath.2015.11.022.

Qiu, M., Zhou, X., Zippi, M., Goyal, H., Basharat, Z., Jagielski, M., Hong, W., 2023. Comprehensive review on the pathogenesis of hypertriglyceridaemia-associated acute pancreatitis. Annals of Medicine 55 (2). https://doi.org/10.1080/07853890.2023.2265939.

Rosales-Muñoz, G.J., Souza-Arroyo, V., Bucio-Ortiz, L., Miranda-Labra, R.U., Gomez-Quiroz, L.E., Gutiérrez-Ruiz, M.C., 2025. Acute pancreatitis experimental models, advantages and disadvantages. Journal of Physiology & Biochemistry 81 (3), 539–556. https://doi.org/10.1007/s13105-025-01091-w.

Ross, 2010. Histology a text and Atlas with correlated cell and molecular biology, Fifth. Wolters Kluwer.

Sah, R.P., Garg, P., Saluja, A.K., 2012. Pathogenic mechanisms of acute pancreatitis. Current Opinion in Gastroenterology 28 (5), 507–515. https://doi.org/10.1097/MOG.0b013e3283567f52.

Samuel, I., Yuan, Z., Meyerholz, D.K., Twait, E., Williard, D.E., Kempuraj, D., 2010. A novel model of severe gallstone pancreatitis: murine pancreatic duct ligation results in systemic inflammation and substantial mortality. Pancreatology 10 (5), 536–544. https://doi.org/10.1159/000320776.

Samuel, I., Toriumi, Y., Yokoo, H., Wilcockson, D.P., Trout, J.J., Joehl, R.J., 1994. Ligation-induced acute pancreatitis in rats and opossums: a comparative morphologic study of the early phase. Journal of Surgical Research 57 (2), 299–311. https://doi.org/10.1006/jsre.1994.1149.

Schanaider, A., Penna de Carvalho, T., Coelho, S.de O., Renteria, J.M., Eleuthério, E.C.A., Teixeira Lima Castelo-Branco, M., Madi, K., Baetas-da-Cruz, W., Pereira de Souza, H. S., 2015. Ischemia–reperfusion rat model of acute pancreatitis: protein carbonyl as a putative early biomarker of pancreatic injury. Clinical and Experimental Medicine 15 (3), 311–320. https://doi.org/10.1007/s10238-014-0289-2.

Silva-Vaz, P., Margarida Abrantes, A., Castelo-Branco, M., Gouveia, A., Botelho, M.F., Guilherme Tralhão, J., 2019. Murine models of acute pancreatitis: a critical appraisal of clinical relevance. International Journal of Molecular Sciences 20 (11), 2794. https://doi.org/10.3390/ijms20112794.

Singh, P., Garg, P.K., 2016. Pathophysiological mechanisms in acute pancreatitis: current understanding. Indian Journal of Gastroenterology 35 (3), 153–166. https://doi.org/10.1007/s12664-016-0647-y.

Unal, E., Atalay, S., Tolan, K.H., Yuksekdag, S., Gunes, P., Bas, G., Memisoglu, K., 2015. Biliopancreatic duct injection of ethanol as an experimental model of acute and chronic pancreatitis in rats. International Journal of Clinical and Experimental Medicine 8 (1), 304–310.

Vaccaro, M.I., Calvo, E.L., Suburo, A.M., Sordelli, D.O., Lanosa, G., Iovanna, J.L., 2000. Lipopolysaccharide directly affects pancreatic acinar cells: implications on acute pancreatitis pathophysiology. Digestive Diseases and Sciences 45 (5), 915–926. https://doi.org/10.1023/A:1005521007609.

Vonlaufen, A., Wilson, J.S., Pirola, R.C., Apte, M.V., 2007. Role of alcohol metabolism in chronic pancreatitis. Alcohol Research & Health 30 (1), 48–54.

Vrolyk, V., Singh, B., 2020. Animal models to study the role of pulmonary intravascular macrophages in spontaneous and induced acute pancreatitis. Cell and Tissue Research 380 (2), 207–222. https://doi.org/10.1007/s00441-020-03211-y.

Wan, J., Wang, J., Wagner, L.E., Wang, O.H., Gui, F., Chen, J., Zhu, X., Haddock, A.N., Edenfield, B.H., Haight, B., Mukhopadhyay, D., Wang, Y., Yule, D.I., Bi, Y., Ji, B., 2021. Pancreas-specific CHRM3 activation causes pancreatitis in mice. JCI Insight 6 (17). https://doi.org/10.1172/jci.insight.132585.

Wan, M.H., Huang, W., Latawiec, D., Jiang, K., Booth, D.M., Elliott, V., Mukherjee, R., Xia, Q., 2012. Review of experimental animal models of biliary acute pancreatitis and recent advances in basic research. HPB 14 (2), 73–81. https://doi.org/10.1111/j.1477-2574.2011.00408.x.

Wang, Y.L., Zheng, Y.J., Zhang, Z.P., Su, J.Y., Lei, R.Q., Tang, Y.Q., Zhang, S.D., 2009. Effects of gut barrier dysfunction and NF-κB activation on aggravating mechanism of severe acute pancreatitis. Journal of Digestive Diseases 10 (1), 30–40. https://doi.org/10.1111/j.1751-2980.2008.00360.x.

Wu, 2000. Current concept of pathogenesis of severe acute pancreatitis. World Journal of Gastroenterology 6 (1). https://doi.org/10.3748/wjg.v6.i1.32.

Wu, X., Yang, Z., Wang, H., Yang, Z., Gao, X., Zang, B., 2021. High-mobility group box protein-1 induces acute pancreatitis through activation of neutrophil extracellular trap and subsequent production of IL-1β. Life Sciences 286, 119231. https://doi.org/10.1016/j.lfs.2021.119231.

Yamamiya, Tominaga, Hoshi, Nagashima, Minaguchi, Haruyama, et al., 2022. The risk factors for progression to chronic pancreatitis in patients with past-history of acute pancreatitis: a retrospective analysis based on mechanistic definition. Journal of Clinical Medicine 11 (8). https://doi.org/10.3390/jcm11082209.

Yanai, H., Matsuda, A., An, J., Koshiba, R., Nishio, J., Negishi, H., Ikushima, H., Onoe, T., Ohdan, H., Yoshida, N., Taniguchi, T., 2013. Conditional ablation of HMGB1 in mice reveals its protective function against endotoxemia and bacterial infection. Proceedings of the National Academy of Sciences 110 (51), 20699–20704. https://doi.org/10.1073/pnas.1320808110.

Zhang, X., Xu, C., Ji, L.J., Zhang, H., 2025. Endoplasmic reticulum stress in acute pancreatitis: exploring the molecular mechanisms and therapeutic targets. Cell Stress & Chaperones 30 (3), 119–129. https://doi.org/10.1016/j.cstres.2025.03.001.

CHAPTER

Epidemiology of pancreatitis

2

Yuichiro Kitagawa
Gifu University Hospital, Advanced Critical Care Center, Gifu, Japan

2.1 Incidence rate

According to a study by Baichuan Zhou et al. using the Global Burden of Disease (GBD) database, the global age-standardized incidence rate of acute pancreatitis in 2021 was 32.806 per 100,000 individuals (28.853–37.377). The GBD database provides a comprehensive analysis of 371 diseases and injuries across 204 countries and territories from 1990 to 2021. During this period, the age-standardized incidence rate of pancreatitis exhibited a downward trend, peaking at 37.760 (32.805–43.501) in 1992 and declining to 32.806 (28.853–37.377) in 2021 (Baichuan Zhou, 2025).

In 2021, the incidence rates for males and females were 36.747 (32.194–42.243) and 28.822 (25.378–32.548) per 100,000 individuals, respectively, yielding a male-to-female ratio of 1.27. Regionally, the highest age-standardized incidence rate was observed in Eastern Europe at 99.349 (82.691–117.934) per 100,000 population, whereas the lowest rate was reported in Tropical Latin America at 16.628 (14.701–18.834). From 1990 to 2021, the global age-standardized incidence rate changed by −0.128 (−0.147 to −0.107), with the largest increase observed in Southern Latin America (0.151, 0.104–0.202) and the greatest decrease in East Asia (−0.318, −0.346 to −0.289) (Baichuan Zhou, 2025).

When regions are classified according to the Socio-demographic Index (SDI), similar trends in age-standardized incidence rates were observed. In high-SDI regions, which exhibited the highest incidence rates in 2021, the rates peaked at 48.727 (42.418–55.981) in 1994 and subsequently declined to 38.919 (34.160–44.029) in 2021 (Baichuan Zhou, 2025).

In Japan, an epidemiological survey using questionnaires conducted by Yuichi Tanaka and colleagues reported an annual incidence rate of 49 per 100,000 population in 2021. Compared with the 2016 survey, this represents a 22% decrease. While previous surveys had shown an increasing trend, the incidence declined for the first time in 2021. The male-to-female ratio among patients with acute pancreatitis in 2021 was 2.0 (Yuichi Tanaka, 2025).

Globally, the age-standardized prevalence of pancreatitis decreased from 93.743 per 100,000 population (68.311–127.077) in 1990 to 68.992 (51.258–91.293) in 2021, corresponding to a percentage change of −0.264 (−0.295 to −0.231). In 2021, the prevalence rates were 78.573 (95% uncertainty interval [UI]:

Advanced Insights into Acute Pancreatitis. https://doi.org/10.1016/B978-0-443-44736-5.00004-8

58.107–105.750) per 100,000 for males and 59.338 (44.233–79.510) per 100,000 for females, resulting in a male-to-female ratio of 1.32 (Baichuan Zhou, 2025).

Regionally, the highest prevalence was observed in Eastern Europe at 723.277 (525.235–1003.430) per 100,000 population, whereas the lowest prevalence was in East Asia at 24.179 (17.355–33.342). Among regions showing a decline, East Asia experienced the largest reduction (−0.309, −0.392 to −0.221), whereas Tropical Latin America exhibited the greatest increase (0.141, 0.111–0.169). From 1990 to 2021, the global age-standardized prevalence of pancreatitis significantly decreased, peaking at 94.032 per 100,000 person-years (68.821–127.124) in 1992 and declining to 68.992 (51.258–91.293) in 2021. In high-middle SDI regions, which had the highest prevalence in 2021, the rate decreased from 214.934 (154.830–296.032) in 1993 to 147.281 (107.933–200.878) in 2021. Conversely, low SDI regions, which had the lowest prevalence in 2021, exhibited a slight increase, from 32.125 (21.842–44.413) in 1990 to 32.803 (22.170–45.413) in 2021 (Baichuan Zhou, 2025).

Age-specific analyses revealed that the incidence rate of pancreatitis per 100,000 population increased with age, peaking in the ≥95-year age group in 2021. In contrast, the prevalence rate initially increased with age, peaking in the 60–64 age group for males and the 80–84 age group for females, before declining in older age groups. Among individuals aged 25–59 years, the incidence rate was significantly higher in males than in females; after age 60, the increase in male incidence slowed, and female incidence slightly exceeded that of males. Similarly, the number of new cases was substantially higher in males than in females between 20 and 59 years of age, whereas this trend reversed in other age groups. The peak number of new cases occurred in the 35–39 age group for males and the 55–59 age group for females (Baichuan Zhou, 2025).

In Japan, the mean age of patients was 61.9 ± 18.4 years, with males averaging 60.1 ± 17.5 years and females 65.4 ± 19.5 years. The most frequently affected age groups were 70–79 years for males and 80–89 years for females. Female patients were significantly older than male patients ($P < .001$) (Yuichi Tanaka, 2025).

2.2 Etiology

2.2.1 Etiology of acute pancreatitis

Although the relative contribution of etiological factors of pancreatitis varies by region, alcohol consumption is globally recognized as the major risk factor (Baichuan Zhou, 2025).

In general, the etiologies of acute pancreatitis (AP) include alcoholic, biliary (gallstone-related), hypertriglyceridemia-induced, post–endoscopic retrograde cholangiopancreatography (post-ERCP), traumatic, drug-induced, and idiopathic causes (Weiss FU, 2019).

An observational study conducted in Europe reported no sex difference in the overall incidence of hospitalized acute pancreatitis; however, marked sex differences were observed with respect to etiology. Alcoholic pancreatitis was more common in men, whereas biliary pancreatitis predominated in women (Weiss FU, 2019).

In Japan, the main causes of AP were alcohol (31.2%), gallstones (22.5%), and idiopathic disease (22.1%). Among male patients, alcohol was the most frequent cause (40.9%), followed by idiopathic (18.6%) and gallstone-related pancreatitis (17.5%). In contrast, gallstones were the leading cause in female patients (32.4%), followed by idiopathic (28.9%) and alcohol-related pancreatitis (12.2%). Alcohol-related AP was most prevalent in the 50–59-year age group, whereas the incidence of gallstone-related and idiopathic pancreatitis increased with advancing age. When stratified by sex and age, gallstone-related AP was more common in elderly women, alcohol-related AP predominated in middle-aged men, and the prevalence of idiopathic AP was nearly equal between sexes (Takeda, 2010).

With regard to race, Black individuals have been reported to have a 2–3-fold higher risk of developing pancreatitis compared with White individuals (Weiss FU, 2019).

2.2.2 Alcoholic acute pancreatitis

Global studies have shown that the proportion of pancreatitis-related deaths attributable to alcohol consumption increased overall from 1990 to 2021; however, a clear decline has been observed since 2010 (Baichuan Zhou, 2025).

The peak incidence of alcoholic acute pancreatitis has been reported to occur at 25–34 years of age in women and 34–45 years in men.

The association between alcohol and acute pancreatitis has been recognized for more than a century. It is now well established that excessive alcohol consumption can trigger the onset of acute pancreatitis and increase susceptibility to chronic pancreatitis.

Although the relationship between alcohol intake and acute pancreatitis is well documented epidemiologically, only a minority of individuals with alcohol dependence develop pancreatitis. This observation suggests that alcohol consumption alone is rarely sufficient to serve as the sole triggering factor. Rather, alcohol appears to sensitize the pancreas, with additional risk factors—such as a high-fat diet, infectious agents, or smoking—contributing to disease onset (Weiss FU, 2019).

2.2.3 Gallstone-related acute pancreatitis

Gallstone-related acute pancreatitis is one of the most common causes of acute pancreatitis, comparable in frequency to alcohol-induced acute pancreatitis. A cross-sectional study of patients with pancreatitis reported the presence of gallbladder stones in 50% of women and 15% of men. It has also been estimated that approximately 8% of patients with gallbladder stones develop acute pancreatitis.

Gallstone-related acute pancreatitis occurs when gallstones become impacted at the duodenal papilla, the outlet of the pancreatic duct, thereby obstructing the outflow of pancreatic juice.

The majority of patients with gallstone-related acute pancreatitis develop mild edematous pancreatitis and subsequently recover completely; however, 15%–30% may progress to severe necrotizing pancreatitis, which requires intensive multidisciplinary management (Weiss FU, 2019).

Unlike other etiologies of acute pancreatitis, gallstone-related acute pancreatitis has a clearly distinct therapeutic approach, underscoring the importance of accurate etiological differentiation.

2.2.4 Hypertriglyceridemia

It is estimated that approximately 10% of cases of acute pancreatitis are associated with elevated serum triglyceride levels.

An increase in serum triglycerides is thought to contribute to the development of acute pancreatitis by increasing blood viscosity and inducing localized tissue ischemia. This ischemia shifts cellular metabolism from aerobic to anaerobic pathways, leading to increased reliance on anaerobic glycolysis, lactate accumulation, and the development of metabolic acidosis. In addition, enhanced toxicity of free fatty acids promotes the autoactivation of trypsinogen, which may serve as a local triggering factor for acute pancreatitis.

Patients with familial hypertriglyceridemia are considered to have a specific risk of recurrent episodes of acute pancreatitis.

Furthermore, secondary hypertriglyceridemia has also been reported to increase the risk of acute pancreatitis. In populations with serum triglyceride levels $\geq$1000 mg/dL, the risk of developing at least one episode of acute pancreatitis is estimated to be approximately 20%. Moreover, each 100 mg/dL increase in serum triglyceride levels has been reported to be associated with a 4% increase in the risk of acute pancreatitis (Weiss FU, 2019).

2.2.5 Post–endoscopic retrograde cholangiopancreatography acute pancreatitis

Lesions of the pancreatic ductal or biliary system, including gallstone disease as described above, are generally indications for ERCP. However, one of the most serious complications associated with ERCP is the development of acute pancreatitis. Although the occurrence of severe acute pancreatitis after ERCP is rare, it tends to be more severe compared with acute pancreatitis of other etiologies, accounting for approximately 0.5% of all acute pancreatitis cases and about 5% of ERCP procedures.

The risk factors for post-ERCP acute pancreatitis include both patient-related and procedure-related factors. Patient-related factors include sex, age, anatomical abnormalities of the duodenal papilla or pancreas, a history of pancreatitis, and

ERCP performed for gallstone pancreatitis. Procedure-related factors primarily include the experience of the endoscopist.

The proposed pathophysiological mechanisms of post-ERCP acute pancreatitis include edema of the duodenal papilla, spasm of the sphincter of Oddi, and obstruction of the pancreatic duct. In addition, the osmolarity, pH, and chemical composition of contrast agents may exert direct chemical effects. Increased intrapancreatic ductal pressure during contrast injection may also activate digestive enzymes, leading to autodigestion of pancreatic tissue (Weiss FU, 2019).

2.2.6 Drug-induced acute pancreatitis

The diagnosis of drug-induced acute pancreatitis is rare, with an estimated incidence ranging from 0.1% to 2% according to previous reports. With the increasing prevalence of polypharmacy associated with population aging, the incidence of drug-induced acute pancreatitis may rise (Weiss FU, 2019).

The proposed mechanisms of drug-induced acute pancreatitis include spasm of the sphincter of Oddi, direct cytotoxic effects, metabolic disturbances, allergic reactions, and localized vascular edema. A wide variety of drugs have been implicated as potential causes. In one report, more than 500 drugs were identified as having a possible risk of inducing acute pancreatitis; however, a definitive causal relationship has not been established for all of them (Sci, 2015).

The diagnosis of drug-induced pancreatitis requires fulfillment of three criteria: onset of pancreatitis during drug administration, resolution of pancreatitis after discontinuation of the suspected drug, and recurrence of pancreatitis upon re-administration of the drug. However, re-challenge is often ethically unacceptable, making definitive confirmation of drug-induced pancreatitis difficult (Fig. 2.1).

Drugs Associated with Drug-Induced Acute Pancreatitis
(Drugs for which at least one positive rechallenge test has been reported and other potential etiologies have been excluded)

α-Methyldopa	Cytosine	Mesalamine	Simvastatin
Olsalazine	Dapsone	Metronidazole	Antimony gluconate
Bezafibrate	Enalapril	Pentamidine	Sulfamethoxazole
Cannabis	arabinofuranosyl cytosine	Pravastatin	Sulindac
Carbimazole	Furosemide	Procainamide	Tetracycline
Codeine	Isoniazid	Pyritinol	Valproic acid

Adapted from: Japanese guidelines for the management of acute pancreatitis 2015.

FIGURE 2.1 Drugs associated with drug-induced acute pancreatitis.

List adapted from Japanese guidelines for the management of acute pancreatitis (2015). Drug-induced acute pancreatitis is suspected when pancreatitis develops during drug exposure and improves after drug discontinuation; recurrence on readministration supports causality, although rechallenge is often ethically unacceptable. This table is intended as a screening aid when evaluating unexplained acute pancreatitis; medication reconciliation is essential.

2.2.7 Traumatic acute pancreatitis

A study conducted in Germany reported that traumatic acute pancreatitis occurred in approximately 20% of patients with abdominal trauma, with 95% of cases resulting from blunt trauma. In general, traumatic pancreatic injury is relatively rare, with an incidence of approximately 6%. Notably, there is no clear correlation between the magnitude of external force and the severity of pancreatic injury (Weiss FU, 2019).

2.2.8 Rare etiologies

Other factors that have been reported to be associated with the development of acute pancreatitis include pregnancy; viral infections (such as mumps, Coxsackie B virus, hepatitis B virus, cytomegalovirus, herpes simplex virus type 2, and varicella-zoster virus); bacterial infections (including Salmonella typhi, Leptospira, and Legionella species); fungal infections (e.g., Aspergillus); parasitic infections (such as Toxoplasma gondii, Cryptosporidium, and Ascaris); Mycoplasma infection; hyperparathyroidism; and end-stage renal disease (Sci, 2015).

2.3 Genetic variants associated with the risk of acute pancreatitis

Genetic variants have been reported to increase the risk of developing acute pancreatitis (AP). A systematic review of 96 studies identified three genetic variants that showed highly credible associations with the risk of AP. These variants were found in the SPINK1 gene (odds ratio [OR] 2.87, 95% confidence interval [CI] 1.89–4.34), IL1B (OR 1.23, 95% CI 1.06–1.42), and IL6 (OR 1.64, 95% CI 1.15–2.32).

Serine protease inhibitor Kazal type 1 (SPINK1) functions as a trypsin inhibitor in the pancreas. Loss-of-function variants in this gene are thought to enhance the autoactivation of trypsinogen, leading to autodigestion of pancreatic tissue and triggering a key early process in the pathogenesis of pancreatitis.

Interleukin-1β and interleukin-6 (IL1B and IL6) are cytokines primarily produced by macrophages and lymphocytes and act as mediators of intercellular communication within the immune system, exerting either proinflammatory or anti-inflammatory effects. IL1B is one of the proinflammatory cytokines that increases during episodes of acute pancreatitis, and genetic variants that enhance its expression have been reported to be associated with disease severity. IL6 is also a proinflammatory cytokine and is known to correlate with the severity of the disease.

In addition, two single-nucleotide polymorphisms showing credible associations with AP risk have been reported in Asian populations: ALDH2 (OR 0.48, 95% CI 0.36–0.64) and IL18 (OR 1.47, 95% CI 1.18–1.82). Mitochondrial aldehyde dehydrogenase (ALDH2) is involved in the ethanol metabolism pathway. Variants in this gene are common in Asian populations and are known to be associated with susceptibility to alcohol dependence and alcohol sensitivity (van den Berg et al., 2020).

2.4 Severity

Based on the Japanese severity classification criteria (Takeda, 2010), 68.9% of patients with acute pancreatitis (AP) were categorized as mild, while 31.1% were classified as severe. Compared with the 2016 survey conducted by the same group, the estimated number of patients with mild AP increased by 29.8%, whereas the estimated number of severe cases remained largely unchanged.

Among patients with severe AP, 75.0% were classified as severe based solely on CT grading, 11.2% were classified based on prognostic factors, and 13.8% met both CT grading and prognostic criteria.

The mean age (± standard deviation) of patients with severe AP was 60.4 ± 19.3 years, with males averaging 57.8 ± 18.4 years and females 65.9 ± 19.9 years.

Regarding etiology, alcohol-related AP was the most common cause of severe disease (37.3%), followed by idiopathic (22.3%) and gallstone-related (21.5%) etiologies (Yuichi Tanaka, 2025).

2.5 Mortality

In 2021, the global age-standardized mortality rate (ASMR) for pancreatitis was 1.452 per 100,000 population (95% uncertainty interval [UI]: 1.303–1.674). The ASMR fluctuated between 1990 and 2005, peaking at 1.758 (1.573–2.010) in 1994, and subsequently showing a declining trend after 2005 (Baichuan Zhou, 2025).

Regionally, in 2021, the highest age-standardized mortality rate was observed in Eastern Europe at 5.599 (5.151–6.066), whereas the lowest was recorded in the high-income Asia–Pacific region at 0.466 (0.405–0.511). From 1990 to 2021, the global ASMR changed by −0.141 (−0.245 to −0.011). Among regions, the greatest increase was observed in Eastern Europe at 0.955 (0.777–1.169), while the largest decrease occurred in the high-income Asia–Pacific region at −0.506 (−0.551 to −0.441) (Baichuan Zhou, 2025).

The ASMR by sex in 2021 was 1.929 (1.662–2.256) for males and 0.992 (0.834–1.253) for females, yielding a male-to-female ratio of 1.94 (Baichuan Zhou, 2025).

In Japan, the mortality rate among patients with acute pancreatitis was 2.1%, and among patients with severe acute pancreatitis, it was 5.3%. Compared with 2016, the mortality rate in patients with severe acute pancreatitis decreased by 13.1% (Yuichi Tanaka, 2025).

When stratified by etiology, mortality rates were as follows: post-therapeutic ERCP pancreatitis, 4.8%; alcohol-induced, 1.8%; gallstone-related, 2.1%; and idiopathic, 2.1%. Mortality was significantly higher in post-therapeutic ERCP pancreatitis compared with other etiologies ($P = .03$) (Yuichi Tanaka, 2025).

2.6 Future projections of pancreatitis

A study using the Bayesian age-period-cohort (BAPC) model demonstrated a consistent decline in the age-standardized incidence rate (ASIR) and age-standardized mortality rate (ASMR) of pancreatitis from 1990 to 2020. Future projections suggest that this downward trend is likely to continue through 2050 (Tianyu Li, 2024).

References

Baichuan Zhou, J.Z., 2025. The global, regional, and National burden of pancreatitis from 1990 to 2021: a systematic analysis for the global burden of disease study 2021. Journal of Gastroenterology and Hepatology.

van den Berg, F.F., et al., 2020. Meta-analysis and field synopsis of genetic variants associated. BJS Open.

Sci, J.H., 2015. Japanese guidelines for the management of acute pancreatitis. Journal of Hepatobiliary Pancreatic Sciences.

Takeda, K.,Y.M., 2010. Assessment of severity of. Journal of Hepatobiliary Pancreatic Sciences.

Tianyu Li, C.Q., 2024. Global and regional burden of pancreatitis:epidemiological trends, risk factors, and projections to 2050 from the global burden of disease study 2021. BMC Gastroenterology.

Weiss FU, L.F., 2019. Etiology and pathogenesis of acute pancreatitis. Nature Reviews Gastroenterology & Hepatology.

Yuichi Tanaka, A.M., 2025. Nationwide epidemiological survey of acute pancreatitis in Japan, 2021: the impact of the COVID-19 pandemic and revised clinical guidelines. Journal of Gastroenterology.

CHAPTER

Acute pancreatitis and infectious diseases 3

Nobuyuki Tetsuka

Gifu University, Graduate School of Medicine, Department of Infection Control, and the Advanced Critical Care Center at Gifu University Hospital, Gifu, Japan

3.1 Introduction

Acute pancreatitis (AP) is an acute inflammatory disorder of the pancreas that typically presents with severe epigastric pain and elevated serum pancreatic enzyme levels. Gallstones and excessive alcohol consumption account for most cases, while metabolic disturbances, anatomic abnormalities, autoimmune disease, ischemia, malignancy, trauma, endoscopic retrograde cholangiopancreatography, and drugs represent less common etiologies (Sakorafas and Tsiotou, 2000; Tenner, 2014). Several antimicrobial agents have been implicated in drug-induced pancreatitis, although such events remain uncommon.

From a clinical standpoint, it is essential to distinguish between infectious causes of pancreatitis and infectious complications arising during its course. While primary pancreatic infection is rare, secondary infection—particularly of necrotic pancreatic tissue—plays a decisive role in determining outcomes in severe disease. Accordingly, this chapter focuses on infections complicating AP, emphasizing contemporary diagnostic and therapeutic strategies rather than rare infectious etiologies.

3.2 Infectious etiologies of acute pancreatitis: Limited clinical relevance

A wide range of viral, bacterial, fungal, and parasitic pathogens has been reported in association with AP, largely on the basis of isolated case reports and small series. Systematic re-evaluations of these reports have demonstrated that only a limited subset fulfills rigorous diagnostic criteria for both pancreatitis and infection (Parenti et al., 1996).

Even in cases with plausible associations—such as certain viral infections or ascariasis in endemic regions (Khuroo et al., 1992)—there is no convincing evidence that identification or treatment of specific pathogens alters clinical outcomes. Consequently, routine investigation for infectious etiologies in patients with idiopathic AP is not recommended. In modern clinical practice, the relevance of infection in AP lies overwhelmingly in its role as a complication rather than a primary cause.

Advanced Insights into Acute Pancreatitis. https://doi.org/10.1016/B978-0-443-44736-5.00002-4

3.3 Acute pancreatitis in immunocompromised hosts

Human immunodeficiency virus infection has been associated with an increased incidence of pancreatitis, although direct viral invasion of pancreatic tissue has not been demonstrated. Common contributing factors include gallstones, alcohol or intravenous drug use, drug toxicity (e.g., pentamidine or didanosine), and opportunistic infections (Dutta et al., 1997). An inverse correlation between CD4 lymphocyte counts and serum pancreatic enzyme levels has been reported.

In advanced disease, pancreatic microabscesses or isolated enzyme elevations may be detected incidentally, often without clinical or radiologic features of AP (Murthy et al., 1992). These observations underscore the importance of cautious interpretation of pancreatic abnormalities in immunocompromised patients.

3.4 Infectious complications of acute pancreatitis: A temporal framework

Severe AP frequently requires intensive, multidisciplinary management, including admission to the intensive care unit, prolonged hospitalization, and the use of invasive devices. As a consequence, these patients are exposed to a substantially increased risk of hospital-acquired infections. Mechanical ventilation, central venous catheterization, urinary catheterization, and exposure to broad-spectrum antimicrobials all contribute to the development of hospital-associated infections such as pneumonia, catheter-related bloodstream infection, and urinary tract infection.

Accordingly, when patients with AP develop fever, recurrent elevation of inflammatory markers, or deterioration in respiratory or hemodynamic status, these changes should not be attributed solely to progression of pancreatitis or pancreatic infection. Clinicians should avoid anchoring on pancreatic pathology and instead perform a comprehensive reassessment of the entire patient, considering not only pancreatic infection but also extrapancreatic and hospital-acquired infections involving other organ systems. This global perspective is particularly critical in severe AP, where nonpancreatic infections may significantly influence outcomes.

From a clinical standpoint, infectious complications of AP can be usefully conceptualized according to the timing of their occurrence: early extrapancreatic infections and later secondary infections of pancreatic or peripancreatic necrosis.

3.4.1 Early extrapancreatic infections

During the early phase of AP, patients are susceptible to extrapancreatic infections, including pneumonia, bacteremia, urinary tract infection, and cholangitis (Fan et al., 1993; Besselink et al., 2009a). These infections often occur in parallel with systemic inflammatory response syndrome and may be difficult to distinguish from inflammation caused by pancreatitis itself.

In severe cases, management in the intensive care setting further increases the risk of ventilator-associated pneumonia and catheter-related bloodstream infection, both of which are important contributors to morbidity and mortality. Therefore, when infection is suspected in patients with AP, evaluation should not be limited to the pancreas but should include a thorough assessment of the respiratory tract, bloodstream, urinary tract, and biliary system.

3.4.2 Late pancreatic and peripancreatic infections

In contrast, most pancreatic and peripancreatic infections develop later in the disease course and represent secondary infection of necrotic tissue. This process is thought to be driven primarily by bacterial translocation from the gastrointestinal tract following disruption of the intestinal barrier.

Infected pancreatic necrosis is among the most severe complications of AP and is strongly associated with systemic toxicity, persistent organ failure, and increased mortality. Because these infections typically occur after an initial period of sterile necrosis, clinicians must maintain a high index of suspicion when patients demonstrate secondary clinical deterioration during the later stages of disease, even if early imaging or clinical findings suggested noninfected necrosis (Besselink et al., 2009a).

3.5 Pathophysiology and clinical impact of infection

Approximately 80% of patients with AP develop interstitial edematous pancreatitis, which usually resolves with supportive care (Ratschko et al., 1999). In 5%–15% of cases, sustained microcirculatory impairment and inflammatory injury result in pancreatic necrosis.

Among patients with necrotizing pancreatitis, secondary infection develops in approximately one-third and accounts for the majority of disease-related deaths (Besselink et al., 2009a; Karimgani et al., 1992). Infection and persistent organ failure exert synergistic effects on mortality, defining a subgroup of patients with extremely severe disease (Petrov et al., 2010). The risk of infection increases in proportion to the extent of necrosis (Beger et al., 1986; Pederzoli et al., 1993).

3.6 Definitions and evolution of pancreatic collections

Terminology for pancreatic and peripancreatic collections is based on the Atlanta classification and its 2012 revision (Bradley, 1993; Banks et al., 2013). The main entities are defined as follows (Banks et al., 2013):

- **Acute pancreatitis**: An acute inflammatory process of the pancreas that may involve surrounding tissues and distant organ systems.
- **Necrotizing pancreatitis**: A form of pancreatitis characterized by necrosis of the pancreatic parenchyma and/or peripancreatic tissues.
- **Acute necrotic collection (ANC)**: A heterogeneous collection containing both fluid and necrotic material, occurring in the early phase of necrotizing pancreatitis before a defined wall has formed.

- **Acute peripancreatic fluid collection (APFC)**: A collection of fluid without necrotic debris, seen within the first 4 weeks of interstitial edematous pancreatitis and lacking a defined capsule.
- **Pancreatic pseudocyst**: A well-defined, encapsulated fluid collection with little or no necrotic content, typically developing more than 4 weeks after the onset of interstitial edematous pancreatitis.
- **Walled-off necrosis (WON)**: A mature, encapsulated collection containing necrotic material, usually forming more than 4 weeks after the onset of necrotizing pancreatitis.

In the early phase, acute necrotic collections may develop without encapsulation. Over time, sterile necrosis may evolve into walled-off necrosis (WON), whereas interstitial edematous pancreatitis may lead to the formation of pancreatic pseudocysts, typically several weeks after disease onset (Bittner et al., 1987; Bohidar et al., 2003; Banks et al., 1995; Gerzof et al., 1987).

In general, infected necrosis is associated with greater systemic toxicity and higher mortality than pancreatic abscess (Bittner et al., 1987; Banks et al., 1995).

3.7 Diagnosis of pancreatic infection

Diagnosing infection in AP is challenging, as systemic inflammatory responses may occur in the absence of infection. Clinical suspicion should increase when patients demonstrate secondary deterioration after initial stabilization, particularly when accompanied by recurrent fever, rising inflammatory markers, or new-onset organ failure (Bohidar et al., 2003).

The presence of gas within necrotic collections on contrast-enhanced computed tomography is highly suggestive but not universally present (van Grinsven et al., 2018). Although CT-guided fine-needle aspiration was historically advocated (Banks et al., 1995; Gerzof et al., 1987), contemporary guidelines no longer recommend its routine use, as results rarely alter management and may delay intervention.

3.8 Microbiology of pancreatic infection

Pancreatic infections typically arise from translocation of gastrointestinal microbiota and involve both aerobic and anaerobic organisms. Historically, Gram-negative enteric bacteria predominated, while Gram-positive organisms were less frequent (Lumsden and Bradley, 1990).

More recent series have demonstrated a shift toward enterococci, multidrug-resistant organisms, and fungi, particularly in patients exposed to broad-spectrum antibiotics or prolonged intensive care (Schmidt et al., 2014; Mowbray et al., 2018). This evolving microbiology underscores the importance of antimicrobial stewardship in AP.

3.9 Antimicrobial exposure and fungal infection

Multiple randomized trials and meta-analyses have demonstrated that prophylactic antibiotic therapy does not reduce the incidence of infected necrosis or mortality in AP (Lim et al., 2015; Vege et al., 2018; Lee et al., 2014). Moreover, unnecessary antimicrobial exposure is associated with increased rates of resistant bacterial and fungal infections (Lee et al., 2014; Kochhar et al., 2011).

Fungal infections are associated with increased systemic complications and may contribute to worse outcomes, particularly in patients with prolonged hospitalization or multiple interventions (Kochhar et al., 2011).

3.10 Management of pancreatic infection

Effective management requires integration of antimicrobial therapy and source control.

3.10.1 Minimally invasive drainage

Percutaneous or endoscopic drainage can successfully manage infected necrosis in a substantial proportion of patients, often obviating the need for surgery (Mortelé et al., 2009; Van Santvoort et al., 2010).

3.10.2 Step-up approach and surgery

When drainage alone is insufficient, treatment is escalated using a step-up approach, favoring minimally invasive necrosectomy over open surgery (Van Santvoort et al., 2010, 2011; Farkas et al., 1996). Open necrosectomy is reserved as a last resort.

3.10.3 Timing of intervention

Delayed intervention, typically after 4 weeks, allows demarcation of necrotic tissue and is associated with improved outcomes compared with early surgery (Van Santvoort et al., 2011).

3.11 Prevention strategies

3.11.1 Enteral nutrition

Enteral feeding may preserve gut barrier function and reduce infectious complications, although results across studies are heterogeneous (Kalfarentzos et al., 1997; Oláah et al., 2002; Bakker et al., 2014). Current guidelines prioritize enteral nutrition when feasible.

3.11.2 Probiotics and selective decontamination

Despite early promise (Oláh et al., 2002), probiotics have been associated with increased mortality and bowel ischemia in severe AP (Besselink et al., 2008; Sand and Nordback, 2008; Besselink et al., 2009b). Selective digestive tract decontamination remains controversial (Luiten et al., 1995, 1997).

3.11.3 Prophylactic antibiotics

Extensive investigation has failed to demonstrate benefit from prophylactic antibiotics, and routine use is not recommended (Dellinger et al., 2007; Xue et al., 2009; Jafri et al., 2009; Ukai et al., 2015; Wittau et al., 2011; Villatoro et al., 2010).

3.12 Real-world practice

Despite strong evidence and guideline recommendations, antibiotics continue to be overused in AP. Observational studies demonstrate frequent antibiotic administration even in mild disease (Baltatzis et al., 2016; Greenberg et al., 2016), highlighting the gap between evidence and practice.

3.13 Conclusions

In AP, clinical outcomes are driven primarily by infectious complications rather than by infectious etiologies. Severe AP requires intensive, multidisciplinary care and is therefore associated with a high risk of hospital-acquired infections. Clinical deterioration in these patients should not be attributed solely to progression of pancreatic disease or pancreatic infection.

Optimal management demands avoidance of anchoring on pancreatic pathology and requires comprehensive reassessment of the entire patient, including consideration of extrapancreatic and hospital-associated infections. Current evidence supports judicious antimicrobial use, early enteral nutrition, and delayed, minimally invasive source control using a step-up approach. Advances in supportive care may have reduced the baseline risk of infected necrosis, whereas unnecessary antimicrobial exposure promotes antimicrobial resistance and fungal infection (Howard, 2007; De Waele et al., 2004).

AI disclosure

During the preparation of this work the author(s) used ChatGPT 5.2(OpenAI) in order to ChatGPT was used to assist with language editing, structural reorganization, and refinement of clarity in the manuscript. All scientific content, interpretations, and conclusions were determined by the author, who takes full responsibility for the accuracy and integrity of the work.

After using this tool/service, the author(s) reviewed and edited the content as needed and take (s) full responsibility for the content of the publication.

References

Bakker, O.J., Van Brunschot, S., Van Santvoort, H.C., Besselink, M.G., Bollen, T.L., Boermeester, M.A., Dejong, C.H., Van Goor, H., Bosscha, K., Ahmed Ali, U., Bouwense, S., Van Grevenstein, W.M., Heisterkamp, J., Houdijk, A.P., Jansen, J.M., Karsten, T.M., Manusama, E.R., Nieuwenhuijs, V.B., Schaapherder, A.F., Van Der Schelling, G.P., Schwartz, M.P., Spanier, B.W.M., Tan, A., Vecht, J., Weusten, B.L., Witteman, B.J., Akkermans, L.M., Bruno, M.J., Dijkgraaf, M.G., Van Ramshorst, B., Gooszen, H.G., 2014. Early versus on-demand nasoenteric tube feeding in acute pancreatitis. New England Journal of Medicine 371 (21), 1983–1993. https://doi.org/10.1056/NEJMoa1404393. Available from: http://www.nejm.org/doi/pdf/10.1056/NEJMoa1404393.

Baltatzis, M., Mason, J.M., Chandrabalan, V., Stathakis, P., McIntyre, B., Jegatheeswaran, S., Jamdar, S., O'Reilly, D.A., Siriwardena, A.K., 2016. Antibiotic use in acute pancreatitis: an audit of current practice in a tertiary centre. Pancreatology 16 (6), 946–951. https://doi.org/10.1016/j.pan.2016.08.012.

Banks, P.A., Gerzof, S.G., Langevin, R.E., Silverman, S.G., Sica, G.T., Hughes, M.D., 1995. CT-guided aspiration of suspected pancreatic infection—bacteriology and clinical outcome. International Journal of Pancreatology 18 (3), 265–270. https://doi.org/10.1007/BF02784951.

Banks, P.A., Bollen, T.L., Dervenis, C., Gooszen, H.G., Johnson, C.D., Sarr, M.G., Tsiotos, G. G., Vege, S.S., Windsor, J.A., Horvath, K.D., Mortele, K.J., Gardner, T.B., Van Santvoort, H., Pelaez-Luna, M., Yadav, D., Stefanidis, G., Delakidis, S., Morgan, D.E., Thoeni, R.F.L., Acosta, J.M., Amann, S.T., Andren-Sandberg, A., Aranha, G.V., Asciutti, S., Barauskas, G., Baron, T.H., Bassi, C., Behrman, S., Behrns, K.E., Belliappa, V., Berzin, T.M., Besselink, M.G.H., Bhasin, D.K., Biankin, A., Bishop, M. D., Bonini, C.J., Bradley, E.L., Buechler, M., Carter, R., Cavestro, G.M., Chari, S.T., Chavez-Rodriguez, J.J., Da Cunha, J.E., D'Agostino, H., De Campos, T., De-Madaria, E., Deprez, P.H., DiSario, J.A., Doria, C., Falconi, M., Fernandez-del Castillo, C., Freeny, P.C., Frey, C.F., Friess, H., Frossard, J.L., Fuchshuber, P., Gallagher, S.F., Garg, P.K., Ghattas, G., Glasgow, R., Gonzalez, J.A., Gress, T.M., Gumbs, A.A., Hallibruton, C., Helton, S., Hill, M.C., Hoyos, S., Imrie, C.W., Isenmann, R., Izbicki, J.R., Karagiannis, J.A., Klar, E., Kolokythas, O., Lau, J., Litvin, A.A., Longnecker, D.S., Lowenfels, A.B., Mackey, R., Mah'Moud, M., Malangoni, M., McFadden, D.W., Mishra, G., Moody, F.G., Morinville, V., Neoptolemos, J.P., Nordback, I., Pap, A., Papachristou, G.I., Parks, R., Pedrazolli, S., Pezzilli, R., Pitt, H.A., Prosanto, C., Ramesh, H., Ramirez, F.C., Raper, S.E., Rasheed, A., Reed, D.N., Romangnuolo, J., Rossaak, J., Sanabria, J., Schaefer, C., Schmidt, J., Schmidt, P.N., Serrablo, A., Senkowski, C.K., Sharma, M., Sigman, K.M., Singh, P., Steinberg, W., Steiner, J., Strasberg, S., Strum, W., Takada, T., Tanaka, M., Vaccaro, M., Villavicencio, R.L., Vrochides, D., Wagner, M., Warshaw, A.L., Wilcox, C.M., Wysocki, P., Zenilman, M.E., Zyromski, N.J., 2013. Classification of acute pancreatitis—2012: revision of the Atlanta classification and definitions by international consensus. Gut 62 (1), 102–111. https://doi.org/10.1136/gutjnl-2012-302779UnitedStates.

Beger, H.G., Bittner, R., Block, S., Büchler, M., 1986. Bacterial contamination of pancreatic necrosis. A prospective clinical study. Gastroenterology 91 (2), 433–438. https://doi.org/10.1016/0016-5085(86)90579-2.

Besselink, M.G., van Santvoort, H.C., Buskens, E., Boermeester, M.A., van Goor, H., Timmerman, H.M., Nieuwenhuijs, V.B., Bollen, T.L., van Ramshorst, B., Witteman, B. J., Rosman, C., Ploeg, R.J., Brink, M.A., Schaapherder, A.F., Dejong, C.H., Wahab, P. J., van Laarhoven, C.J., van der Harst, E., van Eijck, C.H., Cuesta, M.A., Akkermans, L.M., Gooszen, H.G., 2008. Probiotic prophylaxis in predicted severe acute pancreatitis: a randomised, double-blind, placebo-controlled trial. The Lancet 371 (9613), 651–659. https://doi.org/10.1016/S0140-6736(08)60207-X.

Besselink, M.G., van Santvoort, H.C., Boermeester, M.A., Nieuwenhuijs, V.B., van Goor, H., Dejong, C.H.C., Schaapherder, A.F., Gooszen, H.G., 2009a. Timing and impact of infections in acute pancreatitis. British Journal of Surgery 96 (3), 267–273. https://doi.org/10.1002/bjs.6447.

Besselink, M.G., Van Santvoort, H.C., Renooij, W., De Smet, M.B., Boermeester, M.A., Fischer, K., Timmerman, H.M., Ahmed Ali, U., Cirkel, G.A., Bollen, T.L., Van Ramshorst, B., Schaapherder, A.F., Witteman, B.J., Ploeg, R.J., Van Goor, H., Van Laarhoven, C.J., Tan, A.C., Brink, M.A., Van Der Harst, E., Wahab, P.J., Van Eijck, C. H., Dejong, C.H., Van Erpecum, K.J., Akkermans, L.M., Gooszen, H.G., 2009b. Intestinal barrier dysfunction in a randomized trial of a specific probiotic composition in acute pancreatitis. Annals of Surgery 250 (5), 712–719. https://doi.org/10.1097/SLA.0b013e3181bce5bd.

Bittner, R., Block, S., Büchler, M., Beger, H.G., 1987. Pancreatic abscess and infected pancreatic necrosis—different local septic complications in acute pancreatitis. Digestive Diseases and Sciences 32 (10), 1082–1087. https://doi.org/10.1007/BF01300192.

Bohidar, N.P., Garg, P.K., Khanna, S., Tandon, R.K., 2003. Incidence, etiology, and impact of fever in patients with acute pancreatitis. Pancreatology 3 (1), 9–13. https://doi.org/10.1159/000069146.

Bradley, E.L., 1993. A clinically based classification system for acute pancreatitis: summary of the international Symposium on acute pancreatitis, Atlanta, Ga, September 11 through 13, 1992. Archives of Surgery 128 (5), 586–590. https://doi.org/10.1001/archsurg.1993.01420170122019.

De Waele, J.J., Vogelaers, D., Hoste, E., Blot, S., Colardyn, F., 2004. Emergence of antibiotic resistance in infected pancreatic necrosis. Archives of Surgery 139 (12), 1371–1375. https://doi.org/10.1001/archsurg.139.12.1371.

Dellinger, E.P., Tellado, J.M., Soto, N.E., Ashley, S.W., Barie, P.S., Dugernier, T., Imrie, C.W., Johnson, C.D., Knaebel, H.P., Laterre, P.F., Maravi-Poma, E., Kissler, J.J.O., Sanchez-Garcia, M., Utzolino, S., 2007. Early antibiotic treatment for severe acute necrotizing pancreatitis: a randomized, double-blind, placebo-controlled study. Annals of Surgery 245 (5), 674–683. https://doi.org/10.1097/01.sla.0000250414.09255.84.

Dutta, S.K., Ting, C.D., Lai, L.L., 1997. Study of prevalence, severity, and etiological factors associated with acute pancreatitis in patients infected with human immunodeficiency virus. American Journal of Gastroenterology 92 (11), 2044–2048.

Fan, S.T., Lai, E., Mok, F., lo, C.M., Zheng, S.S., Wong, J., 1993. Early treatment of acute biliary pancreatitis by endoscopic papillotomy. New England Journal of Medicine 328 (4), 228–232. https://doi.org/10.1056/NEJM199301283280402.

Farkas, G., Márton, J., Mándi, Y., Szederkényi, E., 1996. Surgical strategy and management of infected pancreatic necrosis. Journal of British Surgery 83 (7), 930–933. https://doi.org/10.1002/bjs.1800830714.

Gerzof, S.G., Banks, P.A., Robbins, A.H., Johnson, W.C., Spechler, S.J., Wetzner, S.M., Snider, J.M., Langevin, R.E., Jay, M.E., 1987. Early diagnosis of pancreatic infection by computed tomography-guided aspiration. Gastroenterology 93 (6), 1315–1320. https://doi.org/10.1016/0016-5085(87)90261-7.

Greenberg, J.A., Hsu, J., Bawazeer, M., Marshall, J., Friedrich, J.O., Nathens, A., Coburn, N., Huang, H., McLeod, R.S., 2016. Compliance with evidence-based guidelines in acute pancreatitis: an audit of practices in University of Toronto Hospitals. Journal of Gastrointestinal Surgery 20 (2), 392–400. https://doi.org/10.1007/s11605-015-3023-9.

Howard, T.J., 2007. As good as it gets: the study of prophylactic antibiotics in severe acute pancreatitis. Annals of Surgery 245 (5), 684–685. https://doi.org/10.1097/01.sla.0000261458.39058.fc.

Jafri, N.S., Mahid, S.S., Idstein, S.R., Hornung, C.A., Galandiuk, S., 2009. Antibiotic prophylaxis is not protective in severe acute pancreatitis: a systematic review and meta-analysis. The American Journal of Surgery 197 (6), 806–813. https://doi.org/10.1016/j.amjsurg.2008.08.016.

Kalfarentzos, F., Kehagias, J., Mead, N., Kokkinis, K., Gogos, C.A., 1997. Enteral nutrition is superior to parenteral nutrition in severe acute pancreatitis: results of a randomized prospective trial. British Journal of Surgery 84 (12), 1665–1669. https://doi.org/10.1002/bjs.1800841207.

Karimgani, I., Porter, K.A., Eugene Langevin, R., Banks, P.A., 1992. Prognostic factors in sterile pancreatic necrosis. Gastroenterology 103 (5), 1636–1640. https://doi.org/10.1016/0016-5085(92)91189-b.

Khuroo, M.S., Zargar, S.A., Yattoo, G.N., Koul, P., Khan, B.A., Dar, M.Y., Alai, M.S., 1992. Ascaris-induced acute pancreatitis. British Journal of Surgery 79 (12), 1335–1338. https://doi.org/10.1002/bjs.1800791231.

Kochhar, R., Noor, M.T., Wig, J., 2011. Fungal infections in severe acute pancreatitis. Journal of Gastroenterology and Hepatology 26 (6), 952–959. https://doi.org/10.1111/j.1440-1746.2011.06685.x.

Lee, H.S., Lee, S.K., Park, D.H., Lee, S.S., Seo, D.W., Kim, M.H., Chong, Y.P., 2014. Emergence of multidrug resistant infection in patients with severe acute pancreatitis. Pancreatology 14 (6), 450–453. https://doi.org/10.1016/j.pan.2014.10.003.

Lim, C.L.L., Lee, W., Liew, Y.X., Tang, S.S.L., Piotr Chlebicki, M., Lay-Hoon Kwa, A., 2015. Role of antibiotic prophylaxis in necrotizing pancreatitis: a meta-analysis. Journal of Gastrointestinal Surgery 19 (3), 480–491. https://doi.org/10.1007/s11605-014-2662-6.

Luiten, E.J.T., Hop, W.C.J., Lange, J.F., Bruining, H.A., 1995. Controlled clinical trial of selective decontamination for the treatment of severe acute pancreatitis. Annals of Surgery 222 (1), 57–65. https://doi.org/10.1097/00000658-199507000-00010.

Luiten, E.J.T., Hop, W.C.J., Lange, J.F., Bruining, H.A., 1997. Differential prognosis of Gram-negative versus Gram-positive infected and sterile pancreatic necrosis: results of a randomized trial in patients with severe acute pancreatitis treated with adjuvant selective decontamination. Clinical Infectious Diseases 25 (4), 811–816. https://doi.org/10.1086/515545.

Lumsden, A., Bradley, E.L., 1990. Secondary pancreatic infections. Surgery Gynecology & Obstetrics 170 (5), 459–467.

Mortelé, K.J., Girshman, J., Szejnfeld, D., Ashley, S.W., Erturk, S.M., Banks, P.A., Silverman, S.G., 2009. CT-guided percutaneous catheter drainage of acute necrotizing pancreatitis: clinical experience and observations in patients with sterile and infected necrosis. American Journal of Roentgenology 192 (1), 110–116. https://doi.org/10.2214/AJR.08.1116UnitedStates.

Mowbray, N.G., Ben-Ismaeil, B., Hammoda, M., Shingler, G., Al-Sarireh, B., 2018. The microbiology of infected pancreatic necrosis. Hepatobiliary and Pancreatic Diseases International 17 (5), 456–460. https://doi.org/10.1016/j.hbpd.2018.08.007.

Murthy, U.K., DeGregorio, F., Oates, R.P., Blair, D.C., 1992. Hyperamylasemia in patients with the acquired immunodeficiency syndrome. American Journal of Gastroenterology 87 (3), 332–336. https://doi.org/10.1111/j.1572-0241.1992.tb02819.x.

Oláah, A., Pardavi, G., Beláagyi, T., Nagy, A., Issekutz, Á.A., Mohamed, G.E., 2002. Early nasojejunal feeding in acute pancreatitis is associated with a lower complication rate. Nutrition 18 (3), 259–262. https://doi.org/10.1016/s0899-9007(01)00755-9.

Oláh, A., Belágyi, T., Issekutz, Á., Gamal, M.E., Bengmark, S., 2002. Randomized clinical trial of specific lactobacillus and fibre supplement to early enteral nutrition in patients with acute pancreatitis. British Journal of Surgery 89 (9), 1103–1107. https://doi.org/10.1046/j.1365-2168.2002.02189.x.

Parenti, D.M., Steinberg, W., Kang, P., 1996. Infectious causes of acute pancreatitis. Pancreas 13 (4), 356–371. https://doi.org/10.1097/00006676-199611000-00005.

Pederzoli, P., Bassi, C., Vesentini, S., Campedelli, A., 1993. A randomized multicenter clinical trial of antibiotic prophylaxis of septic complications in acute necrotizing pancreatitis with imipenem. Surgery Gynecology & Obstetrics 176 (5), 480–483.

Petrov, M.S., Shanbhag, S., Chakraborty, M., Phillips, A.R.J., Windsor, J.A., 2010. Organ failure and infection of pancreatic necrosis as determinants of mortality in patients with acute pancreatitis. Gastroenterology 139 (3), 813–820. https://doi.org/10.1053/j.gastro.2010.06.010.

Ratschko, M., Fenner, T., Lankisch, P.G., 1999. The role of antibiotic prophylaxis in the treatment of acute pancreatitis. Gastroenterology Clinics of North America 28 (3), 641–659. https://doi.org/10.1016/S0889-8553(05)70078-9.

Sakorafas, G.H., Tsiotou, A.G., 2000. Etiology and pathogenesis of acute pancreatitis: current concepts. Journal of Clinical Gastroenterology 30 (4), 343–356. https://doi.org/10.1097/00004836-200006000-00002.

Sand, J., Nordback, I., 2008. Probiotics in severe acute pancreatitis. The Lancet 371 (9613), 634–635. https://doi.org/10.1016/S0140-6736(08)60284-6.

Schmidt, P.N., Roug, S., Hansen, E.F., Knudsen, J.D., Novovic, S., 2014. Spectrum of microorganisms in infected walled-off pancreatic necrosis—impact on organ failure and mortality. Pancreatology 14 (6), 444–449. https://doi.org/10.1016/j.pan.2014.09.001.

Tenner, S., 2014. Drug induced acute pancreatitis: does it exist? World Journal of Gastroenterology 20 (44), 16529–16534. https://doi.org/10.3748/wjg.v20.i44.16529.

Ukai, T., Shikata, S., Inoue, M., Noguchi, Y., Igarashi, H., Isaji, S., Mayumi, T., Yoshida, M., Takemura, Y.C., 2015. Early prophylactic antibiotics administration for acute necrotizing pancreatitis: a meta-analysis of randomized controlled trials. Journal of Hepato-Biliary-Pancreatic Sciences 22 (4), 316–321. https://doi.org/10.1002/jhbp.221.

van Grinsven, J., van Brunschot, S., van Baal, M.C., Besselink, M.G., Fockens, P., Goor, H. van, van Santvoort, H.C., Bollen, T.L., 2018. Natural history of gas configurations and encapsulation in necrotic collections during necrotizing pancreatitis. Journal of Gastrointestinal Surgery 22 (9), 1557–1564. https://doi.org/10.1007/s11605-018-3792-z.

Van Santvoort, H.C., Besselink, M.G., Bakker, O.J., Hofker, H.S., Boermeester, M.A., Dejong, C.H., Van Goor, H., Schaapherder, A.F., Van Eijck, C.H., Bollen, T.L., Van Ramshorst, B., Nieuwenhuijs, V.B., Timmer, R., Laméris, J.S., Kruyt, P.M., Manusama, E.R., Van Der Harst, E., Van Der Schelling, G.P., Karsten, T., Hesselink, E. J., Van Laarhoven, C.J., Rosman, C., Bosscha, K., De Wit, R.J., Houdijk, A.P., Van Leeuwen, M.S., Buskens, E., Gooszen, H.G., 2010. A step-up approach or open necrosectomy for necrotizing pancreatitis. New England Journal of Medicine 362 (16), 1491–1502. https://doi.org/10.1056/NEJMoa0908821.

Van Santvoort, H.C., Bakker, O.J., Bollen, T.L., Besselink, M.G., Ahmed Ali, U., Schrijver, A. M., Boermeester, M.A., Van Goor, H., Dejong, C.H., Van Eijck, C.H., Van Ramshorst, B., Schaapherder, A.F., Van Der Harst, E., Hofker, S., Nieuwenhuijs, V.B., Brink, M.A., Kruyt, P.M., Manusama, E.R., Van Der Schelling, G.P., Karsten, T., Hesselink, E.J., Van Laarhoven, C.J., Rosman, C., Bosscha, K., De Wit, R.J., Houdijk, A.P., Cuesta, M. A., Wahab, P.J., Gooszen, H.G., 2011. A conservative and minimally invasive approach to necrotizing pancreatitis improves outcome. Gastroenterology 141 (4), 1254–1263. https://doi.org/10.1053/j.gastro.2011.06.073.

Vege, S.S., DiMagno, M.J., Forsmark, C.E., Martel, M., Barkun, A.N., 2018. Initial medical treatment of acute pancreatitis: American gastroenterological association Institute Technical review. Gastroenterology 154 (4), 1103–1139. https://doi.org/10.1053/j.gastro.2018.01.031.

Villatoro, E., Mulla, M., Larvin, M., 2010. Antibiotic therapy for prophylaxis against infection of pancreatic necrosis in acute pancreatitis. Cochrane Database of Systematic Reviews 5, CD002941.

Wittau, M., Mayer, B., Scheele, J., Henne-Bruns, D., Patchen Dellinger, E., Isenmann, R., 2011. Systematic review and meta-analysis of antibiotic prophylaxis in severe acute pancreatitis. Scandinavian Journal of Gastroenterology 46 (3), 261–270. https://doi.org/10.3109/00365521.2010.531486.

Xue, P., Deng, L.-H., Zhang, Z.-D., Yang, X.-N., Wan, M.-H., Song, B., Xia, Q., 2009. Effect of antibiotic prophylaxis on acute necrotizing pancreatitis: results of a randomized controlled trial. Journal of Gastroenterology and Hepatology 24 (5), 736–742. https://doi.org/10.1111/j.1440-1746.2008.05758.x.

CHAPTER

Symptoms and complications of pancreatitis

4

Tomotaka Miura
Gifu University Hospital, Advanced Critical Care Center, Gifu, Japan

Acute pancreatitis presents with diverse symptoms, most commonly sudden-onset, severe epigastric or left upper quadrant pain radiating to the back. Nausea, vomiting, fever, ileus, and jaundice may accompany the disease, while elderly patients or those with comorbidities may exhibit atypical or mild presentations, potentially delaying diagnosis. Disease severity and prognosis are primarily determined by systemic complications such as respiratory failure, circulatory collapse, and acute kidney injury, while local complications evolve over time from acute peripancreatic fluid or necrotic collections to pancreatic pseudocysts, walled-off necrosis, or infected necrosis. Understanding these clinical trajectories is essential for risk stratification, early intervention, and improved patient outcomes. This chapter provides an explanation of the symptoms and complications of pancreatitis.

4.1 Symptoms

4.1.1 First symptom of acute pancreatitis

Abdominal pain is the most common initial symptom of acute pancreatitis, occurring in approximately 80%–90% of patients (Masamune et al., 2020; Reid et al., 2017; Tanaka et al., 2025). Abdominal pain is part of the diagnostic criteria for acute pancreatitis, and in clinical practice, the onset of the disease generally coincides with the patient's first perception of pain (Banks et al., 2013). The pain most commonly localizes to the left upper quadrant or epigastrium, may radiate, and can be alleviated by leaning forward or assuming a squatting position (Quinlan, 2014). The criteria for "typical" pancreatitis pain vary across reports. Based on the Revised Atlanta classification (Banks et al., 2013), it is characterized as "acute onset of persistent, severe epigastric pain often radiating to the back." Similarly, the American College of Gastroenterology Guidelines (Tenner et al., 2024) describe the pain as "usually constant, with radiation to the back, chest, or flanks; however, but this description is nonspecific. The intensity of the pain is typically severe but may vary, and neither its intensity nor location correlates with disease severity. Pain described as dull, colicky, or located in the lower abdomen is not consistent with acute pancreatitis and suggests an alternative etiology." Notably, gallstones, a causative factor for

Advanced Insights into Acute Pancreatitis. https://doi.org/10.1016/B978-0-443-44736-5.00013-9

pancreatitis, can cause abdominal pain. Among patients presenting with acute-onset abdominal pain, less than a few percent are ultimately diagnosed with acute pancreatitis (Cervellin et al., 2016). Following abdominal pain, vomiting (27.0%), fever (16.9%), and back pain (16.7%) are the most frequent. Although rare, altered consciousness (2.2%) or shock (1.1%) may represent the initial manifestation (Masamune et al., 2020). A nationwide epidemiological survey conducted by the Japanese Society of Pancreas in 2016 included nearly 3000 patients with acute pancreatitis. The most common initial symptom was abdominal pain, present in 92% of cases. Other symptoms occurred less frequently, including vomiting in 27%, fever in 17%, back pain in 17%, loss of appetite in 16%, abdominal distension in 10%, general fatigue in 10%, jaundice in 6%, diarrhea in 4%, disturbance of consciousness in 2%, and shock in 1% of patients. These findings, subsequently incorporated into the Japanese Clinical Practice Guidelines for Acute Pancreatitis (2021), represent one of the most comprehensive datasets available on the frequency of initial symptoms at disease onset (Masamune et al., 2020; Takada et al., 2022). A Hungarian multicenter cohort (600 patients) reported data on symptoms at admission. In this study, 95.3% of patients presented with abdominal pain, whereas 85.4% experienced nausea and/or vomiting. In addition, a smaller proportion of patients presented with other symptoms, including fever (22%), diarrhea (17%), and constipation (11%) (Párniczky, 2016).

Beyond the initial presentation, the temporal evolution of pain in acute pancreatitis is clinically informative. Pain typically starts abruptly and reaches maximal intensity within hours, often persisting for ≥24 h without substantial fluctuation; transient improvement after emesis or bowel rest does not exclude the diagnosis of pancreatitis (Banks et al., 2013; Boxhoorn et al., 2020). Radiation patterns of pain in acute pancreatitis reflect the retroperitoneal course of the splanchnic nerves. Posterior radiation to the back is common, whereas migration to the lower quadrants suggests an alternative diagnosis (Banks et al., 2013; Quinlan, 2014). In elderly patients and those with diabetes or chronic kidney disease, pain can be attenuated or atypical, thereby increasing reliance on biochemical and imaging criteria (Lankisch et al., 2015; Mederos et al., 2021). Coexisting biliary colic may confound its presentation. Biliary pain is typically colicky and episodic, whereas pancreatitis pain is continuous and more prolonged (Quinlan, 2014; Lankisch et al., 2015).

Therefore, the "typical" first presentation of acute pancreatitis consists of acute-onset epigastric or left upper quadrant pain accompanied by nausea and vomiting, often with radiation. Figure 4.1 illustrates the initial clinical manifestations of acute pancreatitis.

4.1.2 Clinical symptoms of acute pancreatitis

Over the clinical course of acute pancreatitis, the most frequently observed symptoms are abdominal pain, nausea/vomiting, and fever. In addition, reduced intestinal peristalsis often leads to abdominal distension (meteorism) or ileus. Jaundice, shock, and neurological findings may also occur; however, each is observed in less than half

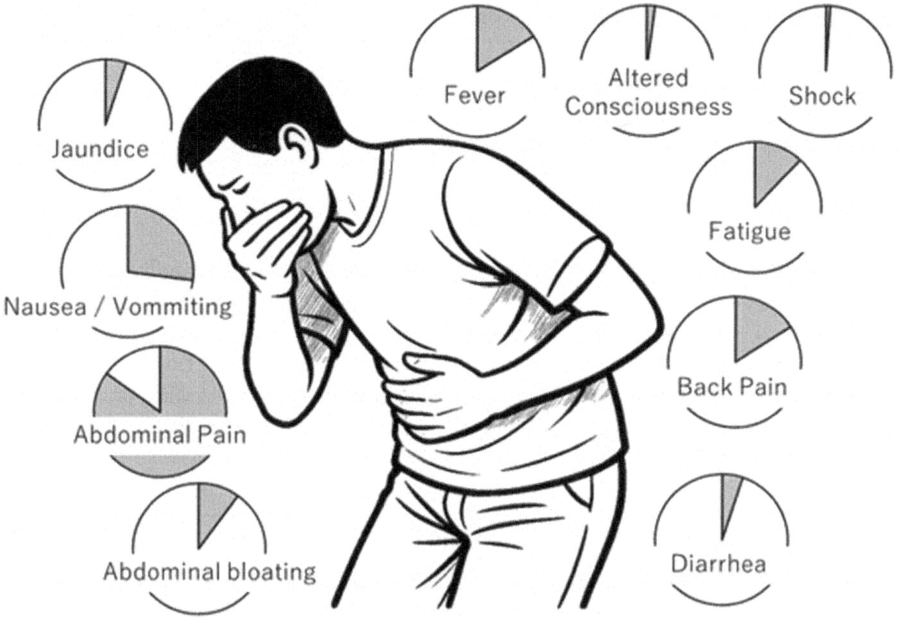

FIGURE 4.1 Initial symptoms of acute pancreatitis.

Abdominal pain is the most common presenting symptom of acute pancreatitis, occurring in approximately 80%–90% of patients. It typically localizes to the epigastrium or left upper quadrant, may radiate to the back, and is continuous in nature. Patients often adopt a forward-leaning or squatting posture to relieve the pain, a characteristic clinical feature of pancreatitis. Nausea and vomiting occur in about one-quarter of patients, whereas fever and back pain are reported in approximately 15%–20%. Altered consciousness or shock is rarely the initial manifestation. The accompanying pie chart illustrates the approximate frequency distribution of these initial symptoms, highlighting abdominal pain as the predominant presentation compared with other, less common manifestations.

of cases. Their frequencies vary by cohort and timing of assessment and may newly arise during the course of pancreatitis (Masamune et al., 2020; Malfertheiner and Kemmer, 1991; Pahuja et al., 2025). Jaundice and pleural or peritoneal effusions tend to be associated with relatively severe disease (Boxhoorn et al., 2020; Patel et al., 2022).

Gastrointestinal dysmotility contributes to nausea, early satiety, and abdominal distension, mediated by inflammatory cytokines and autonomic imbalance. Ileus may be further exacerbated by opioid use for analgesia (Lankisch et al., 2015; Mederos et al., 2021). Systemic inflammatory response frequently produces fever, tachycardia, and tachypnea within the first 24–48 h; however, fever alone has poor specificity for early infection (Boxhoorn et al., 2020). Jaundice generally reflects transient obstruction of the distal common bile duct due to edema or choledocholithiasis. Cholangitis should be suspected when jaundice is accompanied by fever and leukocytosis, especially in gallstone pancreatitis (Quinlan, 2014; Boxhoorn et al., 2020). Neurological symptoms, including agitation or confusion, are uncommon at initial presentation but may develop with worsening systemic illness and metabolic derangements (Masamune et al., 2020; Patel et al., 2022).

Classically described subcutaneous ecchymoses—the Gray–Turner (flank) and Cullen (periumbilical) signs—are recognized manifestations of acute pancreatitis, though they occur in <3% of cases and typically appear several hours or more after onset (Bem and Bradley, 1998; Mookadam and Cikes, 2005; Murthy et al., 2020). Although often considered indicative of severe disease, supporting evidence for this association is limited; moreover, these findings are nonspecific and can be observed in conditions such as ectopic pregnancy or retroperitoneal bleeding (Epperla et al., 2015).

4.1.3 Clinical features of abdominal pain in acute pancreatitis

The temporal evolution of key symptoms is a characteristic feature of acute pancreatitis. Pain typically begins abruptly, reaches maximal intensity within hours, and—unlike biliary colic—remains continuous rather than episodic (Boxhoorn et al., 2020). In mild disease, the overall clinical syndrome is self-limited and usually resolves within approximately 1 week. Nausea and vomiting are the most prominent early symptoms and tend to subside as ileus and inflammation recede (Szatmary et al., 2022). These temporal patterns reflect the biphasic pathophysiology of acute pancreatitis, characterized by an early inflammatory phase (within the first few days) and a later phase, in which local collections and necrosis mature over several weeks. Consequently, early symptom relief does not exclude the development of local complications, whereas prolonged pain beyond several days should prompt reassessment (Tenner et al., 2024; Boxhoorn et al., 2020; Baron et al., 2020).

4.1.3.1 Time from symptom onset to presentation

Most patients present to the emergency department within either 1–6 h or 19–24 h after the onset of abdominal pain. Delayed presentation beyond 4 days is associated with a higher risk of severe disease and mortality (Párniczky, 2016). Independent cohort studies have reported a median pain-to-admission interval of approximately 27 h (interquartile range, 6–72 h), with shorter intervals observed among patients requiring intensive care, consistent with more intense pain at onset (Phillip et al., 2013). These findings contextualize the early, sustained character of pain in acute pancreatitis and underscore that both very delayed and early presentations may signal higher acuity for different reasons (fulminant inflammation vs. progressed complications).

4.1.3.2 Symptom dynamics as prognostic signals

Pain characteristics on admission carry prognostic significance. In a large prospective analysis, intense and sharp pain at presentation correlated with higher odds of severe acute pancreatitis and with both systemic and local complications, underscoring the need for closer monitoring (Földi et al., 2022). Conversely—and importantly, in atypical cases—the absence of abdominal pain at admission can be a risk factor for mortality, reminding clinicians that blunted pain is not equivalent to benign disease. Moreover, nausea and vomiting are frequent early complaints; however, their absence is not associated with increased mortality (Párniczky, 2016). Overall, symptom intensity and the presence or absence of cardinal symptoms should be interpreted alongside objective markers, rather than in isolation.

4.1.3.3 Linking symptom course to severity assessment

Dynamic changes in symptoms during the first 24–48 h meaningfully interact with formal severity assessment. Current guidelines emphasize that this early management window (approximately the first 24–48 h) is decisive for resuscitation and risk stratification. A persistent systemic inflammatory response beyond 48 h strongly predicts persistent organ failure and worse outcomes (Tenner et al., 2024; Boxhoorn et al., 2020). In clinical practice, persistent or escalating pain despite adequate analgesia, ongoing vomiting, inability to tolerate oral intake beyond the first 48–72 h, or new/worsening pain after a transient improvement should trigger re-evaluation for complications such as evolving necrosis, biliary obstruction or cholangitis, and abdominal compartment physiology. Radiologic reassessment is often warranted, as cross-sectional imaging within the first 48–72 h can underestimate necrosis. Symptom trajectories, therefore, complement scoring systems and biomarkers during serial bedside reassessment (Tenner et al., 2024; Boxhoorn et al., 2020; Porter et al., 2019).

4.1.3.4 Practical implications

1) Early, continuous epigastric pain that peaks within hours and persists $\geq$24 h supports the clinical diagnosis and justifies early, guideline-directed care;
2) A markedly delayed presentation after symptom onset indicates accumulated risk and should lower the threshold for higher-level care;
3) Marked admission pain intensity or paradoxically minimal pain in a systemically ill patient should both be treated as red flags;
4) Symptom improvement within the first few days is expected in mild disease; however, a secondary rise in pain or persistent vomiting warrants prompt imaging and multidisciplinary review. These principles integrate patient-reported symptom dynamics into routine severity assessments, complementing, rather than replacing, validated scoring systems.

4.1.4 Differences in symptom presentation by age, sex, and comorbidity

Symptom expression in acute pancreatitis is not uniform across patient groups. Aging, sex, and comorbid conditions—particularly long-standing diabetes—shape the intensity and salience of abdominal pain, prominence of nausea and constitutional complaints, and likelihood of atypical or blunted presentations that delay diagnosis and risk stratification. Recognizing these patterns can reduce missed or late diagnoses and contextualize symptom trajectories alongside objective markers (Boxhoorn et al., 2020; Aziz, 2025).

4.1.4.1 Older adults

In older adults, the cardinal symptoms of abdominal pain may be absent, mild, or described vaguely, and may sometimes be accompanied by acute confusional states. Such atypical presentations are well documented and contribute to diagnostic delay (Rossetti et al., 2009). Contemporary overviews emphasize that the clinical picture of acute pancreatitis in older adults is often less specific and outcomes worsen with increasing age, underscoring the need for a lower threshold to test lipase and pursue imaging when epigastric pain is equivocal or transient (Zhang, 2024).

4.1.4.2 Diabetes and neuropathy

In individuals with diabetes, autonomic and somatic neuropathies can blunt visceral pain perception and alter gastrointestinal symptom reporting. This may diminish the intensity of epigastric pain during acute pancreatitis while amplifying nonspecific complaints, such as early satiety or bloating from gastric dysmotility (Pop-Busui et al., 2017; Kuźnik et al., 2020).

4.1.4.3 Sex-related differences

Direct, head-to-head evidence comparing how women and men report acute pancreatitis symptoms is limited. However, high-quality reviews consistently show sex differences in etiology: gallstone-related pancreatitis is predominant in women, whereas alcohol-related and hypertriglyceridemic pancreatitis are more frequent in men. These differences plausibly shape accompanying symptom constellations (for example, women with gallstone disease may more often highlight nausea, anorexia, or malaise in addition to pain), although abdominal pain remains the central complaint in both sexes (Drake et al., 2021; Wang et al., 2021). Population-based analyses of acute biliary pancreatitis and broader hospitalization datasets also suggest sex-linked differences in management and outcomes, although findings are heterogeneous and do not isolate symptom-reporting patterns specifically. Therefore, clinicians should integrate sex with etiology and comorbidity rather than assume categorical differences in pain description (Shen et al., 2013).

4.1.4.4 Practical implications

In older adults and individuals with diabetes, a lower threshold for serum lipase testing and early repeat assessment is prudent when abdominal pain is muted or short-lived but systemic illness is apparent. Symptom interpretation should be anchored to serial examinations and objective markers rather than to baseline assumptions.

4.1.5 Symptoms of acute pancreatitis in children

The clinical presentation of acute pancreatitis in children overlaps with that in adults but includes age-specific nuances that can delay recognition. Abdominal pain remains the dominant complaint, typically epigastric and sometimes radiating, reported in approximately 85%–95% of pediatric cases. Nausea and vomiting are also common, affecting approximately ≥60% of patients (Wang et al., 2022). Importantly, infants and toddlers may not verbalize pain; irritability, decreased oral intake, and lethargy can be the leading manifestations in this age group, necessitating a high index of suspicion (Abu-El-Haija et al., 2018). Consensus definitions in pediatrics still hinge on the "two-of-three" criteria—characteristic abdominal pain, pancreatic enzyme elevation ≥3× the upper limit of normal, or imaging evidence—which anchors symptom interpretation to objective testing when the history is equivocal (Morinville et al., 2012). Compared with adults, back pain and fever are less frequent in children, and physical findings are often nonspecific (such as epigastric tenderness without peritoneal signs) (Suzuki, 2014). Symptom dynamics may also reflect pediatric-specific etiologies—trauma, drugs, and systemic diseases are more prominent than gallstones or alcohol—therefore, clinicians should contextualize symptom onset and associated features (recent injury, new medications, or intercurrent infection) during triage (Abu-El-Haija et al., 2018; Suzuki, 2014). A recent single-center pediatric series quantifying presenting symptoms (abdominal pain ~95%, vomiting ~60%, nausea ~58%) reinforces that a normal temperature or the absence of jaundice does not exclude acute pancreatitis at first contact. Early hydration and analgesia should proceed in parallel with diagnostic confirmation (Wang et al., 2022). Finally, guideline statements further emphasize that younger children may present with irritability rather than classic pain, highlighting the importance of symptom-triggered reassessment after fluids/analgesia and early consultation with pediatric gastroenterology as practical safeguards against under-recognition (Abu-El-Haija et al., 2018).

4.1.6 Physical findings of acute pancreatitis

Vital signs in acute pancreatitis frequently include fever and associated tachycardia. In severe cases, hypotension and reduced SpO_2 may be observed, reflecting systemic inflammatory response and organ dysfunction (Wang et al., 2025).

Abdominal tenderness is typically localized to the upper abdomen, sometimes with mild tenderness (Wang et al., 2025). Since the pancreas is a retroperitoneal organ, classic peritonitic signs, including marked guarding, rigidity, and rebound tenderness, are comparatively uncommon (Wu and Conwell, 2010). Similar to McBurney's point or Rosenstein's sign in acute appendicitis, several findings have been proposed for pancreatitis. Examples include Koyama's point—tenderness elicited to the left of the midpoint between the xiphoid process and umbilicus along the rectus abdominis (a sign described in Japanese medical textbooks, although rarely reported internationally)—and Mallet–Guy sign—tenderness on deep palpation from the epigastrium to the left upper quadrant in the right lateral decubitus position (originally described in chronic pancreatitis); however, the diagnostic utility of these signs remains uncertain.

With symptom progression, additional findings may include diminished bowel sounds, as well as abdominal distension (due to reduced peristalsis), jaundice (due to biliary obstruction), mucosal dryness, decreased skin turgor (due to dehydration/intravascular volume depletion), and cutaneous ecchymoses (Wang et al., 2025; Oppenlander et al., 2022). Hypocalcemia may cause tetany and positive Chvostek or Trousseau signs, which at the bedside may indicate a poor prognosis in acute pancreatitis (Chhabra et al., 2016).

On examination, tenderness localized to the epigastrium predominates, whereas guarding and rebound are limited by the retroperitoneal location of the pancreas (Wu and Conwell, 2010). Gray–Turner and Cullen signs are late and rare ($<3\%$), representing tracking of hemoperitoneum or enzyme-rich exudate into subcutaneous tissues; their presence should prompt consideration of severe disease or alternative sources of retroperitoneal bleeding (Bem and Bradley, 1998; Mookadam and Cikes, 2005; Epperla et al., 2015). Early signs of intravascular volume depletion, such as dry mucous membranes, decreased skin turgor, and orthostasis, are frequent and correlate with a higher risk of organ dysfunction (Boxhoorn et al., 2020; Oppenlander et al., 2022).

4.1.7 Clinical symptoms of chronic pancreatitis

Historically, chronic pancreatitis was defined as an irreversible inflammatory condition of the pancreas leading to endocrine and exocrine dysfunction. More recently, it has been defined as a "pathological fibro-inflammatory syndrome of the pancreas in individuals with genetic, environmental, and/or other risk factors who develop persistent pathological responses to parenchymal injury or stress" (Gardner et al., 2020). Approximately 10% of patients experience a first episode of acute pancreatitis, and approximately 30% of patients with recurrent pancreatitis progress to chronic pancreatitis (Sankaran et al., 2015). Abdominal pain is the most common symptom in chronic pancreatitis; however, a broad array of manifestations arises from exocrine and endocrine insufficiency, and disease progression can culminate in diabetes mellitus. This corresponds to type 3c (pancreatogenic) diabetes, or diabetes of the exocrine pancreas (Hart et al., 2016; Shimizu et al., 2022). Perineural inflammation and neural remodeling around the celiac plexus foster neuropathic features, including hyperalgesia and allodynia, which help explain why pain intensity correlates imperfectly with morphological severity on imaging (Wu et al., 2023; Luo et al., 2021). Centrally, spinal and supraspinal sensitization—modulated by microglial activation and descending inhibitory pathways—may perpetuate pain after biochemical resolution, a mechanism shared with chronic pancreatitis (Andrén-Sandberg et al., 2002; Olesen et al., 2017).

Abdominal pain occurs in approximately 80%–90% of patients with chronic pancreatitis; however, the pain pattern varies from continuous to intermittent, and intensity differs markedly among individuals (Andrén-Sandberg et al., 2002; Wilcox et al., 2015). Improving the quality of life depends on effective pain control. Symptoms of exocrine pancreatic insufficiency include steatorrhea, bloating/abdominal distension, weight loss, fat-soluble vitamin deficiency, and malnutrition (Whitcomb et al., 2023). Endocrine insufficiency manifests as disordered glucose tolerance, which predisposes to hypoglycemia in patients with type 3c diabetes (Hart et al., 2016). Additional features include painless jaundice and pale stools due to biliary stricture/obstruction (Rodrigues and Boike, 2021), as well as vomiting, anorexia, and early satiety due to duodenal stenosis (Tomioka et al., 2019). Abrupt weight loss and dysglycemia should raise suspicion for pancreatic cancer (Le Cosquer et al., 2023).

4.1.8 Mechanism of abdominal pain in acute pancreatitis

The pancreas is often termed a "silent organ," and pancreatic cancer, in particular, is a silent killer given its paucity of early symptoms. Animal models have demonstrated sparse nociceptive innervation within the pancreatic parenchyma (Lindsay et al., 2006). Although the precise mechanisms underlying severe pain in acute pancreatitis remain incompletely elucidated, three interrelated processes—visceral, inflammatory, and neuropathic pains—converge; therefore, the severity of pain does not necessarily correlate with disease severity (Wu et al., 2023; Luo et al., 2021). Visceral pain can be triggered by acinar cell injury, increased intraductal pressure, capsular stretch, edema, and ischemia, with afferent signals transmitted via the celiac plexus, splanchnic nerves, and thoracic spinal cord (T5–T12). Within the pancreas, intrapancreatic trypsin activation and NF-κB signaling are induced, amplifying inflammatory mediators, such as bradykinin and cytokines, including TNF-α, IL-1β, and IL-6. Infiltration of inflammatory cells into the peripancreatic nerve plexus and consequent neural injury contribute to both inflammatory and neuropathic pain. Mechanistically, trypsin activation and NF-κB–driven inflammation sensitize pancreatic afferents, whereas bradykinin, prostanoids, and cytokines (TNF-α, IL-1β, and IL-6) lower nociceptor thresholds (Luo et al., 2021). Perineural inflammation and neural remodeling around the celiac plexus foster neuropathic features, including hyperalgesia and allodynia, which help explain why pain intensity correlates imperfectly with morphological severity on imaging (Wu et al., 2023; Luo et al., 2021). Centrally, spinal and supraspinal sensitization—modulated by microglial activation and descending inhibitory pathways—may perpetuate pain after biochemical resolution, a mechanism shared with chronic pancreatitis (Andrén-Sandberg et al., 2002; Olesen et al., 2017).

Abdominal pain is also the most common symptom of chronic pancreatitis; however, its mechanisms differ in several aspects from those in acute disease. These mechanisms are multifactorial, encompassing peripheral and central sensitization, neural remodeling, psychosocial contributors, and, in part, sustained inflammatory signaling pathways that overlap with those implicated in acute disease (Andrén-Sandberg et al., 2002; Olesen et al., 2017; Atsawarungruangkit, 2015; Lin et al., 2025).

4.1.9 Differential diagnosis of acute pancreatitis

A wide spectrum of disorders may present with upper abdominal pain resembling that of acute pancreatitis; representative examples are listed below (Quinlan, 2014).

A. Hepatobiliary disorders.

Acute hepatitis, perihepatitis, acute cholecystitis, cholangitis, biliary colic, and common bile duct stones. *Note:* Biliary diseases, including gallstones, may cause acute pancreatitis.

B. Gastrointestinal disorders.

Gastrointestinal perforation (especially upper gastrointestinal perforation), peptic ulcer disease, acute gastritis/duodenitis, spontaneous esophageal rupture (Boerhaave syndrome), gastric volvulus, intestinal obstruction, incarcerated/strangulated hernia, infectious enteritis, diverticulitis, and early appendicitis.

C. Cardiovascular disorders.

Acute myocardial infarction (particularly of the inferior wall), pulmonary embolism/infarction, aortic dissection, ruptured aortic aneurysm, celiac or mesenteric artery dissection, mesenteric ischemia, and splenic/renal/adrenal infarction.

D. Genitourinary and gynecologic disorders.

Ureteral stones, acute pyelonephritis/renal abscess, ruptured ectopic pregnancy, ovarian torsion, ovarian hemorrhage, and tubo-ovarian abscess.

E. Others.

Pleuritis/pericarditis, lower-lobe pneumonia, empyema, splenomegaly, splenic rupture, splenic abscess, intraspinal hemorrhage (including spinal epidural, subdural, and intramedullary hemorrhage), iliopsoas hematoma, psoas abscess, paraspinal (erector spinae) muscle hematoma/abscess, rectus sheath hematoma, diabetic ketoacidosis, herpes zoster (including the pre-eruptive phase), porphyria, and heavy metal poisoning.

4.2 Complications

Acute pancreatitis is triggered by acute inflammation of the pancreas and can produce various complications—pancreatic, peripancreatic, and systemic—that vary in severity and duration (Yegneswaran et al., 2011). These complications directly influence prognosis and therapeutic decision-making and are categorized as systemic or local complications.

4.2.1 Systemic complications

Approximately 20% of patients with acute pancreatitis develop organ failure, which is a key determinant of prognosis (Garg and Singh, 2019). In the 2012 Revised Atlanta classification, organ failure was assessed using the Modified Marshall scoring system across three organ systems: respiratory, circulatory, and renal (Banks et al., 2013; Marshall et al., 1995). More recently, the Sequential Organ Failure Assessment (SOFA) score, a widely adopted measure in intensive care settings (ICU), has been increasingly applied to evaluate organ dysfunction in acute pancreatitis. Numerous studies have demonstrated its association with mortality; trajectories of the SOFA score after ICU admission have been correlated with subsequent death (Juneja et al., 2010; Para et al., 2021; Vincent et al., 1996; Tee et al., 2018). Advanced age, comorbidities, obesity (particularly visceral), hypertriglyceridemia, and extensive pancreatic necrosis are recognized risk factors for progression to organ failure in acute pancreatitis (Garg and Singh, 2019; Garg et al., 2005). Although early fluid resuscitation is essential for managing severe disease, excessive fluid administration may worsen systemic complications (de-Madaria et al., 2022).

From a temporal perspective, systemic complications are divided into two phases: an early (primary) phase—occurring within the first several days (1–2 weeks), and a late (secondary) phase, spanning several weeks to months (Banks et al., 2013; Garg and Singh, 2019).

For severity assessment, the Modified Marshall score remains the Revised Atlanta standard for defining organ failure (Banks et al., 2013; Marshall et al., 1995). The SOFA score and its trajectories after ICU admission correlate with outcomes and can complement bedside reassessments when dynamic risk stratification is needed (Juneja et al., 2010; Para et al., 2021; Tee et al., 2018). Risk modifiers include advanced age, comorbid burden, visceral obesity, hypertriglyceridemia, and extensive necrosis (Garg and Singh, 2019; Garg et al., 2005).

During the early phase (first few days to 1–2 weeks), widespread capillary leakage and microcirculatory dysfunction—driven by cytokine release and endothelial injury—cause hypovolemia, interstitial edema, and tissue hypoxia (Boxhoorn et al., 2020; Garg and Singh, 2019). Over-resuscitation can worsen pulmonary edema and abdominal compartment physiology, whereas under-resuscitation perpetuates shock and renal ischemia; a goal-directed approach targeting hemodynamic stability without a positive fluid balance is advisable (de-Madaria et al., 2022; Van Brunschot et al., 2014). During the late phase (weeks to months), secondary infection of necrosis and immune dysregulation dominate the trajectory and are closely associated with mortality risk (Kylänpää et al., 2010; Zhang et al., 2023).

In contrast to the early phase, where the systemic inflammatory response syndrome (SIRS) predominates, sepsis is the principal driver of pathology in the late phase. Progression to sepsis is influenced by an immunosuppressed state, and the late phase is further characterized by compensated anti-inflammatory syndrome (CARS), mixed antagonistic response syndrome (MARS), and persistent inflammation, immunosuppression, and catabolism syndrome (PICS) arising from the dynamic interplay between SIRS, CARS, and MARS (Kylänpää et al., 2010; Zhang et al., 2023). In critical care, endothelial glycocalyx injury has been implicated in the progression of multiorgan failure; in pancreatitis-associated multiorgan failure, the glycocalyx is increasingly recognized as a potential therapeutic target (Suzuki et al., 2023; Li et al., 2025).

4.2.1.1 Pathophysiologic framework

Acute pancreatitis triggers an intense innate immune response driven by intrapancreatic trypsin activation, damage-associated molecular patterns, and downstream NF-κB signaling. Early cytokine surges (such as TNF-α, IL-1β, and IL-6) propagate endothelial activation and capillary leak, causing third-spacing, tissue edema, and microcirculatory impairment (Garg and Singh, 2019; Kylänpää et al., 2010). Endothelial glycocalyx shedding further increases permeability, promotes leukocyte adhesion, and enhances microthrombosis, thereby amplifying organ dysfunction (Suzuki et al., 2023; Li et al., 2025). Concomitantly, CARS may ensue, predisposing patients to secondary infections and a protracted immune-dysregulated state (Kylänpää et al., 2010; Zhang et al., 2023).

4.2.1.2 Gut-organ axis

Splanchnic hypoperfusion, ileus, and inflammatory mediators disrupt mucosal tight junctions, increasing intestinal permeability and bacterial translocation. Translocated organisms and pathogen-associated molecular patterns can seed areas of pancreatic and peripancreatic necrosis, thereby amplifying the SIRS and establishing a link between local injury and systemic organ failure (Boxhoorn et al., 2020; Mederos et al., 2021).

4.2.1.3 Coagulation–microcirculation

Contact pathway activation, platelet–neutrophil interactions, and microthrombus formation contribute to patchy ischemia in the lungs, kidneys, and gut. Clinically, elevated D-dimer levels and thrombocytopenia may accompany severe disease courses; however, routine anticoagulation therapy for this indication alone has not been established (Boxhoorn et al., 2020; Garg and Singh, 2019).

4.2.1.4 Mitochondrial dysfunction and bioenergetics

Systemic inflammation impairs cellular respiration and calcium handling, lowering ATP reserves and rendering organs vulnerable to "supply-dependent" failure despite normalized macrohemodynamics (Garg and Singh, 2019).

4.2.2 Organ failure manifestations

Based on the 2012 Revised Atlanta classification, organ failure is defined as a Modified Marshall score $\geq$2 in any system. Persistent organ failure, lasting $\geq$48 h, constitutes the defining criterion for severe acute pancreatitis (Banks et al., 2013). Practically, this corresponds to: [I] Respiratory system: $PaO_2/FiO_2 \leq 300$, or the need for mechanical ventilation. [II] Cardiovascular system: Systolic BP $<$ 90 mmHg, unresponsive to fluid resuscitation and requiring vasopressor support. [III] Renal system: Serum creatinine $\geq$1.9 mg/dL ($\geq$170 μmol/L). These thresholds are applied when describing system-specific complications. Figure 4.2 shows the systemic complications associated with acute pancreatitis.

4.2.2.1 Pulmonary complications

Pulmonary complications are among the most frequent organ complications in acute pancreatitis (Iyer et al., 2020; Browne and Pitchumoni, 2006). Increased vascular permeability can lead to non-cardiogenic pulmonary edema, acute respiratory

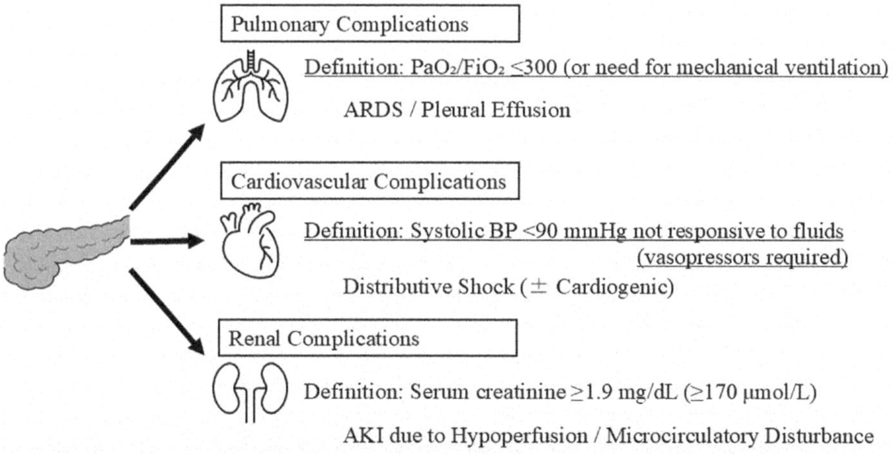

FIGURE 4.2 Systemic complications of acute pancreatitis.

Systemic complications of acute pancreatitis include respiratory failure, manifested as acute respiratory distress syndrome (ARDS) and/or pleural effusion; cardiovascular failure, typically distributive shock (sometimes with a cardiogenic component); and renal failure, presenting as acute kidney injury (AKI) due to hypoperfusion and microcirculatory disturbance. Persistent organ failure in any of these systems defines severe acute pancreatitis based on the 2012 Revised Atlanta Classification. Abbreviations: *AKI*, acute kidney injury; *ARDS*, acute respiratory distress syndrome;.

- Banks, P.A., Bollen, T.L., Dervenis, C., Gooszen, H.G., Johnson, C.D., Sarr, M.G., Tsiotos, G.G., Vege, S.S., Windsor, J.A., Horvath, K.D., Mortele, K.J., Gardner, T.B., Van Santvoort, H., Pelaez-Luna, M., Yadav, D., Stefanidis, G., Delakidis, S., Morgan, D.E., Thoeni, R.F.L., ... Zyromski, N.J., 2013. Classification of acute pancreatitis—2012: Revision of the Atlanta classification and definitions by international consensus. Gut, 62 (1), 102–111. https://doi.org/10.1136/gutjnl-2012-302779.

distress syndrome (ARDS), and pleural effusions. Atelectasis, pneumonia, and pulmonary embolism may also occur due to prolonged bed rest or mechanical ventilation (Zeng et al., 2023).

ARDS in pancreatitis results from non-cardiogenic permeability edema with alveolar damage. Pro-inflammatory mediators and pancreatic enzymes increase endothelial leak, whereas immobility and ventilation increase the risk of atelectasis and pneumonia (Iyer et al., 2020; Browne and Pitchumoni, 2006; Zeng et al., 2023). Pleural effusions—left-sided predominance—correlate with disease severity and may reflect transdiaphragmatic lymphatic flow or fistulization in necrotizing pancreatitis (Boxhoorn et al., 2020).

4.2.2.2 Cardiovascular complications

Shock commonly develops in patients with severe acute pancreatitis, most often due to distributive physiology secondary to increased vascular permeability. Inadequate fluid administration can readily precipitate hypovolemia and shock. Depressed cardiac contractility and arrhythmias have also been reported and may contribute to circulatory failure (Luo et al., 2021; Yegneswaran et al., 2011). Early hemodynamic reassessment is essential for distinguishing fluid-responsive hypovolemia from vasoplegia.

4.2.2.3 Renal complications

Reduced renal perfusion resulting from circulatory failure, along with microcirculatory disturbances driven by inflammatory cytokines and nephrotoxicity exposures, renders patients susceptible to acute kidney injury (AKI) (Nassar and Qunibi, 2019). AKI is associated with increased mortality and greater utilization of healthcare resources in acute pancreatitis (Chen et al., 2025). AKI stems from renal hypoperfusion, microvascular injury, and tubular toxicity; even transient AKI leads to higher mortality and prolonged hospitalization (Nassar and Qunibi, 2019; Chen et al., 2025). Avoidance of nephrotoxins and attention to abdominal compartment physiology are key supportive measures (Van Brunschot et al., 2014).

4.2.2.4 Other systemic complications

Beyond the respiratory, circulatory, and renal systems, acute pancreatitis may also lead to acute liver failure and central nervous system dysfunction (such as encephalopathy), gastrointestinal injury (including bleeding), disseminated intravascular coagulation (DIC), and electrolyte and metabolic derangements (Lankisch et al., 2015; Van Santvoort et al., 2010). Coagulopathy and DIC, hepatic dysfunction, and encephalopathy may occur in severe disease and often reflect the severity of systemic inflammation or evolving sepsis (Lankisch et al., 2015; Van Santvoort et al., 2010). Neurologic dysfunction.

Encephalopathy is uncommon at onset but may develop with hypoxemia, sepsis, or metabolic derangements such as uremia, hyponatremia, or hypocalcemia (Masamune et al., 2020; Patel et al., 2022). Hematologic and coagulation abnormalities.

Leukocytosis and hemoconcentration reflect the SIRS and fluid third-spacing, whereas declining platelet counts and rising D-dimer levels suggest consumptive coagulopathy in severe courses. Prophylactic venous thromboembolism prevention remains indicated unless the risk of bleeding is prohibitive (Boxhoorn et al., 2020; Garg and Singh, 2019).

Severe acute pancreatitis is characterized by a hypermetabolic, catabolic, and insulin-resistant state. Common metabolic derangements include stress hyperglycemia, hypocalcemia due to saponification, hypophosphatemia, hypomagnesemia, hypoalbuminemia related to capillary leakage, and a negative nitrogen balance. Early initiation of enteral nutrition helps preserve gut-barrier function and is associated with lower rates of infectious complications and organ failure than parenteral feeding (Tenner et al., 2024; Boxhoorn et al., 2020; Mederos et al., 2021). Micronutrient depletion may occur with prolonged fasting or biliary obstruction. Following recovery, persistent steatorrhea or weight loss should be promptly evaluated for secondary pancreatic exocrine insufficiency (Gardner et al., 2020; Whitcomb et al., 2023).

4.2.3 Local complications

Numerous local complications arise within the pancreatic parenchyma and peripancreatic tissues, materially affecting disease severity and prognosis. The Revised Atlanta classification distinguishes peripancreatic collections into two major categories: fluid collections, composed of liquid contents, and necrotic collections, which include nonliquid necrotic debris (Banks et al., 2013). Fluid collections associated with interstitial edematous pancreatitis are classified as acute peripancreatic fluid collections (APFC) when they occur within 4 weeks of onset and pancreatic pseudocysts (PPC) thereafter; necrotic collections associated with necrotizing pancreatitis are classified as acute necrotic collections (ANC) within 4 weeks and walled-off necrosis (WON) thereafter. Infected pancreatic necrosis (IPN) denotes bacterial or fungal superinfection of ANC or WON and represents one of the most important predictors of mortality in acute pancreatitis, being consistently associated with an increased risk of death (Garg et al., 2005; Sion and Davis, 2019; Tran et al., 2022).

Thus, the key elements of classification are: (i) the morphological type of acute pancreatitis (interstitial vs. necrotizing), (ii) time from onset (particularly before vs. after 4 weeks), (iii) contents (fluid vs. necrotic material), and (iv) encapsulation. Contrast-enhanced computed tomography (CT), magnetic resonance imaging (MRI), and ultrasonography are useful for evaluation, although early-phase imaging may underestimate the extent of disease (Bollen, 2016).

Imaging distinctions help operationalize the Revised Atlanta scheme: APFCs are homogeneous, fluid-attenuation collections without a definable wall within the first $\leq$4 weeks, whereas ANCs contain non-liquid components (such as fat stranding, hemorrhage, and parenchymal debris). After approximately 4 weeks, PPC and

WON exhibit well-formed enhancing walls on contrast-enhanced CT or MRI (Bollen, 2016). Figure 4.3 depicts the local complications of acute pancreatitis.

4.2.3.1 Natural history and intervention

Most APFCs resolve spontaneously; drainage is reserved for persistent or symptomatic collections causing pain, gastric outlet obstruction, or biliary obstruction (Tenner et al., 2024). PPCs generally require intervention only when infected, enlarging, or causing complications (such as rupture or bleeding) (Habashi and Draganov, 2009; Vitas and Sarr, 1992). WON often represents the infectable substrate in necrotizing disease; when infection or a significant mass effect occurs, step-up strategies—such as drainage followed by minimally invasive necrosectomy as needed—reduce morbidity compared to primary open necrosectomy (Baron et al., 2020; Van Santvoort et al., 2010). Early cross-sectional imaging may underestimate the extent of necrosis; repeating imaging after clinical evolution clarifies the compartmentalization and maturity of collections (Bollen, 2016).

4.2.3.2 APFCs

APFCs—defined in the 2012 Revised Atlanta classification as early ($\leq$4 weeks after onset), non-encapsulated, fluid-only collections associated with interstitial edematous pancreatitis—are observed in approximately one-third of patients with acute pancreatitis, although incidence varies by cohort and imaging timing (Solakoglu et al., 2023; Molla et al., 2022). APFCs are frequently observed in mild to moderately severe pancreatitis and represent the most common local complication (Tenner et al., 2024). Many resolve spontaneously without symptoms; however, a subset persists, becomes encapsulated, and evolves into a pseudocyst (Zheng et al., 2021).

4.2.3.3 PPC

An encapsulated fluid collection that develops $\geq$4 weeks after onset, most often arising when an APFC fails to resolve (Bollen, 2016; Habashi and Draganov, 2009). Although some patients remain asymptomatic, larger lesions may cause gastric outlet or biliary obstruction. Complications such as rupture and hemorrhage have also been reported (Vitas and Sarr, 1992). Asymptomatic lesions are typically managed conservatively, and drainage is considered when symptoms develop or the collection enlarges.

4.2.3.4 ANC

ANCs are single or multiple collections associated with necrotizing pancreatitis, appearing over several days to weeks and involving necrosis of the pancreatic parenchyma and/or peripancreatic fat (Van Santvoort et al., 2010; Thoeni, 2012). Approximately 10%–20% of patients develop necrotizing pancreatitis, and a substantial proportion of these form ANC (Bollen, 2016). With persistent inflammation, patients may develop fever and clinical deterioration; after 4 weeks, ANC evolves into WON.

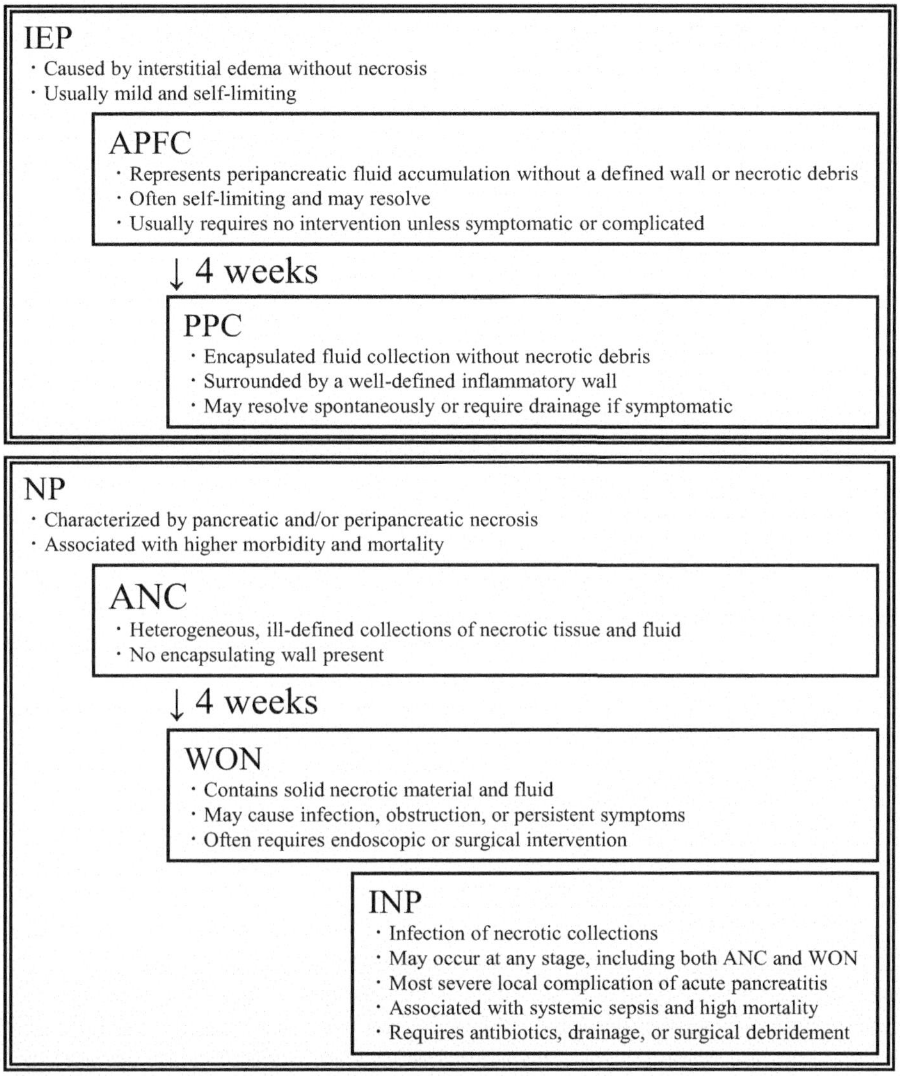

FIGURE 4.3 Local complications of acute pancreatitis.

Local complications of acute pancreatitis include interstitial edematous pancreatitis (IEP), which is characterized by interstitial edema without necrosis, and acute peripancreatic fluid collection (APFC), which represents early peripancreatic fluid accumulation without a defined wall or necrotic debris. When persistent, APFC may evolve into a pancreatic pseudocyst (PPC), an encapsulated fluid collection surrounded by a well-defined inflammatory wall. In contrast, necrotizing pancreatitis (NP) involves pancreatic and/or peripancreatic necrosis and is associated with high morbidity and mortality rates. In its early phase, NP presents as an acute necrotic collection (ANC), which is ill-defined and non-encapsulated, whereas in the late phase, these collections may mature into

4.2.3.5 WON

A well-circumscribed encapsulated necrotic collection that appears ≥4 weeks after disease onset and often causes symptoms due to infection or mass effect (Boxhoorn et al., 2020). The severity of pancreatitis correlates with the likelihood of WON development (Tanaka et al., 2025; Bollen, 2016). Patients may present with fever, leukocytosis, and worsening organ failure; gas may be visible within the collection on imaging. Management centers on antibiotic therapy and drainage, with necrosectomy performed as needed. Whenever feasible, intervention should be delayed (Baron et al., 2020).

4.2.4 Other complications

Vascular complications include pseudoaneurysms (commonly in the splenic artery), hemorrhage into pancreatic or peripancreatic collections, and intra-abdominal bleeding. These events typically occur during the intermediate to late course of the disease due to vascular invasion of necrotic areas or cyst walls, and can cause gastrointestinal or intraperitoneal bleeding. Although infrequent (less than several percent of acute pancreatitis cases), they can be fatal. Early diagnosis using contrast-enhanced CT or angiography and prompt endovascular management (embolization) are crucial (Barge and Lopera, 2012; Maatman et al., 2020; Hoilat et al., 2025). Visceral artery pseudoaneurysms—most commonly involving the splenic, gastroduodenal, and left gastric arteries—arise from enzymatic and inflammatory erosion of peripancreatic vessels. Despite their low overall incidence, rupture carries a high mortality risk; therefore, clinicians should maintain a low threshold for CT angiography when gastrointestinal bleeding, a "sentinel bleed" from a cyst cavity, or a sudden fall in hemoglobin is observed (Barge and Lopera, 2012; Maatman et al., 2020). In necrotizing pancreatitis managed with transmural drainage and endoscopic necrosectomy using lumen-apposing metal stents, early access to the necrotic cavity has been associated with increased incidence of early bleeding and a higher risk of pseudoaneurysm formation. Coordinated surveillance and early consultation with interventional radiology are advisable when debridement is anticipated (Maatman et al., 2020; Abdallah et al., 2022). First-line management is generally endovascular, but rebleeding can occur; therefore, cross-sectional reassessment and readiness for repeat interventions are important (Maatman et al., 2020).

walled-off necrosis (WON), an encapsulated collection containing both necrotic tissue and fluid. Infection of these necrotic collections leads to infected necrotizing pancreatitis (INP), the most severe local complication that often requires antibiotics, drainage, or surgical intervention. Abbreviations: *ANC*, acute necrotic collection; *APFC*, acute peripancreatic fluid collection; *IEP*, interstitial edematous pancreatitis; *INP*, infected necrotizing pancreatitis *NP*, necrotizing pancreatitis; *PPC*, pancreatic pseudocyst; *WON*, walled-off necrosis.

Abdominal compartment syndrome (ACS) is defined as an intra-abdominal pressure (IAP) $\geq$20 mmHg accompanied by new-onset organ failure, most often due to massive third-spacing, ileus-related gaseous distension, and large peripancreatic collections. Intravesical pressure measurement is the standard method for assessment (Kirkpatrick et al., 2013). Management progresses from sedation and nasogastric/rectal decompression to percutaneous drainage of fluid collections, and, in refractory cases, to decompressive laparotomy, acknowledging the high morbidity associated with surgical decompression (Van Brunschot et al., 2014). In acute pancreatitis, intense inflammation leads to exudation and edema within the abdominal cavity, predisposing the patient to an elevated IAP (De Waele and Leppäniemi, 2009). Ascites, perinecrotic fluid collections, gaseous distension from paralytic ileus, and third-space fluid accumulation from aggressive resuscitation are the principal drivers of increasing IAP, which can trigger ACS, thereby worsening overall clinical status (Van Brunschot et al., 2014).

4.2.5 Complications of chronic pancreatitis

Chronic pancreatitis gives rise to a characteristic spectrum of complications that influence long-term outcomes and may underline late presentations after progression from recurrent acute pancreatitis. Exocrine pancreatic insufficiency leads to steatorrhea, weight loss, and fat-soluble vitamin deficiency. The diagnosis is primarily clinical, supported by fecal elastase testing, and pancreatic enzyme replacement therapy improves nutrition and quality of life (Gardner et al., 2020; Whitcomb et al., 2023). Endocrine failure manifests as type 3c (pancreatogenic) diabetes, reflecting combined insulin deficiency and resistance, predisposing affected individuals to brittle glycemia (Hart et al., 2016; Shimizu et al., 2022). Inflammatory fibrosis, calcifications, and mass effects from pseudocysts can cause distal biliary strictures (painless jaundice and cholangitis) and duodenal obstructions (vomiting, early satiety, and weight loss) (Shimizu et al., 2022; Rodrigues and Boike, 2021; Tomioka et al., 2019). Pancreatic duct stones and pseudocysts contribute to pain, obstruction, and superinfection (Gardner et al., 2020; Shimizu et al., 2022). Vascular sequelae include splenic vein thrombosis, leading to left-sided portal hypertension and an increased risk of gastric variceal bleeding (Gardner et al., 2020). Chronic pancreatitis is also associated with an increased risk of pancreatic ductal adenocarcinoma. Reported cumulative risks include approximately 4% after 15–20 years in alcohol-related disease, with higher risks in hereditary pancreatitis (for example, PRSS1 (approximately 19%) and SPINK1 (approximately 12%) mutations by age 60) (Le Cosquer et al., 2023). Notably, approximately 10% of patients after a first episode of acute pancreatitis and approximately 30% with recurrent attacks progress to chronic pancreatitis, underscoring why these complications often manifest late in the disease course (Sankaran et al., 2015; Nair et al., 2007).

4.2.6 Pancreatitis and pancreatic cancer

Several studies have reported an association between pancreatitis and the subsequent development of pancreatic cancer. Chronic pancreatitis has been consistently linked to a markedly increased risk of pancreatic ductal adenocarcinoma, likely due to persistent inflammation, cycles of injury and regeneration, fibrosis, and a pro-tumorigenic microenvironment (Le Cosquer et al., 2023; Lowenfels et al., 1993; Gandhi et al., 2022). Meta-analyses have confirmed the association between chronic pancreatitis and pancreatic cancer, with standardized incidence ratios often exceeding 20, even after excluding cases of pancreatic cancer diagnosed shortly after the onset of chronic pancreatitis to minimize reverse causation (Gandhi et al., 2022).

Acute pancreatitis has also been implicated in an increased risk of pancreatic cancer, although the relationship is less clearly defined. A nationwide cohort study from Korea demonstrated that patients with acute pancreatitis, especially those with recurrent episodes or progression to chronic disease, had a significantly elevated hazard of pancreatic cancer compared to controls (Park et al., 2023). A pooled analysis further suggested that a history of pancreatitis was associated with approximately a six-fold higher odds of developing pancreatic cancer (Duell et al., 2012). The risk appears greatest within the first few years after the diagnosis of pancreatitis and gradually decreases thereafter, although it remains higher than that of the general population (Park et al., 2023; Kirkegård et al., 2018).

Notably, the absolute risk of pancreatic cancer remains low, and many individuals with pancreatitis do not develop malignancies. Shared risk factors, such as tobacco use, heavy alcohol consumption, and genetic predisposition, contribute to both conditions and complicate causal inference. In addition, early pancreatic cancer can sometimes present as acute pancreatitis, underscoring the importance of appropriate lag-time analysis to avoid misclassification (Kirkegård et al., 2018). Nevertheless, although formal guidelines have not yet been established, patients with recurrent or long-standing pancreatitis may benefit from increased clinical vigilance and potentially targeted surveillance strategies.

4.3 Summary of symptoms and complications of pancreatitis

In this chapter, the symptoms and complications of acute pancreatitis have been described. Recognizing the presenting symptoms is clinically important, and monitoring their trajectory provides key insight into disease progression and treatment response. The typical initial manifestation is acute-onset, severe, and persistent abdominal pain, which remains a cornerstone for diagnosis; however, its expression varies with age, sex, and comorbidities. Nausea, vomiting, fever, ileus, and jaundice may appear as initial findings or emerge during the clinical course. Atypical or mild presentations require close observation and confirmation through biochemical testing and imaging. As the disease progresses, systemic complications such as respiratory failure, circulatory failure, and AKI are decisive determinants of severity and prognosis, reflecting the central role of persistent organ failure in defining severe acute pancreatitis. In contrast, local complications exhibit a time-dependent progression: acute peripancreatic fluid and acute necrotic collections in the early phase may evolve into pseudocysts or WON after 4 weeks, sometimes complicated by infection. These features underscore the heterogeneity of clinical presentations and the dynamic nature of disease progression. For clinicians, integrating an understanding of symptom trajectories with awareness of both systemic and local complications is indispensable for timely severity assessment and individualized management, ultimately contributing to improved long-term outcomes.

AI disclosure

During the preparation of this work the author(s) used ChatGPT (OpenAI) in order to The tool was used to assist with English language editing and proofreading. Following assistance from ChatGPT with the English translation, the manuscript was subsequently subjected to professional English proofreading by a native speaker at Editage. After using this tool/service, the author(s) reviewed and edited the content as needed and take(s) full responsibility for the content of the publication.

References

Abdallah, M., Vantanasiri, K., Young, S., Azeem, N., Amateau, S.K., Mallery, S., Freeman, M. L., Trikudanathan, G., 2022. Visceral artery pseudoaneurysms in necrotizing pancreatitis: risk of early bleeding with lumen-apposing metal stents. Gastrointestinal Endoscopy 95 (6), 1150–1157. https://doi.org/10.1016/j.gie.2021.11.030.

Abu-El-Haija, M., Kumar, S., Quiros, J.A., Balakrishnan, K., Barth, B., Bitton, S., Eisses, J.F., Foglio, E.J., Fox, V., Francis, D., Freeman, A.J., Gonska, T., Grover, A.S., Husain, S.Z., Kumar, R., Lapsia, S., Lin, T., Liu, Q.Y., Maqbool, A., Sellers, Z.M., Szabo, F., Uc, A., Werlin, S.L., Morinville, V.D., 2018. Management of acute pancreatitis in the pediatric population: a clinical report from the north American society for pediatric gastroenterology, hepatology and nutrition pancreas committee. Journal of Pediatric Gastroenterology and Nutrition 66 (1), 159–176. https://doi.org/10.1097/MPG.0000000000001715.

Andrén-Sandberg, A., Hoem, D., Gislason, H., 2002. Pain management in chronic pancreatitis. European Journal of Gastroenterology and Hepatology 14 (9), 957–970. https://doi.org/10.1097/00042737-200209000-00006.

Atsawarungruangkit, A., 2015. Current understanding of the neuropathophysiology of pain in chronic pancreatitis. World Journal of Gastrointestinal Pathophysiology 6 (4), 193. https://doi.org/10.4291/wjgp.v6.i4.193.

Aziz, A.A., 2025. Gender-related differences in hospitalization outcomes of acute pancreatitis: a retrospective analysis. Cureus 17 (6), e86371.

Banks, P.A., Bollen, T.L., Dervenis, C., Gooszen, H.G., Johnson, C.D., Sarr, M.G., Tsiotos, G. G., Vege, S.S., Windsor, J.A., Horvath, K.D., Mortele, K.J., Gardner, T.B.,

Van Santvoort, H., Pelaez-Luna, M., Yadav, D., Stefanidis, G., Delakidis, S., Morgan, D. E., Thoeni, R.F.L., Acosta, J.M., Amann, S.T., Andren-Sandberg, A., Aranha, G.V., Asciutti, S., Barauskas, G., Baron, T.H., Bassi, C., Behrman, S., Behrns, K.E., Belliappa, V., Berzin, T.M., Besselink, M.G.H., Bhasin, D.K., Biankin, A., Bishop, M. D., Bonini, C.J., Bradley, E.L., Buechler, M., Carter, R., Cavestro, G.M., Chari, S.T., Chavez-Rodriguez, J.J., Da Cunha, J.E., D'Agostino, H., De Campos, T., De-Madaria, E., Deprez, P.H., DiSario, J.A., Doria, C., Falconi, M., Fernandez-del Castillo, C., Freeny, P.C., Frey, C.F., Friess, H., Frossard, J.L., Fuchshuber, P., Gallagher, S.F., Garg, P.K., Ghattas, G., Glasgow, R., Gonzalez, J.A., Gress, T.M., Gumbs, A.A., Hallibruton, C., Helton, S., Hill, M.C., Hoyos, S., Imrie, C.W., Isenmann, R., Izbicki, J.R., Karagiannis, J.A., Klar, E., Kolokythas, O., Lau, J., Litvin, A.A., Longnecker, D.S., Lowenfels, A.B., Mackey, R., Mah'Moud, M., Malangoni, M., McFadden, D.W., Mishra, G., Moody, F.G., Morinville, V., Neoptolemos, J.P., Nordback, I., Pap, A., Papachristou, G.I., Parks, R., Pedrazolli, S., Pezzilli, R., Pitt, H.A., Prosanto, C., Ramesh, H., Ramirez, F.C., Raper, S.E., Rasheed, A., Reed, D.N., Romangnuolo, J., Rossaak, J., Sanabria, J., Schaefer, C., Schmidt, J., Schmidt, P.N., Serrablo, A., Senkowski, C.K., Sharma, M., Sigman, K.M., Singh, P., Steinberg, W., Steiner, J., Strasberg, S., Strum, W., Takada, T., Tanaka, M., Vaccaro, M., Villavicencio, R.L., Vrochides, D., Wagner, M., Warshaw, A.L., Wilcox, C.M., Wysocki, P., Zenilman, M.E., Zyromski, N.J., 2013. Classification of acute pancreatitis—2012: Revision of the Atlanta classification and definitions by international consensus. Gut 62 (1), 102–111. https://doi.org/10.1136/gutjnl-2012-302779.

Barge, J.U., Lopera, J.E., 2012. Vascular complications of pancreatitis: role of interventional therapy. Korean Journal of Radiology 13 (Suppl. 1), S45. https://doi.org/10.3348/kjr.2012.13.s1.s45.

Baron, T.H., DiMaio, C.J., Wang, A.Y., Morgan, K.A., 2020. American gastroenterological association clinical practice update: management of pancreatic necrosis. Gastroenterology 158 (1), 67. https://doi.org/10.1053/j.gastro.2019.07.064.

Bem, J., Bradley, E.L., 1998. Subcutaneous manifestations of severe acute pancreatitis. Pancreas 16 (4), 551–555. https://doi.org/10.1097/00006676-199805000-00015.

Bollen, T.L., 2016. Acute pancreatitis: international classification and nomenclature. Clinical Radiology 71 (2), 121–133. https://doi.org/10.1016/j.crad.2015.09.013.

Boxhoorn, L., Voermans, R.P., Bouwense, S.A., Bruno, M.J., Verdonk, R.C., Boermeester, M. A., van Santvoort, H.C., Besselink, M.G., 2020. Acute pancreatitis. The Lancet 396 (10252), 726–734. https://doi.org/10.1016/s0140-6736(20)31310-6.

Browne, G.W., Pitchumoni, C.S., 2006. Pathophysiology of pulmonary complications of acute pancreatitis. World Journal of Gastroenterology 12 (44), 7087–7096. https://doi.org/10.3748/wjg.v12.i44.7087.

Cervellin, G., Mora, R., Ticinesi, A., Meschi, T., Comelli, I., Catena, F., Lippi, G., 2016. Epidemiology and outcomes of acute abdominal pain in a large urban Emergency Department: retrospective analysis of 5,340 cases. Annals of Translational Medicine 4 (19). https://doi.org/10.21037/atm.2016.09.10, 362–362.

Chen, F., Xu, K., Han, Y., Ding, J., Ren, J., Cao, F., Wang, Y., Qian, W., Wang, Z., Wu, Z., Ma, Z., 2025. Risk factors for acute kidney injury in patients with acute pancreatitis and construction of nomogram model: a single-center study and external validation. Frontiers of Medicine 12. https://doi.org/10.3389/fmed.2025.1626664.

Chhabra, P., Rana, S.S., Sharma, V., Sharma, R., Bhasin, D.K., 2016. Hypocalcemic tetany: a simple bedside marker of poor outcome in acute pancreatitis. Annals of Gastroenterology 29 (2), 214–220. https://doi.org/10.20524/aog.2016.0015.

De Waele, J.J., Leppäniemi, A.K., 2009. Intra-abdominal hypertension in acute pancreatitis. World Journal of Surgery 33 (6), 1128–1133. https://doi.org/10.1007/s00268-009-9994-5.

de-Madaria, E., Buxbaum, J.L., Maisonneuve, P., de Paredes, A.G.G., Zapater, P., Guilabert, L., Vaillo-Rocamora, A., Rodríguez-Gandía, M.Á., Donate-Ortega, J., Lozada-Hernández, E.E., Moreno, A.J.R.C., Lira-Aguilar, A., Llovet, L.P., Mehta, R., Tandel, R., Navarro, P., Sánchez-Pardo, A.M., Sánchez-Marin, C., Cobreros, M., Fernández-Cabrera, I., Casals-Seoane, F., Deza, D.C., Lauret-Braña, E., Martí-Marqués, E., Camacho-Montaño, L.M., Ubieto, V., Ganuza, M., Bolado, F., 2022. Aggressive or moderate fluid resuscitation in acute pancreatitis. New England Journal of Medicine 387 (11), 989–1000. https://doi.org/10.1056/nejmoa2202884.

Drake, M., Dodwad, Shah-J.M., Davis, J., Kao, L.S., Cao, Y., Ko, T.C., 2021. Sex-related differences of acute and chronic pancreatitis in adults. Journal of Clinical Medicine 10 (2), 300. https://doi.org/10.3390/jcm10020300.

Duell, E.J., Lucenteforte, E., Olson, S.H., Bracci, P.M., Li, D., Risch, H.A., Silverman, D.T., Ji, B.T., Gallinger, S., Holly, E.A., Fontham, E.H., Maisonneuve, P., Bueno-de-Mesquita, H.B., Ghadirian, P., Kurtz, R.C., Ludwig, E., Yu, H., Lowenfels, A.B., Seminara, D., Petersen, G.M., La Vecchia, C., Boffetta, P., 2012. Pancreatitis and pancreatic cancer risk: a pooled analysis in the international pancreatic cancer case-control consortium (PanC4). Annals of Oncology 23 (11), 2964–2970. https://doi.org/10.1093/annonc/mds140.

Epperla, N., Mazza, J.J., Yale, S.H., 2015. A review of clinical signs related to ecchymosis. Wisconsin Medical Journal 114 (2), 61–65.

Földi, M., Gede, N., Kiss, S., Vincze, Á., Bajor, J., Szabó, I., Szepes, Z., Izbéki, F., Gervain, J., Hamvas, J., Vitális, Z., Fehér, E., Crai, S., Sallinen, V., Maldonado, E.R., Meczker, Á., Varjú, P., Poropat, G., Stimac, D., Faluhelyi, N., Miseta, A., Nagy, T., Márton, Z., Vereczkei, A., Hegyi, P.J., Párniczky, A., Hegyi, P., Szentesi, A., 2022. The characteristics and prognostic role of acute abdominal on-admission pain in acute pancreatitis: a prospective cohort analysis of 1432 cases. European Journal of Pain 26 (3), 610–623. https://doi.org/10.1002/ejp.1885.

Gandhi, S., de la Fuente, J., Hassan Murad, M., Majumder, S., 2022. Chronic pancreatitis is a risk factor for pancreatic cancer, and incidence increases with duration of disease: a systematic review and meta-analysis. Clinical and Translational Gastroenterology 13 (3), e00463. https://doi.org/10.14309/ctg.0000000000000463.

Gardner, T.B., Adler, D.G., Forsmark, C.E., Sauer, B.G., Taylor, J.R., Whitcomb, D.C., 2020. ACG clinical guideline: chronic pancreatitis. American Journal of Gastroenterology 115 (3), 322–339. https://doi.org/10.14309/ajg.0000000000000535.

Garg, P.K., Singh, V.P., 2019. Organ failure due to systemic injury in acute pancreatitis. Gastroenterology 156 (7), 2008–2023. https://doi.org/10.1053/j.gastro.2018.12.041.

Garg, P.K., Madan, K., Pande, G.K., Khanna, S., Sathyanarayan, G., Bohidar, N.P., Tandon, R. K., 2005. Association of extent and infection of pancreatic necrosis with organ failure and death in acute necrotizing pancreatitis. Clinical Gastroenterology and Hepatology 3 (2), 159–166. https://doi.org/10.1016/S1542-3565(04)00665-2.

Habashi, S., Draganov, P.V., 2009. Pancreatic pseudocyst. World Journal of Gastroenterology 15 (1), 38–47. https://doi.org/10.3748/wjg.15.38.

Hart, P.A., Bellin, M.D., Andersen, D.K., Bradley, D., Cruz-Monserrate, Z., Forsmark, C.E., Goodarzi, M.O., Habtezion, A., Korc, M., Kudva, Y.C., Pandol, S.J., Yadav, D., Chari, S. T., 2016. Type 3c (pancreatogenic) diabetes mellitus secondary to chronic pancreatitis and pancreatic cancer. The Lancet Gastroenterology & Hepatology 1 (3), 226–237. https://doi.org/10.1016/S2468-1253(16)30106-6.

Hoilat, G.J., Mathew, G., Ahmad, H., 2025. Pancreatic Pseudoaneurysm. StatPearls. https://www.ncbi.nlm.nih.gov/books/.

Iyer, H., Elhence, A., Mittal, S., Madan, K., Garg, P.K., 2020. Pulmonary complications of acute pancreatitis. Expert Review of Respiratory Medicine 14 (2), 209–217. https://doi.org/10.1080/17476348.2020.1698951.

Juneja, D., Gopal, P.B., Ravula, M., 2010. Scoring systems in acute pancreatitis: which one to use in intensive care units? Journal of Critical Care 25 (2), 358.e9. https://doi.org/10.1016/j.jcrc.2009.12.010.

Kirkegård, J., Cronin-Fenton, D., Heide-Jørgensen, U., Mortensen, F.V., 2018. Acute pancreatitis and pancreatic cancer risk: a nationwide matched-cohort study in Denmark. Gastroenterology 154 (6), 1729–1736. https://doi.org/10.1053/j.gastro.2018.02.011.

Kirkpatrick, A.W., Roberts, D.J., De Waele, J., Jaeschke, R., Malbrain, M.L.N.G., De Keulenaer, B., Duchesne, J., Bjorck, M., Leppaniemi, A., Ejike, J.C., Sugrue, M., Cheatham, M., Ivatury, R., Ball, C.G., Reintam Blaser, A., Regli, A., Balogh, Z.J., D'Amours, S., Debergh, D., Kaplan, M., Kimball, E., Olvera, C., 2013. Canada Intra-abdominal hypertension and the abdominal compartment syndrome: updated consensus definitions and clinical practice guidelines from the World Society of the Abdominal Compartment Syndrome. Intensive Care Medicine 7, 1190–1206. https://doi.org/10.1007/s00134-013-2906-z, 39.

Kuźnik, E., Dudkowiak, R., Adamiec, R., Poniewierka, E., 2020. Diabetic autonomic neuropathy of the gastrointestinal tract. Gastroenterology Review 15 (2), 89–93. https://doi.org/10.5114/pg.2020.95554.

Kylänpää, M.L., Repo, H., Puolakkainen, P.A., 2010. Inflammation and immunosuppression in severe acute pancreatitis. World Journal of Gastroenterology 16 (23), 2867–2872. https://doi.org/10.3748/wjg.v16.i23.2867.

Lankisch, P.G., Apte, M., Banks, P.A., 2015. Acute pancreatitis. The Lancet 9988, 85–96. https://doi.org/10.1016/S0140-6736(14)60649-8.

Le Cosquer, G., Maulat, C., Bournet, B., Cordelier, P., Buscail, E., Buscail, L., 2023. Pancreatic cancer in chronic pancreatitis: pathogenesis and diagnostic approach. Cancers 15 (3), 761. https://doi.org/10.3390/cancers15030761.

Li, H., Wen, H., Liu, J., Luo, X., Pei, B., Peng, G., Sun, Z., Liu, J., Wang, J., Chen, H., 2025. The glycocalyx: a key target for treatment of severe acute pancreatitis-associated multiple organ dysfunction syndrome. Human Cell 38 (4). https://doi.org/10.1007/s13577-025-01227-6.

Lin, Z., Pandol, S., Apte, M., Jiang, Y., 2025. Navigating chronic pancreatitis pain: a pathophysiological and therapeutic overview. Frontiers in Physiology 16. https://doi.org/10.3389/fphys.2025.1622845.

Lindsay, T.H., Halvorson, K.G., Peters, C.M., Ghilardi, J.R., Kuskowski, M.A., Wong, G.Y., Mantyh, P.W., 2006. A quantitative analysis of the sensory and sympathetic innervation of the mouse pancreas. Neuroscience 137 (4), 1417–1426. https://doi.org/10.1016/j.neuroscience.2005.10.055.

Lowenfels, A.B., Maisonneuve, P., Cavallini, G., Ammann, R.W., Lankisch, P.G., Andersen, J. R., Dimagno, E.P., Andren-Sandberg, A., Domellof, L., 1993. Pancreatitis and the risk of pancreatic cancer. New England Journal of Medicine 328 (20), 1433–1437. https://doi.org/10.1056/NEJM199305203282001.

Luo, Y., Li, Z., Ge, P., Guo, H., Li, L., Zhang, G., Xu, C., Chen, H., 2021. Comprehensive mechanism, novel markers and multidisciplinary treatment of severe acute pancreatitis-associated cardiac injury—a narrative review. Journal of Inflammation Research 14, 3145–3169. https://doi.org/10.2147/JIR.S310990.

Maatman, T.K., Heimberger, M.A., Lewellen, K.A., Roch, A.M., Colgate, C.L., House, M.G., Nakeeb, A., Ceppa, E.P., Max Schmidt, C., Zyromski, N.J., 2020. Visceral artery pseudoaneurysm in necrotizing pancreatitis: incidence and outcomes. Canadian Journal of Surgery 63 (3), E272. https://doi.org/10.1503/cjs.009519.

Malfertheiner, P., Kemmer, T.P., 1991. Clinical picture and diagnosis of acute pancreatitis. Hepato-Gastroenterology 38 (2), 97–100.

Marshall, J.C., Cook, D.J., Christou, N.V., Bernard, G.R., Sprung, C.L., Sibbald, W.J., 1995. Multiple organ dysfunction score: a reliable descriptor of a complex clinical outcome. Critical Care Medicine 23 (10), 1638–1652. https://doi.org/10.1097/00003246-199510000-00007.

Masamune, A., Kikuta, K., Hamada, S., Tsuji, I., Takeyama, Y., Shimosegawa, T., Okazaki, K., Matsumoto, R., Kume, K., Miura, S., Takikawa, T., Hongo, S., Tanaka, Y., Sakai, T., Ueno, M., Unno, J., Toki, M., Nishimura, M., Kitamura, K., Kobashigawa, K., Ashida, K., Takahashi, K., Watanabe, T., Sato, Y., Satoh, A., Yoshida, N., Tanaka, K., Kondo, H., Suzuki, N., Sato, H., Mitoro, A., Minami, K., Iwasaki, E., Okuwaki, K., Mukai, T., Kudo, Y., Azuma, S., Takeda, Y., Setsuda, M., Tsuchida, K., Irisawa, A., Motoya, D., Shirahata, N., Iwata, K., Oza, N., Kawaguchi, Y., Hoshino, A., Hirano, A., Sakakihara, I., Watanabe, N., Funayama, H., Nakamura, Y., Yamamoto, T., Uchida, H.H., Iizawa, Y., Imamura, M., Sekine, M., Inui, K., Yamamoto, S., Sato, S., Ishii, S., Yakushijin, T., Yamamoto, K., Masaka, T., Yoshida, M., Arizumi, T., Goto, M., Takenaka, M., Miyakawa, H., Otsuka, Y., Nishi, M., Suwa, K., Sakagami, J., Satoh, M., Koga, H., Takagi, T., Nagahama, M., Hatayama, K., Kaneko, N., Mori, M., Nakai, Y., Iwashita, Y., Iwatsuka, K., Hayashi, K., Terai, S., Kaneto, H., Nishikawa, T., Uchi, M., Honjo, M., Takagi, T., Mizukami, K., Tamura, T., Gushima, T., Sato, A., Shimizu, K., Aruga, Y., Nonogaki, K., Kaino, M., Iwashita, T., Tsuruoka, M., Miura, M., Hasegawa, H., Takai, T., Yokota, H., Fujisawa, T., Uetsuki, K., Kobayashi, T., yamaji, N., Itoi, T., Muraki, T., Watanabe, T., Inatomi, O., Ushijima, T., Takehara, Y., Kawamura, H., Fukasawa, M., Hagiwara, H., Tamura, H., Moriya, T., Kuwatani, M., Kawaji, Y., Kamisawa, T., Abue, M., Fukunishi, M., Onoyama, H., Mochizuki, S., Sumimoto, K., Koyabu, M., Nishio, A., Sakao, M., Imamura, Y., Ohta, H., Nishimura, N., Shiihara, M., Anan, H., Motoya, M., Chiba, M., Deguchi, T., Kofunato, Y., Sawada, N., Katada, K., Matsumoto, S., Higashide, S., Okano, H., Ishida, M., Yamazaki, M., Tamura, N., Muraoka, T., Uza, N., Kodama, Y., Muro, S., Fujimori, N., Ikeda, T., Shimokawahara, N., Otsu, N., Yano, Y., Sugawara, S., Takaura, K., Ogura, Y., Furuuchi, T., Shibasaki, M., Kishimoto, M., Ishida, M., 2020. Clinical practice of acute pancreatitis in Japan: an analysis of nationwide epidemiological survey in 2016. Pancreatology 20 (4), 629–636. https://doi.org/10.1016/j.pan.2020.04.013.

Mederos, M.A., Reber, H.A., Girgis, M.D., 2021. Acute pancreatitis: a review. JAMA, the Journal of the American Medical Association 325 (4), 382–390. https://doi.org/10.1001/jama.2020.20317.

Molla, N.W., Alsergani, A.H., Alyamani, A.A., Aljohani, M.A., Aljohani, A.A., Alfaiz, F.A., Alomar, M.O., BinMayouf, M.S., 2022. Incidence of peripancreatic fluid collections in patients presenting with acute pancreatitis. Saudi Medical Journal 43 (12), 1341–1346. https://doi.org/10.15537/smj.2022.43.12.20220508. Available from: https://smj.org.sa/lookup/doi/10.15537/smj.2022.43.12.20220508.

Mookadam, Cikes, 2005. Images in clinical medicine. Cullen's and Turner's signs. New England Journal of Medicine 353 (13).

Morinville, V.D., Husain, S.Z., Bai, H., Barth, B., Alhosh, R., Durie, P.R., Freedman, S.D., Himes, R., Lowe, M.E., Pohl, J., Werlin, S., Wilschanski, M., Uc, A., 2012. Definitions of pediatric pancreatitis and survey of present clinical practices. Journal of Pediatric Gastroenterology and Nutrition 55 (3), 261–265. https://doi.org/10.1097/MPG.0b013e31824f1516.

Murthy, R.N., Viswanathan, S., Jain, D., 2020. Multiple ecchymotic signs in acute pancreatitis. BMJ Case Reports 13 (4), e235147. https://doi.org/10.1136/bcr-2020-235147.

Nair, R.J., Lawler, L., Miller, M.R., 2007. Chronic pancreatitis. American Family Physician 76 (11), 1679–1694.

Nassar, T.I., Qunibi, W.Y., 2019. AKI associated with acute pancreatitis. Clinical Journal of the American Society of Nephrology 14 (7), 1106–1115. https://doi.org/10.2215/CJN.13191118.

Olesen, S.S., Krauss, T., Demir, I.E., Wilder-Smith, O.H., Ceyhan, G.O., Pasricha, P.J., Drewes, A.M., 2017. Towards a neurobiological understanding of pain in chronic pancreatitis: mechanisms and implications for treatment. PAIN Reports 2 (6), e625. https://doi.org/10.1097/pr9.0000000000000625.

Oppenlander, K.E., Chadwick, C., Carman, K., 2022. Acute pancreatitis: rapid evidence review. American Family Physician 106 (1), 44–50.

Pahuja, V., Rani, N., Singla, G., Dugg, P., Singh, S., 2025. A prospective cohort study of clinical presentation and outcomes of acute pancreatitis at tertiary care center. International Surgery Journal 12 (3), 339–343. https://doi.org/10.18203/2349-2902.isj20250565.

Para, O., Caruso, L., Savo, M.T., Antonielli, E., Blasi, E., Capello, F., Ciarambino, T., Corbo, L., Curto, A., Giampieri, M., Maddaluni, L., Zaccagnini, G., Nozzoli, C., 2021. The challenge of prognostic markers in acute pancreatitis: internist's point of view. Journal of Genetic Engineering and Biotechnology 19 (1), 77. https://doi.org/10.1186/s43141-021-00178-3.

Park, S.M., Bae Kim, K., Han, J.-H., Kim, N., Kang, T.U., Swan, H., Jung Kim, H., 2023. Incidence and risk of pancreatic cancer in patients with acute or chronic pancreatitis: a population-based cohort study. Scientific Reports 13 (1). https://doi.org/10.1038/s41598-023-45382-y.

Párniczky, A., 2016. Prospective, multicenter, nationwide clinical data from 600 cases of acute pancreatitis. PLoS One 11 (10), e0165309.

Patel, M.L., Shyam, R., Atam, V., Bharti, H., Sachan, R., Parihar, A., 2022. Clinical profile, etiology, and outcome of acute pancreatitis: experience at a tertiary care center. Annals of African Medicine 21 (2), 118–123. https://doi.org/10.4103/aam.aam_83_20.

Phillip, V., Schuster, T., Hagemes, F., Lorenz, S., Matheis, U., Preinfalk, S., Lippl, F., Saugel, B., Schmid, R.M., Huber, W., 2013. Time Period from onset of pain to hospital admission and patients' awareness in acute pancreatitis. Pancreas 42 (4), 647–654. https://doi.org/10.1097/mpa.0b013e3182714565.

Pop-Busui, R., Boulton, A.J.M., Feldman, E.L., Bril, V., Roy, F., Malik, R.A., Sosenko, J.M., Ziegler, D., 2017. Diabetic neuropathy: a position statement by the American diabetes association. Diabetes Care 40 (1), 136–154. https://doi.org/10.2337/dc16-2042.

Porter, K.K., Zaheer, A., Kamel, I.R., Horowitz, J.M., Arif-Tiwari, H., Bartel, T.B., Bashir, M. R., Camacho, M.A., Cash, B.D., Chernyak, V., Goldstein, A., Grajo, J.R., Gupta, S., Hindman, N.M., Kamaya, A., McNamara, M.M., Carucci, L.R., 2019. ACR appropriateness Criteria® acute pancreatitis. Journal of the American College of Radiology 16 (11), S316. https://doi.org/10.1016/j.jacr.2019.05.017.

Quinlan, J.D., 2014. Acute pancreatitis. American Family Physician 90 (9), 632–639.

Reid, G.P., Williams, E.W., Francis, D.K., Lee, M.G., 2017. Acute pancreatitis: a 7 year retrospective cohort study of the epidemiology, aetiology and outcome from a tertiary hospital in Jamaica. Annals of Medicine and Surgery 20, 103–108. https://doi.org/10.1016/j.amsu.2017.07.014.

Rodrigues, T., Boike, J.R., 2021. Biliary strictures: etiologies and medical management. Seminars in Interventional Radiology 38 (3), 255–262. https://doi.org/10.1055/s-0041-1731086.

Rossetti, B., Spizzirri, A., Migliaccio, C., Mura, F.La, Cattorini, L., Trastulli, S., Cirocchi, R., Giustozzi, G., Sciannameo, F., 2009. Acute pancreatitis in the elderly: our experience. BMC Geriatrics 9 (S1). https://doi.org/10.1186/1471-2318-9-s1-a47.

Sankaran, S.J., Xiao, A.Y., Wu, L.M., Windsor, J.A., Forsmark, C.E., Petrov, M.S., 2015. Frequency of progression from acute to chronic pancreatitis and risk factors: a meta-analysis. Gastroenterology 149 (6), 1490. https://doi.org/10.1053/j.gastro.2015.07.066.

Shen, H.-N., Wang, W.-C., Lu, C.-L., Li, C.-Y., Einwaechter, H., 2013. Effects of gender on severity, management and outcome in acute biliary pancreatitis. PLoS One 8 (2), e57504. https://doi.org/10.1371/journal.pone.0057504.

Shimizu, K., Ito, T., Irisawa, A., Ohtsuka, T., Ohara, H., Kanno, A., Kida, M., Sakagami, J., Sata, N., Takeyama, Y., Tahara, J., Hirota, M., Fujimori, N., Masamune, A., Mochida, S., Enomoto, N., Shimosegawa, T., Koike, K., 2022. Evidence-based clinical practice guidelines for chronic pancreatitis 2021. Journal of Gastroenterology 57 (10), 709–724. https://doi.org/10.1007/s00535-022-01911-6.

Sion, M.K., Davis, K.A., 2019. Step-up approach for the management of pancreatic necrosis: a review of the literature. Trauma Surgery and Acute Care Open 4 (1), e000308. https://doi.org/10.1136/tsaco-2019-000308.

Solakoglu, T., Turkel Kucukmetin, N., Akar, M., Koseoglu, H., 2023. Acute peripancreatic fluid collection in acute pancreatitis: incidence, outcome, and association with inflammatory markers. Saudi Journal of Gastroenterology 29 (4), 225–232. https://doi.org/10.4103/sjg.sjg_443_22.

Suzuki, M., 2014. Acute pancreatitis in children and adolescents. World Journal of Gastrointestinal Pathophysiology 5 (4), 416. https://doi.org/10.4291/wjgp.v5.i4.416.

Suzuki, K., Miura, T., Okada, H., 2023. The endothelial glycocalyx-All the same? No, it is not. Acute Med Surg 10 (1), e896.

Szatmary, P., Grammatikopoulos, T., Cai, W., Huang, W., Mukherjee, R., Halloran, C., Beyer, G., Sutton, R., 2022. Acute pancreatitis: diagnosis and treatment. Drugs 82 (12), 1251–1276. https://doi.org/10.1007/s40265-022-01766-4.

Takada, T., Isaji, S., Mayumi, T., Yoshida, M., Takeyama, Y., Itoi, T., Sano, K., Iizawa, Y., Masamune, A., Hirota, M., Okamoto, K., Inoue, D., Kitamura, N., Mori, Y., Mukai, S., Kiriyama, S., Shirai, K., Tsuchiya, A., Higuchi, R., Hirashita, T., 2022. JPN clinical practice guidelines 2021 with easy-to-understand explanations for the management of acute pancreatitis. Journal of Hepato-Biliary-Pancreatic Sciences 29 (10), 1057–1083. https://doi.org/10.1002/jhbp.1146.

Tanaka, Y., Masamune, A., Matsumoto, R., Takikawa, T., Tanaka, Y., Hamada, S., Miura, S., Kume, K., Takeyama, Y., Kikuta, K., Abe, T., Abe, T., Abue, M., Akiyama, T., Arizumi, T., Asada, S., Azuma, S., Chiba, K., Dohmen, T., Doi, H., Eguchi, T., Endo, H., Fujimori, N., Fujimoto, Y., Fujiwara, J., Fukuda, A., Fukuda, N., Fukuda, Y., Fukue, M., Funayama, H., Gushima, T., Hagiwara, S.-I., Hamano, H., Hamaya, Y., Hashizume, K., Hattori, K., Hattori, N., Hayasaki, A., Hayashi, H., Hayashi, K., Hayashi, K., Hayashi, K., Hirose, S., Hirota, M., Hirota, M., Hisamatsu, T., Hori, Y., Hoshi, T., Hyodo, N., Ibusuki, K., Ida, A., Iemoto, T., Igarashi, T., Iida, T., Iio, S., Iizawa, Y., Ikeura, T., Imamura, T., Inagaki, T., Inatomi, O., Inoue, H., Inoue, K., Inoue, T., Irie, K., Irie, S., Ishida, M., Ishige, K., Ishii, A., Ishii, J., Isomoto, H., Ito, S., Ito, T., Ito, T., Iwasaki, E., Iwasaki, J., Iwashita, T., Iwata, K., Iwata, T., Izumi, T., Jinno, N., Kado, T., Kaino, S., Kajiwara, T., Kaku, T., Kamiga, T., Kanamori, A., Kaneko, K., Kanemaru, K., Kanno, A., Kasugai, S., Kataoka, F., Kato, S., Kawaguchi, J., Kawaguchi, S., Kawaguti, K., Kawahara, R., Kawahara, Y., Kawai, T., Kawaji, Y., Kawakami, H., Kawamoto, T., Kido, O., Kishimoto, Y., Kitada, H., Kitaguchi, T., Kitamura, K., Kitano, R., Kobashigawa, K., Kobayashi, M., Kobori, I., Kodama, Y., Kohi, S., Koike, K., Koizumi, M., Komaki, T., Komoto, S., Komura, T., Konno, N., Koya, Y., Kubota, K., Kuramochi, M., Kurata, H., Kurihara, H., Kuroki, M., Kuwatani, M., Maeda, T., Makino, I., Manaka, T., Mandai, K., Maruo, T., Maruta, A., Maruyama, Y., Masuda, Y., Masuo, T., Matsubara, S., Matsukuma, S., Matsumoto, H., Matsumoto, K., Matsumoto, R., Matsumura, Y., Matsuo, S., Mihara, K., Mitsuhashi, K., Miyagahara, T., Miyajima, I., Miyazawa, M., Mizuno, S., Mizuno, S., Mori, K., Motoya, M., Mukai, S., Murakami, Y., Murawaki, Y., Murayama, Y., Nagamura, R., Nagase, H., Nagata, K., Nakahara, K., Nakahira, T., Nakai, Y., Nakai, Y., Nakamura, A., Nakaoka, K., Nakase, H., Nakazawa, A., Nawa, Y., Niina, Y., Nishida, T., Nishino, T., Noge, S., Oaku, T., Ogawa, T., Ogiyama, H., Oguchi, T., Ogura, Y., Ohata, A., Ohtani, M., Ohyama, H., Oka, H., Oka, S., Okada, T., Okada, Y., Okamoto, H., Okamoto, K., Okamoto, T., Okazaki, A., Okazaki, K., Okino, T., Okochi, O., Okumura, F., Okuwaki, K., Oshima, K., Otani, K., Oyama, T., Saeki, A., Saito, K., Saito, M., Saito, N., Sakai, T., Sakai, Y., Sakano, M., Sano, T., Sasaki, Y., Sasakura, Y., Satake, M., Sato, M., Sato, S., Sato, Y., Satoh, A., Sawai, Y., Sekine, M., Sekino, Y., Setsuda, M., Shigekawa, M., Shiihara, M., Shimatani, M., Shimizu, T., Shinkai, T., Shinohara, N., Shiomi, H., Shirahata, N., Sho, M., Sogame, Y., Suda, T., Suenaga, S., Sumi, H., Sumida, Y., Suzuki, K., Suzuki, K., Suzuki, M., Suzuki, N., Suzuki, Y., Suzumura, K., Tabaru, A., Takaesu, R., Takagaki, K., Takagi, T., Takahashi, K., Takahashi, K., Takahashi, S., Takano, S., Takano, Y., Takasaki, Y., Takenaka, M., Tamagawa, H., Tamoto, E., Tanaka, Y., Tanoue, S., Terai, S., Terajima, H., Toki, M., Torisu, Y., Tsujikawa, T., Tsukamoto, K., Tsukamoto, Y., Tsuruga, Y., Tsushima, K., Tuda, H., Uchida, K., Ueno, M., Ueno, Y., Unno, J., Urata, T., Watanabe, S., Yamada, K., Yamada, R., Yamagami, H., Yamamiya, A., Yamamoto, S., Yamanouchi, K., Yamao, K., Yamashita, S., Yashika, J., Yasuda, H., Yasuda, I., Yogi, T., Yokota, T., Yoneda, K., Yoshida, H., Yoshida, H., Yoshida, N., Yoshimura, H., Yoshioka, R., 2025. Nationwide epidemiological survey of acute pancreatitis in Japan, 2021: the impact of the COVID-19 pandemic and revised clinical guidelines. Journal of Gastroenterology. https://doi.org/10.1007/s00535-025-02284-2.

Tee, Y.-S., Fang, H.-Y., Kuo, I.-M., Lin, Y.-S., Huang, S.-F., Yu, M.-C., 2018. Serial evaluation of the SOFA score is reliable for predicting mortality in acute severe pancreatitis. Medicine 97 (7), e9654. https://doi.org/10.1097/md.0000000000009654.

Tenner, S., Swaroop Vege, S., Sheth, S.G., Sauer, B., Yang, A., Conwell, D.L., Yadlapati, R. H., Gardner, T.B., 2024. American College of gastroenterology guidelines: management of acute pancreatitis. American Journal of Gastroenterology 119 (3), 419–437. https://doi.org/10.14309/ajg.0000000000002645.

Thoeni, R.F., 2012. The revised atlanta classification of acute pancreatitis: its importance for the radiologist and its effect on treatment. Radiology 262 (3), 751–764. https://doi.org/10.1148/radiol.11110947.

Tomioka, A., Shimizu, T., Asakuma, M., Inoue, Y., Taniguchi, K., Hirokawa, F., Hayashi, M., Uchiyama, K., 2019. Duodenal obstruction due to chronic pancreatitis of the pancreas tail treated by surgical intervention: a case report. Medicine 98 (27), e15856. https://doi.org/10.1097/MD.0000000000015856.

Tran, A., Fernando, S.M., Rochwerg, B., Inaba, K., Bertens, K.A., Engels, P.T., Balaa, F.K., Kubelik, D., Matar, M., Lenet, T.I., Martel, G., 2022. Prognostic factors associated with development of infected necrosis in patients with acute necrotizing or severe pancreatitis—a systematic review and meta-analysis. Journal of Trauma and Acute Care Surgery 92 (5), 940–948. https://doi.org/10.1097/ta.0000000000003502.

Van Brunschot, S., Schut, A.J., Bouwense, S.A., Besselink, M.G., Bakker, O.J., Van Goor, H., Hofker, S., Gooszen, H.G., Boermeester, M.A., Van Santvoort, H.C., 2014. Abdominal compartment syndrome in acute pancreatitis: a systematic review. Pancreas 43 (5), 665–674. https://doi.org/10.1097/MPA.0000000000000108.

Van Santvoort, H.C., Besselink, M.G., Bakker, O.J., Hofker, H.S., Boermeester, M.A., Dejong, C.H., Van Goor, H., Schaapherder, A.F., Van Eijck, C.H., Bollen, T.L., Van Ramshorst, B., Nieuwenhuijs, V.B., Timmer, R., Laméris, J.S., Kruyt, P.M., Manusama, E.R., Van Der Harst, E., Van Der Schelling, G.P., Karsten, T., Hesselink, E. J., Van Laarhoven, C.J., Rosman, C., Bosscha, K., De Wit, R.J., Houdijk, A.P., Van Leeuwen, M.S., Buskens, E., Gooszen, H.G., 2010. A step-up approach or open necrosectomy for necrotizing pancreatitis. New England Journal of Medicine 362 (16), 1491–1502. https://doi.org/10.1056/NEJMoa0908821.

Vincent, J.-L., Moreno, R., Takala, J., Willatts, S., De Mendonça, A., Bruining, H., Reinhart, C.K., Suter, P.M., Thijs, L.G., 1996. The SOFA (Sepsis-related Organ Failure Assessment) score to describe organ dysfunction/failure. Intensive Care Medicine 22 (7), 707–710. https://doi.org/10.1007/bf01709751.

Vitas, G.J., Sarr, M.G., 1992. Selected management of pancreatic pseudocysts: operative versus expectant management. Surgery 111 (2), 123–130.

Wang, M., Gorelick, F., Bhargava, A., 2021. Sex differences in the exocrine pancreas and associated diseases. Cellular and Molecular Gastroenterology and Hepatology 12 (2), 427–441. https://doi.org/10.1016/j.jcmgh.2021.04.005.

Wang, C., Fu, B., Su, De, Huang, P., Fu, X., Dipasquale, V., 2022. Acute pancreatitis and recurrent acute pancreatitis in children: a 10-year retrospective study. Gastroenterology Research and Practice 2022, 1–6. https://doi.org/10.1155/2022/5505484.

Wang, C.F., Tariq, A., Chandra, S., 2025. Acute Pancreatitis. StatPearls.

Whitcomb, D.C., Buchner, A.M., Forsmark, C.E., 2023. AGA clinical practice update on the epidemiology, evaluation, and management of exocrine pancreatic insufficiency: expert review. Gastroenterology 165 (5), 1292–1301. https://doi.org/10.1053/j.gastro.2023.07.007.

Wilcox, C.M., Yadav, D., Ye, T., Gardner, T.B., Gelrud, A., Sandhu, B.S., Lewis, M.D., Al-Kaade, S., Cote, G.A., Forsmark, C.E., Guda, N.M., Conwell, D.L., Banks, P.A., Muniraj, T., Romagnuolo, J., Brand, R.E., Slivka, A., Sherman, S., Wisniewski, S.R., Whitcomb, D.C., Anderson, M.A., 2015. Chronic pancreatitis pain pattern and severity are independent of abdominal imaging findings. Clinical Gastroenterology and Hepatology 13 (3), 552–560. https://doi.org/10.1016/j.cgh.2014.10.015.

Wu, B.U., Conwell, D.L., 2010. Acute pancreatitis Part I: approach to early management. Clinical Gastroenterology and Hepatology 8 (5), 410–416. https://doi.org/10.1016/j.cgh.2009.10.033.

Wu, Y., Han, C., Luo, R., Cai, W., Xia, Q., Jiang, R., Ferdek, P.E., Liu, T., Huang, W., 2023. Molecular mechanisms of pain in acute pancreatitis: recent basic research advances and therapeutic implications. Frontiers in Molecular Neuroscience 16. https://doi.org/10.3389/fnmol.2023.1331438.

Yegneswaran, B., Kostis, J.B., Pitchumoni, C.S., 2011. Cardiovascular manifestations of acute pancreatitis. Journal of Critical Care 26 (2), 225.e11. https://doi.org/10.1016/j.jcrc.2010.10.013.

Zeng, T., An, J., Wu, Y., Hu, X., An, N., Gao, L., Wan, C., Liu, L., Shen, Y., 2023. Incidence and prognostic role of pleural effusion in patients with acute pancreatitis: a meta-analysis. Annals of Medicine 55 (2). https://doi.org/10.1080/07853890.2023.2285909.

Zhang, S., 2024. The clinical characteristics and outcomes of acute pancreatitis are different in elderly patients: a single-center study over a 6-year period. Journal of Clinical Medicine 13 (16), 4829.

Zhang, B., Xiao, Q., Ma, Q., Han, L., 2023. Clinical treatment for persistent inflammation, immunosuppression and catabolism syndrome in patients with severe acute pancreatitis (Review). Experimental and Therapeutic Medicine 26 (4). https://doi.org/10.3892/etm.2023.12194.

Zheng, Z., Ding, Y.-X., Qu, Y.-X., Cao, F., Li, F., 2021. A narrative review of acute pancreatitis and its diagnosis, pathogenetic mechanism, and management. Annals of Translational Medicine 9 (1). https://doi.org/10.21037/atm-20-4802, 69–69.

CHAPTER

Diagnostic criteria for pancreatitis

5

Yoshinori Kakino

Gifu University Graduate School of Midicine, Emergency and Disaster Medicine, Gifu, Japan

5.1 Historical evolution of diagnostic criteria

Acute pancreatitis (AP) has been recognized as a distinct clinical entity for over a century. The diagnostic pathways have advanced significantly with the progress in clinical practice, laboratory markers, and imaging modalities. Understanding this historical evolution is necessary to gaining insight into the rationale behind modern classification systems, including the Revised Atlanta Classification (RAC).

5.1.1 Early clinical descriptions

In the late 19th and early 20th centuries, AP was diagnosed primarily based on clinical presentation. Surgeons and internists made the diagnosis based on the characteristic symptom of acute epigastric pain radiating to the back, often accompanied by nausea and vomiting. The early descriptions were predominantly case series that emphasized the severe pain, rapid clinical progression, and high mortality rates in untreated patients (The Etiology of Acute Hemorrhagic Pancreatitis, 1901). Clinicians also recognized the associations with alcohol consumption and gallstones, despite the pathophysiology remaining unclear (Opie, 1904; Comfort et al., 1946). Some studies reported mortality rates higher than 50%, underscoring the importance of diagnostic criteria to enable early intervention (Bradley, 1993).

5.1.2 Introduction of laboratory markers

Serum amylase was first introduced into clinical practice in the early 20th century as a marker of pancreatic injury. Amylase, a key pancreatic enzyme, is released into the circulation when the pancreas is damaged. The seminal work of Elman in 1937 demonstrated dynamic elevations of serum amylase during acute pancreatic disease, thereby establishing its diagnostic relevance (Elman, 1937). However, its specificity was limited, as elevated levels were also observed in conditions unrelated to the pancreas, such as salivary gland disorders. Subsequently, the measurement of lipase, introduced in the 1970s, was specific for pancreatic injury and provided a longer diagnostic window. Ultimately, it became an indispensable component of the diagnostic evaluation in patients with suspected AP (Lott et al., 1986).

Advanced Insights into Acute Pancreatitis. https://doi.org/10.1016/B978-0-443-44736-5.00014-0

5.1.3 The role of imaging diagnostics

Historically, inadequate visualization of the pancreas often resulted in delayed diagnosis of AP (Rifkind et al., 1976). Following the introduction of ultrasound (US) in the 1970s, gallstones and bile duct dilatation could be detected, and it played an important role in assessing the etiology. However, US was frequently restricted by overlying intestinal gas and required highly operator dependency (Block et al., 1986). The advent of contrast-enhanced computed tomography (CECT) in the 1980 and 1990s revolutionized the diagnosis of AP by direct visualization of pancreatic edema, necrosis and peripancreatic fluid collections. In particular, the role of CT imaging in grading disease severity and predicting complications became more critical along with laboratory markers and clinical assessment (Raghuwanshi et al., 2016).

5.1.4 Formal diagnostic criteria

Original Atlanta Classification (1992): Provided standardized terminology for the severity of AP, defining mild, moderately severe and severe forms. Diagnosis was defined as clinical features and biochemical or imaging evidence, however early CT was not universally recommended (Bradley, 1993). Revised Atlanta Classification (2012): Added flexibility, requiring fulfillment of two out of three criteria: (1) typical abdominal pain, (2) serum amylase or lipase levels at least three times the upper limit of normal ($\geq 3\times$ ULN), or (3) imaging finding compatible with AP. This revision was based on observations that early imaging could be equivocal and emphasized clinical judgment supported by laboratory and imaging data. The RAC also introduced clear definitions of local and systemic complications, associating diagnosis with prognosis and treatment decisions. Early diagnosis of pancreatic necrosis or peripancreatic fluid collections enables focused treatment, with less morbidity and mortality (Banks et al., 2013).

5.1.5 Regional variations

Epidemiological studies have demonstrated significant regional differences in diagnostic practices for AP. In Japan, early efforts emphasized serum amylase and lipase measurements, with imaging reserved for selected cases (Takeda, 2007). European guidelines generally favored early contrast-enhanced CT, particularly in patients with severe presentations, while ultrasound remained the first-line modality when gallstone etiology was suspected (Leppäniemi et al., 2019). In North America, integrated scoring systems such as the Ranson and BISAP scores became widely used alongside biochemical testing (Eachempati et al., 2002).

5.1.6 Summary and clinical implications

The evolution of diagnostic criteria for acute pancreatitis (AP) reflects a shift from descriptive symptom-based recognition to a structured, evidence-driven framework.

- **Clinical foundation**: Characteristic abdominal pain remains central, though past reliance on symptoms alone led to misdiagnosis in atypical or mild cases.
  - Laboratory refinement: Serum amylase and later lipase enabled earlier, objective diagnosis, yet both require cautious interpretation due to limited specificity.
  - Imaging advancements: The advent of ultrasonography and CECT revolutionized diagnosis and management, providing severity stratification and guiding timely intervention.
  - Standardization: The Atlanta and Revised Atlanta Classifications unified terminology, linked diagnosis to prognosis, and facilitated global research and practice alignment.
  - Regional adaptation: Differences among Japan, Europe, and North America reflect variations in resource availability and diagnostic emphasis.

Modern AP diagnosis is thus a dynamic synthesis of clinical judgment, laboratory data, and imaging findings—contextualized within regional frameworks. Recognizing this historical progression helps clinicians apply the Revised Atlanta Classification effectively, avoid diagnostic pitfalls, and optimize early management.

5.2 Pathophysiological basis of diagnostic criteria

The formulation of diagnostic criteria for AP is not arbitrary; rather, it is grounded in decades of mechanistic research and clinical validation. Understanding the disease's pathophysiology clarifies why abdominal pain, enzyme elevation, and imaging abnormalities have become the cornerstones of diagnosis. This section explores the major biological processes of AP, linking them to diagnostic hallmarks and clinical markers that inform bedside decision-making.

5.2.1 Initiating event: Premature enzyme activation

At the core of AP lies the premature conversion of trypsinogen to trypsin within pancreatic acinar cells. Normally, zymogen activation is confined to the duodenum via enterokinase, but under pathological conditions, intracellular activation occurs. Studies have demonstrated that disrupted calcium oscillations lead to mitochondrial permeability transition, ATP depletion, and subsequent cell death (Habtezion et al., 2019; Petersen and Tepikin, 2008).

Etiology-specific pathways converge on this mechanism.

- **Gallstone-induced AP**: Obstruction increases ductal pressure, while bile salt reflux directly destabilizes calcium handling, leading to intracellular enzyme activation (Tran et al., 2021).
- **Alcohol-induced AP**: Non-oxidative metabolism of ethanol generates fatty acid ethyl esters, which impair mitochondrial energy supply and sensitize cells to injury (Huang et al., 2014).

- **Hypertriglyceridemia-induced AP**: Hydrolysis of triglycerides produces excessive free fatty acids, which are cytotoxic and promote endothelial injury and microcirculatory collapse (Kiss et al., 2023).

These experimental insights are corroborated by clinical evidence. In a prospective investigation of patients presenting with acute abdominal pain, Keim and colleagues demonstrated that virtually all individuals with imaging-confirmed AP exhibited a rise in serum lipase activity within the first 24 h of symptom onset, with sensitivity exceeding 95% (Keim et al., 1998). These findings highlight serum lipase as a robust and early biochemical correlate of acinar cell injury and premature enzyme activation, thereby reinforcing its pivotal role in the diagnostic evaluation of AP. Importantly, in hypertriglyceridemia, excessively elevated triglycerides (e.g., > 500 mg/dL) can interfere with the colorimetric amylase assay, yielding falsely normal values despite underlying pancreatic injury (Melnick et al., 2016). Therefore, measurement of triglycerides should be performed as a diagnostic adjunct in suspected cases of HTG-induced pancreatitis.

5.2.2 Local inflammation and amplification of tissue injury

Once initiated, enzyme activation triggers a robust local inflammatory response. Activated trypsin catalyzes further conversion of proenzymes such as pro-elastase and pro-phospholipase A2, generating a destructive milieu. Neutrophils and macrophages infiltrate pancreatic tissue, releasing reactive oxygen species and pro-inflammatory cytokines (TNF-α, IL-1β, IL-6, IL-8) (Cosen-Binker and Gaisano, 2007). In animal models, blockade of cytokine signaling reduces tissue injury, supporting their central role in disease amplification (Norman et al., 1995). Histological studies of human pancreas have confirmed fat necrosis, hemorrhage, and extensive acinar destruction in severe cases (Klöppel et al., 1986). Microvascular changes further exacerbate injury. Capillary leak leads to interstitial edema, while microthrombi obstruct pancreatic perfusion. Contrast-enhanced CT studies have shown that non-enhancing pancreatic regions correspond to ischemic necrosis, with strong correlation to histopathological findings (Balthazar et al., 1990).

5.2.3 Systemic inflammatory response and organ dysfunction

The transition from local pancreatic injury to systemic involvement is mediated by "spillover" of inflammatory mediators into the circulation. Elevated serum cytokines, including IL-6, IL-8, and MCP-1, have been consistently associated with persistent organ failure in prospective studies (Ishibashi et al., 2008; Chen et al., 1999). This systemic inflammatory response syndrome (SIRS) manifests clinically with fever, tachycardia, tachypnea, and leukocytosis or leukopenia. If unchecked, it precipitates acute respiratory distress syndrome (ARDS), acute kidney injury (AKI), and cardiovascular instability (Banks et al., 2013).

Some biomarkers can assess the degree of systemic inflammation.

- **C-reactive protein (CRP)**: In a prospective cohort study, levels exceeding 170–190 mg/L at 48 h were strongly predictive of severe AP, pancreatic necrosis, and in-hospital mortality (Cardoso et al., 2013).
- **Procalcitonin (PCT)**: Rises earlier (12–24 h) and discriminates between sterile and infected necrosis. Randomized studies confirm that PCT-guided management reduces unnecessary antibiotic use without compromising outcomes (Rau et al., 1997).
- **IL-6**: Meta-analyses show that early IL-6 elevation predicts persistent organ failure with greater sensitivity than CRP (Chen et al., 1999).

Together, these markers demonstrate how laboratory evaluation mirrors the systemic consequences of local pancreatic injury.

5.2.4 Necrosis and secondary complications

Approximately 20% of patients develop pancreatic necrosis, which significantly alters prognosis (Banks et al., 2013). Necrosis arises from a combination of autodigestion, ischemia, and microvascular obstruction. Perfusion CT studies reveal that early reductions in pancreatic blood flow predict necrosis before it becomes radiologically apparent (Tsuji et al., 2007). The natural history of necrosis underscores the importance of timing. Sterile necrosis may remain clinically stable, whereas secondary bacterial infection—often via translocation from the gut—marks a turning point toward sepsis and multi-organ failure (Banks et al., 2013). Procalcitonin and gas bubbles on imaging serve as practical indicators of infection (Rau et al., 1997; Balthazar, 2002). This progression provides the pathophysiological rationale for why the Revised Atlanta Classification separates local complications (necrosis, collections) from systemic ones (organ failure), as each reflects distinct underlying mechanisms.

5.2.5 Temporal integration into diagnostic criteria

The temporal dynamics of AP illustrate why multiple diagnostic elements are necessary.

- **Early phase (0–24 h)**: Pain and enzyme elevation capture the immediate effects of acinar injury (Pandol et al., 2007).
- **Intermediate phase (24–72 h)**: CRP, BUN, and early imaging findings reflect inflammation and volume status (Balthazar et al., 1990; Wu et al., 2011; Cardoso et al., 2013).
- **Late phase (>72 h)**: Necrosis, infection, and vascular complications emerge, best detected by imaging and infection-specific biomarkers (Balthazar et al., 1990; Rau et al., 1997; Boudghène et al., 1993).

This sequence validates the RAC's "two-out-of-three" rule. Pain corresponds to clinical expression, enzyme elevation to biochemical leakage, and imaging to structural and vascular sequelae. No single marker suffices, but in combination they provide a robust and time-sensitive diagnostic framework (Banks et al., 2013).

5.3 Differential diagnosis

AP typically presents with acute abdominal pain, nausea, and vomiting; however, these symptoms are nonspecific. A broad spectrum of intra-abdominal and systemic conditions can present with similar clinical features, making accurate differential diagnosis essential to avoid misdiagnosis and inappropriate management.

5.3.1 Mimics and confounding conditions

- **Cholelithiasis and Acute Cholecystitis** (Yokoe et al., 2013) Clinical Features: Patients typically present with right upper quadrant or epigastric pain, often postprandial, which may radiate to the back. Nausea and vomiting are frequently observed. Laboratory Findings: Mild elevations of liver enzymes (ALT, AST) and bilirubin are common; amylase may be slightly increased in some cases. Imaging Findings: Ultrasound can detect gallstones, gallbladder wall thickening, and pericholecystic fluid collections. Differential Diagnosis: In contrast, AP usually manifests as more severe epigastric pain radiating to the back, accompanied by a more pronounced elevation of serum lipase.
- **Peptic Ulcer Disease (PUD) and Perforation** (Lanas and Chan, 2017) Clinical Features: Patients present with upper abdominal pain, nausea, and vomiting. Laboratory Findings: Serum pancreatic enzyme levels are usually within the normal range. Imaging/Endoscopy: Endoscopy confirms mucosal ulceration. In cases of perforation, CT may demonstrate localized inflammation or the presence of free intraperitoneal air. Differential Diagnosis: A distinguishing feature of PUD is that pain often improves with food intake or antacid administration. The absence of elevated pancreatic enzymes further differentiates it from AP.
- **Myocardial Ischemia** (Thygesen et al., 2019) Clinical Features: Patients may present with epigastric or upper abdominal pain, often accompanied by nausea and vomiting. Laboratory Findings: Cardiac troponins are elevated, whereas serum amylase and lipase levels are normal. Imaging Findings: Electrocardiogram (ECG) typically demonstrates changes characteristic of ST-segment elevation myocardial infarction (STEMI). Differential Diagnosis: Pain radiating to the left arm or jaw, frequently associated with diaphoresis or dyspnea, helps distinguish myocardial ischemia from AP.
- **Intestinal Obstruction** (Catena et al., 2019) Clinical Features: Colicky abdominal pain, vomiting, and progressive abdominal distension. Laboratory Findings: Leukocytosis may be present; however, serum amylase and lipase levels are usually within the normal range. Imaging Studies: Abdominal X-ray often demonstrates multiple air–fluid levels, while CT scan more precisely identifies the site and cause of obstruction. Differential Diagnosis: Characterized by colicky pain, progressive distension, and the absence of significant lipase elevation, distinguishing it from AP.
- **Mesenteric Ischemia** (Dos Santos Pereira et al., 2025) Clinical Features: Sudden onset of severe abdominal pain, often accompanied by nausea, vomiting, and occasionally diarrhea. Laboratory Findings: Elevated lactate levels and metabolic acidosis are typical. Serum amylase may be mildly elevated in some cases.

Table 5.1 Key laboratory and imaging findings distinguishing acute pancreatitis from other acute abdominal conditions.

Condition	Amylase	Lipase	Liver enzymes	CRP	Distinguishing feature
AP	↑↑ (≥3× ULN)	↑↑ (≥3× ULN)	Normal/↑	↑	Severe epigastric pain radiating to back
Cholelithiasis/ Acute cholecystitis	Mild ↑	Mild ↑	↑	Mild ↑	US: Gallstones, biliary changes
PUD/ perforation	Normal	Normal	Normal	Mild ↑	Endoscopy: ulcer, CT: free air
Myocardial ischemia	Normal	Normal	Normal	Mild ↑	Troponin ↑, ECG changes (STEMI)
Intestinal obstruction	Normal	Normal	Normal	Mild ↑	X-ray: air–fluid levels
Mesenteric ischemia	Mild ↑	Mild ↑	Mild ↑	↑	Lactate ↑, CTA: Arterial occlusion
Acute hepatitis	Mild ↑	Mild ↑	↑↑	Mild ↑	Marked liver enzyme elevation
Pancreatic neoplasms	Mild ↑	Mild ↑	Mild ↑ ↑↑ if biliary obstruction	Mild ↑	Chronic symptoms, CT/ MRI: Mass lesion

Imaging Studies: Contrast-enhanced CT can identify occlusion of major mesenteric vessels, such as the celiac artery or superior mesenteric artery. Differential Diagnosis: Pain is often disproportionate to physical examination findings, with rapid progression to multi-organ failure, distinguishing it from AP.

- **Acute Hepatitis** (Ryder and Beckingham, 2001) Clinical Features: Right upper quadrant abdominal pain, fatigue, and nausea. Laboratory Findings: Significant elevations in liver enzymes (AST, ALT) may occur, while pancreatic enzyme levels may show mild increases. Imaging Findings: Ultrasound may reveal hepatomegaly and increased liver echogenicity, with no pancreatic enlargement. Differential Diagnosis: The predominance of elevated liver enzymes and frequent systemic jaundice help distinguish acute hepatitis from AP.
- **Pancreatic and Periampullary Neoplasms** (Lee and Lee, 2014; Elbanna et al., 2020; Tempero et al., 2021) Clinical Features: Abdominal pain with gradual onset, which may become acute if complicated by biliary obstruction. Laboratory Findings: Serum amylase and lipase levels may be mildly elevated. Imaging Studies: CT or MRI can reveal a pancreatic mass or associated bile duct obstruction. Differential Diagnosis: The chronicity of symptoms and the presence of a distinct mass help distinguish pancreatic neoplasms from AP (Table 5.1).

5.3.2 Summary of mimics

Many intra-abdominal and systemic diseases present with abdominal pain, nausea, and vomiting like AP. Lipase levels exceeding three times the upper limit of normal ($\geq 3\times$ ULN) are the most specific biochemical differential indicator but should be interpreted in conjunction with clinical and imaging findings. Imaging studies (Ultrasound, CT, MRCP, EUS) are crucial for differentiating AP from its mimics and for identifying the etiology and complications. A systematic evaluation of medical history + laboratory data + imaging studies can prevent misdiagnosis and inappropriate interventions.

5.4 Time course of diagnostic markers in AP

AP is a dynamic condition, and accurate diagnosis requires not only identifying specific markers but also understanding their temporal behavior. The dynamics of enzyme markers and inflammatory markers provide essential information for early detection, differentiation between AP and other abdominal emergencies, and severity prediction. Recognizing these dynamics reduces false negatives and ensures timely treatment.

5.4.1 Serum amylase

Serum amylase was the first biochemical marker adopted for diagnosing pancreatitis. Its levels typically rise within 6–12 h after the onset of symptoms, peaks at around 24 h, and normalizes within 48–72 h even in untreated cases. While historically valuable, amylase lacks specificity. Its levels also increase in conditions such as salivary gland disease, renal impairment, intestinal obstruction, and PUD/perforation. Furthermore, in patients presenting more than 48 h after symptom onset, amylase levels may have already normalized despite ongoing inflammation, leading to potential false-negative results. In mild pancreatitis, the peak amylase level tends to be lower, complicating early diagnosis. Recent case reports have similarly described patients with AP presenting with characteristic abdominal pain but normal serum amylase, in whom the diagnosis was later confirmed by imaging or clinical course (Ibrahim et al., 2024; Tenner et al., 2013; Yadav and Lowenfels, 2013).

5.4.2 Serum lipase

Serum lipase has gradually replaced amylase as a more selective enzyme. It rises 4–8 h after symptom onset, peaks within 24–48 h, and remains elevated for 8–14 days. This extended diagnostic window makes lipase particularly useful in patients with delayed diagnosis. The RAC explicitly lists a lipase level exceeding three times the upper limit of normal ($\geq 3\times$ ULN) as one diagnostic criterion. Clinical studies confirm that lipase demonstrates higher specificity than amylase, proving particularly useful in differentiating AP from cholecystitis, and intestinal obstruction (Banks et al., 2013; Spanier et al., 2008).

5.4.3 C-reactive protein

Early investigations identified CRP, together with antiproteases and complement factors, as an objective indicator of the systemic inflammatory response and disease severity in AP (Wilson et al., 1989). Building on these findings, subsequent studies demonstrated that CRP concentrations measured within 48 h of symptom onset are closely associated with the development of severe disease. A cutoff value of approximately 150 mg/L has been repeatedly validated and remains the most widely accepted threshold across both research and clinical guidelines. Although lower values, such as 120 mg/L, have occasionally been proposed to predict severe AP, these results have been inconsistent and have not supplanted the 150 mg/L benchmark (Lee and Cho, 2022). More recently, prospective cohort analyses have shown that incorporating CRP into the Ranson score enhances its ability to predict severe outcomes and organ dysfunction (Basak et al., 2017). It should be emphasized that CRP, while not useful for establishing the diagnosis of AP, is a practical and readily available marker for severity assessment. Sequential measurements are particularly informative, allowing clinicians to monitor disease trajectory and to make timely decisions regarding escalation of care, including the need for transfer to an intensive care unit.

5.4.4 Hematocrit and blood urea nitrogen

Hematocrit (HCT) and blood urea nitrogen (BUN) are indirect indicators of intravascular volume status and blood concentration. An increase in either parameter within 24 h of onset strongly correlates with pancreatic necrosis and persistent organ failure. Temporal monitoring of HCT and BUN provides valuable insights regarding the adequacy of fluid resuscitation and the risk of clinical deterioration. For example, an increase in BUN despite ongoing fluid administration has been consistently shown to predict mortality, underscoring the need for intensified therapeutic intervention (Wu et al., 2009). Beyond BUN, evidence supporting the prognostic utility of HCT values is growing. A retrospective cohort study demonstrated that for every 5% increase in HCT at admission, the risk of severe pancreatitis (odds ratio = 2.8; $P = .001$), necrosis, and intensive care unit admission increased (Gan and Romagnuolo, 2004). Findings suggest that elevated HCT at admission and its temporal changes may serve as early prognostic indicators in AP.

5.4.5 Temporal patterns and clinical applications

Marker dynamics vary by disease stage. Serum lipase rises within 24 h of symptom onset, peaks at 24 h, and remains elevated for 1–2 weeks (Basnayake and Ratnam, 2015). During this period, definitive imaging studies may be limited. Between 24 and 72 h, amylase peaks (Alter and Koch, 2021), CRP begins to rise with the best prognostic accuracy around 48 h (Cardoso et al., 2013), and imaging reveals pancreatic edema and peripancreatic fluid collection (Greenberg et al., 2016). After 72 h, amylase often normalizes, necessitating reliance on lipase, CRP, and imaging findings for definitive diagnosis (Alter and Koch, 2021). This pattern supports the rationale for the RAC guidelines allowing diagnosis based on two of three criteria: characteristic pain, elevated enzymes, and imaging findings (Banks et al., 2013).

5.5 Emerging biomarkers and novel laboratory tools

Conventional biomarkers such as amylase, lipase, and CRP remain central to the diagnosis and monitoring of AP. However, their limitations—low specificity, narrow diagnostic window, and variable expression in atypical cases—have driven the search for novel biomarkers. The aims are threefold: (1) to facilitate earlier and more accurate diagnosis, (2) to enable better prediction of severity and complications, and (3) to clarify disease mechanisms, particularly in subtypes such as hypertriglyceridemia-induced AP. Collectively, these new tools support the transition toward precision medicine in AP.

5.5.1 Trypsinogen activation peptide

TAP is a short peptide released during the conversion of trypsinogen to trypsin, representing one of the earliest steps in AP pathogenesis. It can be detected in plasma or urine within 2–6 h of symptom onset—well before amylase or lipase rise—and correlates with necrosis and multiple organ failure (Neoptolemos et al., 2000; Tenner et al., 1997; Khan et al., 2002). Despite its strong mechanistic relevance, routine testing remains limited by assay availability and lack of standardization (Allemann et al., 2024). Wider clinical adoption will depend on the development of rapid, validated, and affordable test systems.

5.5.2 Urinary trypsinogen-2

Urinary trypsinogen-2 offers a rapid, non-invasive diagnostic option using simple test strips, showing high sensitivity (89%) and specificity (95%) (Baillie, 1997). It is especially valuable in pediatric and atypical presentations and feasible for point-of-care use. Nonetheless, interstudy variability, reduced reliability in renal impairment, and the widespread availability of serum lipase have limited its adoption (Hedström et al., 1996; Sato et al., 2024). Its main value may lie in emergency or resource-limited settings rather than routine hospital practice.

5.5.3 Procalcitonin

Among newer markers, PCT has gained notable attention. Rising within 12–24 h, earlier than CRP, it reflects both systemic inflammation and bacterial infection (Rau et al., 2007). Persistent elevation predicts infected necrosis and sepsis, while declining levels help avoid unnecessary antibiotic exposure (Alberti et al., 2022). PCT thus provides actionable information for antimicrobial stewardship. When combined with cytokines such as IL-6 or TNF-α, its predictive power for complications, organ dysfunction, and mortality further improves (Tarján et al., 2024).

5.5.4 Cytokines

Cytokines including TNF-α, IL-6, and IL-8 play central roles in the inflammatory cascade and correlate strongly with disease severity. IL-6 rises early (within 24–48 h) and has been shown to outperform conventional markers such as CRP in predicting progression to severe disease (Leser et al., 1991). Serial cytokine monitoring offers dynamic insight into systemic inflammation and potential complications (Frossard et al., 2008).

5.5.5 MicroRNAs and genetic markers

MicroRNAs (miR-216a, miR-551b-5p) are small non-coding RNAs released from injured pancreatic cells, serving as extremely early and minimally invasive biomarkers that often precede enzyme elevation (Zhang et al., 2018; Goodwin et al., 2014). Genetic predisposition further influences susceptibility to recurrent and hereditary AP (Kusnierz-Cabala et al., 2015). Mutations in PRSS1, SPINK1, and CFTR alter enzyme activation and ductal function, while combined variants across multiple loci (PRSS1, SPINK1, CTRC, CFTR) confer polygenic risk (Whitcomb et al., 1996; Hegyi and Sahin-Tóth, 2017; LaRusch et al., 2012). These findings suggest that personalized prevention and treatment strategies require polygenic risk assessment.

5.5.6 Lipid-based biomarkers

Hypertriglyceridemia (HTG) accounts for up to 10% of AP cases and can mask enzyme elevations due to lipemic interference (Fallat et al., 1973; Soleimani et al., 2020; de Pretis et al., 2018). Lipolysis within the pancreas releases toxic free fatty acids, promoting acinar and endothelial injury (de Oliveira et al., 2020). In rare genetic disorders such as apoC-II or LPL deficiency (Burnett et al., 1993), aggressive triglyceride lowering—via insulin infusion or plasma exchange—can be lifesaving (Gubensek, 2023). For long-term management, lifestyle modification, fibrates, and omega-3 fatty acids remain standard, while emerging therapies targeting apoC-III and ANGPTL3 show promise for refractory or familial HTG (Berglund et al., 2012; Gaudet et al., 2015).

5.5.7 Multi-marker scoring systems and artificial intelligence models

Traditional scoring systems (Ranson, Glasgow, BISAP) remain useful but limited by delayed parameter availability and moderate accuracy (Mounzer et al., 2012). Multi-marker panels integrating enzymes, inflammatory mediators (CRP, PCT, IL-6), metabolic indices (BUN, triglycerides, hematocrit), and emerging tools (miRNA) enhance early risk stratification (Silva-Vaz et al., 2020). Artificial intelligence (AI) models extend this approach: machine-learning algorithms trained on large clinical; laboratory, imaging, and genetic datasets identify complex patterns beyond human capacity (López Gordo et al., 2025). Convolutional neural networks (CNNs) enable automated CT-based severity assessment (Liang et al., 2023), while integration with electronic health records allows real-time prediction of organ failure and generation of risk alerts (Luo et al., 2023). While challenges remain—such as data standardization, interpretability, and multi-center validation—AI-assisted models embody a paradigm shift toward precision risk stratification in AP.

5.6 Advanced imaging modalities (EUS, perfusion CT, MRI/MRCP, AI analysis)

Imaging remains the foundation of diagnosis, contributing to etiologic clarification, severity grading, and detection of complications in AP. While abdominal ultrasound (US) and contrast-enhanced CT (CECT) are the first-line methods, advances in imaging technology have expanded diagnostic tools. New technologies provide functional and quantitative insights into pancreatic perfusion status, microvascular integrity, and luminal anatomy, complementing traditional morphological assessment (Banks et al., 2013; Tsuji et al., 2007; Tenner et al., 2013). These techniques are particularly beneficial for integrating into clinical practice in early or atypical cases where disease burden is often underestimated by conventional diagnosis.

5.6.1 Endoscopic ultrasound

EUS employs high-frequency ultrasound from within the gastrointestinal tract, offering superior resolution for the pancreas and peripancreatic tissues. It is particularly effective in identifying occult biliary causes such as microcalculi and sludge, which account for a large proportion of idiopathic AP (Umans et al., 2020, 2023). Prospective studies show that EUS reveals etiologic factors in up to two-thirds of "idiopathic" cases, often altering management strategies (Umans et al., 2023). EUS also contributes to evaluation of recurrent disease, sphincter of Oddi dysfunction, and ductal anomalies. Therapeutically, EUS allows fine needle aspiration (EUS-FNA) for differentiating sterile from infected necrosis, and cyst fluid analysis for distinguishing pseudocysts from neoplasms. Thus, it functions as both a diagnostic and interventional tool. Limitations include operator dependency, need for sedation, and limited availability (Muthusamy et al., 2016).

5.6.2 Perfusion CT

Perfusion CT extends CECT by quantifying pancreatic blood flow, blood volume, and permeability. In AP, where microvascular dysfunction drives necrosis, perfusion CT provides early functional biomarkers of severity (Tsuji et al., 2007; Pieńkowska et al., 2016). Reduced perfusion predicts necrosis and systemic complications including multi-organ failure (Tsuji et al., 2017). T Clinically, perfusion CT helps identify high-risk patients within 48 h, supporting timely ICU admission and optimized resuscitation. Despite concerns about radiation and nephrotoxicity, its diagnostic value justifies selective use (Tenner et al., 2013; Pieńkowska et al., 2016; Tsuji et al., 2017). With technological advances, dose reduction strategies and machine learning-based post-processing may further enhance its clinical applicability (Xie et al., 2013; Mossa-Basha et al., 2024).

5.6.3 Magnetic resonance imaging and MRCP

MRI offers excellent soft tissue contrast and no ionizing radiation, making it ideal for detecting subtle necrosis, hemorrhage, and edema (Sun et al., 2019; Porter et al., 2019). MRCP provides non-invasive visualization of the pancreatic and biliary ducts, identifying stones, strictures, and congenital anomalies without ERCP-related risks (Griffin et al., 2012). MRI surpasses CT in evaluating peripancreatic fluid collections and luminal disruptions such as duodenal fistulas, and is preferred in patients with renal dysfunction or pregnancy (Morgan et al., 1997). Limitations include longer acquisition time, motion artifacts in unstable patients, and limited emergency availability. In the subacute or chronic phase, MRI serves as the modality of choice for follow-up and procedural planning.

5.6.4 AI and radiomics

AI-driven image analysis is transforming AP evaluation. Radiomics extracts quantitative features such as texture and heterogeneity that correlate with necrosis and systemic complications (Zhong et al., 2022). Deep learning models trained on CT data can predict ICU admission and organ failure with accuracy surpassing traditional severity scores (Lin et al., 2024). Integrated into clinical workflows, AI may enable real-time, automated severity stratification and early prediction of necrosis, prompting timely interventions (Kiss et al., 2022). Furthermore, combining AI-derived image biomarkers with laboratory markers such as C-reactive protein and procalcitonin may yield multimodal prognostic tools (Yao et al., 2025). Despite these advances, standardized datasets, population-based validation, and resolution of regulatory challenges remain prerequisites for widespread adoption.

Table 5.2 Imaging for complication detection.

Complication	Preferred modality	Key findings	Optimal timing
Pancreatic necrosis	CECT, perfusion CT	Non-enhancing parenchyma; perfusion deficits	48–72 h
Infected necrosis	CECT, MRI, EUS-FNA	Gas within necrotic tissue; positive aspirate	≥5–7 days
Pseudocysts	MRI, CECT, EUS	Well-circumscribed fluid collection	>4 weeks
Walled-off necrosis	CECT, MRI	Encapsulated collection with solid/liquid mix	>4 weeks
Vascular complications	CT angiography, MR angiography	Pseudoaneurysm; thrombosis; venous occlusion	Any stage
Disconnected duct syndrome	MRCP, EUS, ERCP	Ductal disruption with persistent collections	Subacute–chronic

5.6.5 Imaging for complication detection

Imaging is essential for defining complications that influence treatment decisions. CECT and perfusion CT remain the gold standard for assessing necrosis, while MRI and MRCP excel at detailed characterization of cystic lesions and luminal abnormalities. EUS is useful for both diagnosis and treatment in confirming infectious necrosis or suppurative pseudocysts. Vascular complications such as pseudoaneurysms and venous thrombosis are best visualized by CT angiography or MR angiography (Umans et al., 2020, 2023; Muthusamy et al., 2016; Pieńkowska et al., 2016; Tsuji et al., 2017; Xie et al., 2013; Mossa-Basha et al., 2024; Sun et al., 2019; Porter et al., 2019; Griffin et al., 2012; Morgan et al., 1997; O'Connor et al., 2011) (Table 5.2).

5.7 Etiology-specific diagnostic features

AP represents a heterogeneous entity with diverse causes. Although the diagnostic framework remains uniform, identifying etiology at presentation is crucial for targeted therapy, timely intervention, and prevention of recurrence. Epidemiological data indicate that gallstones account for roughly one third of cases, alcohol for about a quarter, and hypertriglyceridemia for a smaller but clinically significant fraction. Drug-induced, post-endoscopic, genetic, and idiopathic forms are less common but carry distinct prognostic implications (Banks et al., 2013; Melnick et al., 2016; Tenner et al., 2013; Zilio et al., 2019; Hung, 2014).

5.7.1 Gallstone-induced (biliary) AP (Banks et al., 2013; Tenner et al., 2013; Tenner et al., 1994; da Costa et al., 2016)

Clinical: Abrupt epigastric pain radiating to the back, often postprandial, with a history of gallstones or biliary colic. Laboratory: Early ALT >150 IU/L within 48 h suggests a biliary cause, with specificity >80%. Mild hyperbilirubinemia may occur if obstruction is present. Imaging: Ultrasound is first line for detecting gallstones, sludge, or ductal dilatation. MRCP offers non-invasive duct visualization; ERCP is reserved for therapy. CECT defines local complications and may indirectly indicate obstruction. Prognosis: Outcome is favorable when cholecystectomy and duct clearance are promptly performed; delays increase recurrence risk.

5.7.2 Alcohol-induced AP (Ocskay et al., 2022; Ahmed Ali et al., 2016; Rana and Vilmann, 2015; Busireddy et al., 2014; Peterson, 2004; Ismail and Bhayana, 2017)

Clinical: Most frequent in middle-aged men with chronic heavy drinking; binge intake can also trigger attacks. Pain resembles biliary AP but recurs more often. Laboratory: Moderate enzyme elevation; lipase > amylase. AST > ALT and elevated GGT suggest alcohol exposure. Imaging: CT/MRI show interstitial edema acutely and fibrosis or calcification with recurrence. EUS detects early parenchymal and ductal changes. Prognosis: Recurrence is common with continued alcohol use; abstinence markedly lowers recurrence and chronic progression risk.

5.7.3 Hypertriglyceridemia-induced AP (Banks et al., 2013; Berglund et al., 2012; Garg and Rustagi, 2018; Bálint et al., 2020)

Clinical: Often occurs in patients with metabolic disorders such as diabetes or dyslipidemia. Laboratory: Triglycerides >1000 mg/dL are diagnostic. Lipase is more reliable than amylase, which may appear normal in lipemic serum. Imaging: No specific findings; CECT typically shows diffuse edema or necrosis correlating with severity. Prognosis: Generally, more severe, with higher necrosis and organ failure rates if lipids remain uncontrolled. Plasmapheresis lowers triglycerides rapidly, though mortality benefit remains uncertain.

5.7.4 Drug-induced AP (Kaurich, 2008; Badalov et al., 2007; Nitsche et al., 2010; Schneider et al., 2021)

Clinical: Diagnosis depends on temporal relation to recent medication. Common agents include azathioprine, valproate, thiazides, and immune checkpoint inhibitors. Laboratory/Imaging: Findings mirror other AP types; a detailed drug history and exclusion of alternatives are essential. Prognosis: Most cases resolve after discontinuation. Severe disease is rare but reported with immunotherapies.

5.7.5 Post-endoscopic and traumatic AP (Moore et al., 1990; Moschetta, 2015; Coccolini et al., 2019; Elmunzer et al., 2012; Buxbaum et al., 2023)

Clinical: Post-ERCP pancreatitis occurs in 3%–10% of procedures. Traumatic AP follows blunt injury, especially in younger patients. Imaging: CECT reveals focal injury, edema, or associated trauma. Prognosis: Post-ERCP cases are usually mild but may prolong hospitalization. Rectal NSAIDs and prophylactic pancreatic stents reduce incidence in high-risk patients. Traumatic AP prognosis depends on damage extent and concomitant injuries.

5.7.6 Genetic and idiopathic AP (Whitcomb et al., 1996; Witt et al., 2000; Cohn et al., 1998; Catalano et al., 2009; Lowenfels et al., 1997; Dermine et al., 2024)

Clinical: Mutations in PRSS1, SPINK1, and CFTR predispose to recurrent or chronic disease. Early-onset and family history raise suspicion. Imaging: Subtle ductal irregularities; labs mirror other etiologies. Prognosis: Hereditary forms have high recurrence and increased pancreatic cancer risk. Next-generation sequencing has reduced the proportion of "true" idiopathic cases (Table 5.3).

5.7.7 Clinical implications

Across etiologies, several diagnostic themes emerge:

- Biochemical pitfalls—Normal amylase in hypertriglyceridemia; blunted enzyme response in the elderly.
- Population modifiers—Pregnancy, trauma, and age alter diagnostic interpretation.
- Iatrogenic and genetic factors—Procedural and hereditary causes require tailored evaluation. These insights affirm that while the Revised Atlanta Classification provides a robust base, practical diagnosis demands flexibility. Integrating laboratory kinetics, evolving imaging findings, and patient-specific variables into dynamic algorithms can bridge the gap between standardized criteria and real-world complexity.

5.8 Diagnostic challenges in special populations

Diagnosing AP in special populations poses unique difficulties owing to atypical clinical symptoms, physiological alterations, and constraints in available diagnostic methods. Careful awareness of these population-specific features is essential to prevent diagnostic delays and to guide appropriate management strategies (Banks et al., 2013; Tenner et al., 2013).

Table 5.3 Etiology-specific diagnostic features and prognosis.

Etiology	Clinical features	Laboratory clues	Imaging	Prognosis and evidence
Gallstones	Postprandial epigastric pain; prior biliary colic	ALT >150 IU/L, bilirubin ↑	US: Gallstones, CBD dilatation; MRCP	Good if early cholecystectomy; recurrence if untreated (Banks et al., 2013; da Costa et al., 2016; Tenner et al., 1994, 2013)
Alcohol	Chronic or binge use; recurrent pain	AST > ALT, GGT ↑	CT/MRI: Edema; chronic calcifications	High recurrence; abstinence ↓ risk (Ahmed Ali et al., 2016; Busireddy et al., 2014; Ismail and Bhayana, 2017; Ocskay et al., 2022; Peterson, 2004; Rana and Vilmann, 2015)
Hypertriglyceridemia	Diabetes, dyslipidemia	TG > 1000 mg/dL; lipase reliable	CECT: Edema, necrosis	Often severe; recurrent unless lipids controlled (Bálint et al., 2020; Banks et al., 2013; Berglund et al., 2012; Garg and Rustagi, 2018)
Drug-induced	Recent exposure	Nonspecific	Non-specific	Resolves with withdrawal; rare severe cases (Badalov et al., 2007; Kaurich, 2008; Nitsche et al., 2010; Schneider et al., 2021)
Post-ERCP/Trauma	Temporal link with procedure/trauma	Typical AP labs	Focal edema or injury	Usually mild; NSAIDs/stents ↓ risk (Buxbaum et al., 2023; Coccolini et al., 2019; Elmunzer et al., 2012; Moore et al., 1990; Moschetta, 2015)
Genetic	Early onset; family history	Non-specific	Subtle ductal changes	High recurrence; ↑ cancer risk (Catalano et al., 2009; Cohn et al., 1998; Dermine et al., 2024; Lowenfels et al., 1997; Whitcomb et al., 1996; Witt et al., 2000)

5.8.1 Pediatric patients (Keim et al., 1998; Uc and Husain, 2019; Chen et al., 2025)

AP in children is relatively uncommon, with an estimated incidence of 1–5 cases per 100,000 annually. Nonetheless, timely recognition is particularly important because of distinctive etiological and clinical characteristics. The most frequently identified causes include biliary tract disease, abdominal trauma, drug-related toxicity (such as valproic acid or asparaginase), systemic illness, and genetic predispositions, including PRSS1, SPINK1, and CFTR mutations. Clinical manifestations in pediatric patients are often nonspecific—irritability, abdominal pain, or vomiting are common—and younger children frequently struggle to localize pain, which may contribute to delayed recognition. Biochemical evaluation usually demonstrates elevations in amylase and lipase similar to those observed in adults, although pediatric reference ranges differ. Because of its superior specificity, serum lipase is generally regarded as the more reliable marker. In recurrent or idiopathic cases, genetic testing should be considered to identify hereditary pancreatitis syndromes. For imaging, ultrasonography is typically the first-line modality due to its safety and absence of radiation. When repeated imaging is required, MRI is preferable to CT, whereas CT should be reserved for severe or complicated disease where precise anatomical definition is necessary.

5.8.2 Elderly patients (Wu et al., 2009; Baeza-Zapata et al., 2021)

In older adults, AP often presents with atypical or muted features. Rather than acute abdominal pain and nausea, patients may describe only mild discomfort or, in some cases, lack gastrointestinal symptoms altogether. The diagnostic process is further complicated by the frequent coexistence of comorbidities such as cardiovascular or renal disease, which can obscure both symptom interpretation and laboratory findings. Amylase concentrations may be deceptively low despite clinically significant disease. By contrast, lipase usually retains its diagnostic utility, and the addition of markers such as CRP, BUN, and HCT has been shown to improve early severity assessment in this group. Ultrasound is often limited by bowel gas or unfavorable body habitus, making early CT or MRI necessary in many cases. Epidemiological evidence further indicates that elderly patients experience higher rates of severe disease, complications, and mortality, highlighting the need for prompt recognition and timely intervention.

5.8.3 Pregnant patients (Copel et al., 2017; Johnson et al., 2015; Mądro, 2022; Zhang et al., 2022)

Although rare, AP during pregnancy is a serious condition that threatens both maternal and fetal outcomes, with most cases occurring in the third trimester. Clinical presentation is especially challenging, as symptoms overlap with pregnancy-related complaints such as nausea, epigastric discomfort, and back pain. Cholelithiasis is the leading cause, although hypertriglyceridemia may also play an important role. Biochemically, both amylase and lipase remain useful diagnostic markers. Mild physiologic elevations can occur during pregnancy, but lipase generally demonstrates greater specificity. Concomitant rises in liver enzymes or bilirubin may support a biliary etiology. Ultrasound is widely accepted as the preferred initial imaging technique because of its safety profile. When sonography is inconclusive, MRI repress without gadolinium enhancement represents a reliable alternative. CT, on the other hand, should be avoided whenever possible and reserved only for life-threatening situations in which diagnostic certainty is urgently required.

5.8.4 Traumatic pancreatitis (Shah et al., 2016; Mitra et al., 2014; Kumar et al., 2016)

Pancreatitis following blunt or penetrating abdominal trauma presents a diagnostic challenge. Symptoms may be overshadowed by concomitant injuries, and biochemical abnormalities may not appear until 24–48 h post-trauma. For this reason, a high index of clinical suspicion together with serial laboratory testing is essential. Serum lipase is generally more sensitive than amylase in detecting traumatic pancreatic injury, although neither is entirely specific, and repeated measurements are often required. Imaging plays a central role, with contrast-enhanced CT as the first-line modality to detect pancreatic lacerations, contusions, or peripancreatic collections. MRI can serve as a useful adjunct when CT findings are equivocal or contraindicated. Ultrasonography is generally of limited utility in trauma due to interference from bowel gas, hematoma, or dressings.

5.8.5 Practical considerations

Taken together, these special populations highlight how diagnostic challenges arise from atypical symptomatology, delayed biochemical responses, and restrictions in imaging choice. In children, symptoms are often vague, and localization is poor. In the elderly, classical features may be absent and comorbidities complicate interpretation. Pregnant patients pose dual concerns of diagnostic overlap and radiation risk, while trauma patients frequently exhibit delayed or masked signs. Recognizing these pitfalls and judiciously combining lipase measurement with severity markers and imaging tailored to the specific population appears crucial for improving diagnostic accuracy. Establishing diagnostic pathways adapted to these groups not only facilitates earlier recognition but also reduces delays in management, thereby improving clinical outcomes and mitigating morbidity and mortality.

5.9 Early detection of complications

Complications of AP are central determinants of mortality, morbidity, and therapeutic strategy. Their early recognition is not a matter of academic precision but a clinical necessity, as delayed detection often translates into preventable deterioration. In practice, clinicians must integrate bedside evaluation, laboratory trajectories, and imaging findings, using each not in isolation but as complementary tools (Banks et al., 2013; Balthazar, 2002).

5.9.1 Pancreatic necrosis

- Pathophysiology Necrosis arises when the inflammatory cascade disrupts pancreatic microcirculation, producing ischemia and parenchymal death. This tissue may remain sterile, in which case supportive care suffices, or it may become infected, a development that dramatically increases morbidity and mortality (Banks et al., 2013; Balthazar et al., 1990; Bhatia et al., 2005; Werge et al., 2016).
- Clinical and Laboratory Indicators Clinical suspicion should rise when patients demonstrate persistent fever, tachycardia, hypotension, or new-onset organ dysfunction despite initial resuscitation (Hamesch et al., 2025). Laboratory markers are imperfect but informative. A CRP $\geq$150 mg/L at 48 h is a well-established predictor of severity (Lee and Cho, 2022), while PCT has superior specificity for infected necrosis (Chen and Jiang, 2022). Persistent hemoconcentration (HCT $>$44%) and rising BUN signal ongoing hypovolemia and correlate with adverse outcome (Wu et al., 2009). More recently, interleukin-6 and the neutrophil-to-lymphocyte ratio (NLR) have been proposed as adjunctive prognostic markers, although their role remains investigational (Rao and Kunte, 2017).
- Imaging Cross-sectional imaging provides the most reliable assessment, but timing is crucial. Contrast-enhanced CT at 72–96 h after onset delineates necrotic areas with high accuracy; non-enhancing regions affecting at least one-third of the gland strongly suggest necrosis. MRI offers equivalent sensitivity and is particularly useful in patients with renal impairment. Gas bubbles within necrotic tissue are virtually pathognomonic for infection. Although EUS-guided fine-needle aspiration can provide microbiological confirmation, the widespread adoption of minimally invasive drainage has reduced the necessity of this procedure (Porter et al., 2019).
- Management Principles The therapeutic distinction lies between sterile and infected necrosis. Sterile necrosis should be managed conservatively. Infected necrosis, however, requires intervention (Leppäniemi et al., 2019). The PANTER trial established that a minimally invasive "step-up" approach—beginning with percutaneous drainage—was safer than primary open necrosectomy (Van Santvoort et al., 2010). The TENSION trial later demonstrated that an endoscopic-first strategy was equally effective, with fewer pancreatic fistulas

and shorter hospital stays (van Brunschot et al., 2018). Meta-analyses have since confirmed that minimally invasive drainage reduces morbidity compared with surgical approaches. Current consensus thus advocates staged, minimally invasive management, reserving surgery for refractory cases (Leppäniemi et al., 2019).

5.9.2 Peripancreatic fluid collections

Peripancreatic fluid collections represent another important complication. Within the first 4 weeks, acute peripancreatic fluid collections are typically homogeneous, non-encapsulated, and often resolve spontaneously. When encapsulation occurs beyond 4 weeks, pseudocysts may develop. These are clinically relevant, as they can cause pain, infection, or obstruction of the gastric or biliary tract. Imaging with CT or MRI confirms their well-defined nature, and EUS is invaluable when drainage is required. The ESGE guideline (2018) emphasizes that intervention should be reserved for symptomatic or complicated cystic lesions (Arvanitakis et al., 2018).

5.9.3 Vascular complications

Although less frequent, vascular events can be catastrophic. Venous complications include splenic, portal, or superior mesenteric vein thrombosis (Buter et al., 2002). Arterial lesions include pseudoaneurysm formation and hemorrhage. Pseudoaneurysms may rupture with devastating consequences (Boudghène et al., 1993). Contrast-enhanced CT, MR angiography, and Doppler ultrasound are complementary diagnostic tools. Management has shifted toward endovascular therapy as first-line treatment, with surgery reserved for cases unamenable to embolization (Hyare et al., 2007).

5.9.4 Algorithm for early complication detection

A practical algorithm highlights the decision points most relevant to daily practice.

- **0–48 h**: clinical and laboratory monitoring; US or non-contrast CT if diagnosis is uncertain (Banks et al., 2013).
- **48–96 h**: CECT or MRI to detect necrosis and early collections; rising CRP or PCT should heighten suspicion of infection (Leppäniemi et al., 2019; Porter et al., 2019).
- **>4 weeks**: imaging reassessment for pseudocysts or walled-off necrosis; intervention only if symptomatic or infected (Banks et al., 2013).
- **At any stage**: vascular suspicion warrants CTA/MRA or Doppler; prompt endovascular management is essential if bleeding is confirmed (Leppäniemi et al., 2019; Boudghène et al., 1993).

5.9.5 Practical recommendations

Several principles are mentioned. First, early CT is rarely justified in mild disease and may expose patients to unnecessary radiation. Second, laboratory trajectories, rather than static thresholds, should guide imaging decisions (Porter et al., 2019). Third, EUS is most valuable in selected cases for drainage planning or microbiological sampling. Finally, when infection is suspected, clinicians should avoid premature open surgery and instead adopt the minimally invasive, step-up approach that has transformed outcomes in recent years (Baron et al., 2020).

The mechanisms underlying pancreatic necrosis and organ failure are outlined in Section 2.5, providing the pathophysiological basis for these complications.

5.10 Future perspectives and research directions

The diagnosis of AP has advanced considerably, yet further refinement is needed to improve early detection, etiology-specific diagnosis, and prediction of severity. Future research is expected to leverage multi-omics data, AI, novel biomarkers, and digital health strategies to establish more precise and dynamic diagnostic frameworks.

5.10.1 Integration of multi-omics approaches

Genomic discoveries, such as mutations in *PRSS1*, *SPINK1*, *CFTR*, and *CTRC*, have clarified genetic predispositions to recurrent and hereditary AP. Large-scale genome-wide association studies continue to identify novel susceptibility loci, offering opportunities for precision risk stratification (Rosendahl et al., 2013). Beyond genomics, proteomic profiling of serum and urine samples has demonstrated potential for detecting biomarkers prior to enzyme elevation, while metabolomic signatures may distinguish between alcoholic and biliary etiologies. Integration of multi-omics platforms is therefore likely to underpin a personalized approach to AP diagnosis and monitoring (Ge et al., 2024).

5.10.2 Advances in imaging and AI integration

Recent advances in radiomics and AI highlight the potential of imaging for diagnostic and prognostic prediction in AP. In a recent multicenter study, applying radiomics analysis to early contrast-enhanced CT scans achieved an AUC of approximately 0.90 for disease severity prediction, outperforming conventional clinical scoring systems (Qi et al., 2024). Similarly, a deep convolutional neural network (DenseNet) was used to classify AP severity on contrast-enhanced CT images, demonstrating an AUC of 0.8 or higher in distinguishing mild, moderate, and severe disease morphologies (Liang et al., 2023). These findings underscore the potential of computational image analysis to complement or even surpass conventional radiological evaluation. However, current research remains constrained by small sample sizes, methodological heterogeneity, and a lack of external validation, underscoring the urgent need for standardization and multicenter prospective trials (Zhong et al., 2022).

5.10.3 Novel laboratory biomarkers

Beyond imaging, novel biomarkers are emerging as tools for ultra-early diagnosis. TAP and urinary trypsinogen-2 demonstrate diagnostic value for early detection of pancreatic injury (Yasuda et al., 2019), while circulating microRNA (miRNA) is increasingly recognized as potentially reflecting inflammatory and necrotic activity (Zhou et al., 2022). Recent studies have also identified phospholipase D2 (PLD2) as a potential molecular indicator of disease severity (Wang et al., 2025). Meanwhile, advances in immunoassays and molecular detection technologies are expanding the potential for rapid point-of-care biomarker testing, bringing these diagnostic methods closer to bedside application (Hao et al., 2025).

5.10.4 Population-specific diagnostic research

Special populations may require tailored diagnostic strategies. In pediatric patients, variations in enzyme activity, clinical presentation, and genetic susceptibility suggest that age-specific reference ranges and broader genetic screening could improve diagnostic precision, although their routine use remains under evaluation (Uc and Husain, 2019). During pregnancy, AP is uncommon but diagnostically complex; current evidence indicates that conventional imaging and laboratory criteria may not always be reliable, and safe, pregnancy-adapted imaging protocols and potential pregnancy-specific biomarkers are areas of ongoing investigation (Zhang et al., 2022). In older adults, atypical or subtle presentations and multiple comorbidities often obscure classic biochemical and imaging findings, implying that modified diagnostic thresholds and individualized interpretation may be warranted (Baeza-Zapata et al., 2021). For trauma-associated pancreatitis, diagnostic delays remain a persistent challenge. Serial clinical, biochemical, and imaging reassessment is often required to detect evolving pancreatic injury, as early imaging or enzyme measurements may appear deceptively normal (Kumar et al., 2016). Future studies focusing on these special populations will be essential to refine diagnostic accuracy and promote equitable evaluation across diverse patient groups.

5.10.5 Digital health and telemedicine

Digital health innovations are recognized as opportunities for continuous monitoring and remote management of patients at risk for AP. Recent telemedicine models enable remote tracking of laboratory value, vital signs, and patient-reported symptoms, potentially allowing early identification of disease progression (Sorribas et al., 2023). Simultaneously, AI-based decision support systems have been developed, integrating medical history, imaging, and laboratory data to offer potential for risk stratification and intervention strategies (López Gordo et al., 2025). These digital tools have the potential to enhance the timeliness and equity of care by bridging geographical and infrastructural gaps in pancreatitis management in both high-resource and resource-constrained settings (Tan et al., 2025).

5.10.6 Toward a unified and dynamic diagnostic algorithm

Current diagnostic frameworks, such as the Revised Atlanta Classification, still rely largely on static thresholds for enzyme levels and imaging findings. Emerging evidence suggests that diagnostic accuracy could be improved by incorporating temporal dynamics of clinical and laboratory parameters rather than single-time-point values. For example, serial assessment of inflammatory and biochemical markers within the first 48 h has been shown to better predict disease progression than baseline measurements alone (Librero-Jiménez et al., 2024). More recently, AI–driven models such as the Adaptive Dynamic AP Tracker (ADAPT) have been developed to integrate real-time clinical data and fluid management variables, exemplifying the shift toward continuously adaptive decision frameworks (Kong et al., 2025). Although these technologies remain in early phases of validation, they collectively point toward a future in which diagnosis and monitoring are continuously updated in response to evolving patient data, enabling earlier and more individualized therapeutic intervention.

5.10.7 Conclusion

Future directions in AP diagnostics will likely be multidimensional, combining omics-based biomarkers for early detection, AI-driven imaging for precise severity assessment, and dynamic, individualized diagnostic algorithms. If these strategies are successfully implemented, they have the potential to minimize diagnostic delays, reduce complications, and improve outcomes across all patient populations.

AI disclosure

During the preparation of this work the author(s) used Chat GPT in order to correct grammatical errors and refine the wording by removing redundant expressions, thereby improving the overall clarity and fluency of the manuscript. After using this tool/service, the author(s) reviewed and edited the content as needed and take(s) full responsibility for the content of the publication.

References

Ahmed Ali, U., Issa, Y., Hagenaars, J.C., Bakker, O.J., van Goor, H., Nieuwenhuijs, V.B., Bollen, T.L., van Ramshorst, B., Witteman, B.J., Brink, M.A., Schaapherder, A.F.,

Dejong, C.H., Spanier, B.W.M., Heisterkamp, J., van der Harst, E., van Eijck, C.H., Besselink, M.G., Gooszen, H.G., van Santvoort, H.C., Boermeester, M.A., 2016. Risk of recurrent pancreatitis and progression to chronic pancreatitis after a first episode of acute pancreatitis. Clinical Gastroenterology and Hepatology 14 (5), 738–746. https://doi.org/10.1016/j.cgh.2015.12.040.

Alberti, P., Pando, E., Mata, R., Cirera, A., Fernandes, N., Hidalgo, N., Gomez-Jurado, M.J., Vidal, L., Dopazo, C., Blanco, L., Gómez, C., Caralt, M., Balsells, J., Charco, R., 2022. The role of procalcitonin as a prognostic factor for acute cholangitis and infections in acute pancreatitis: a prospective cohort study from a European single center. HPB 24 (6), 875–884. https://doi.org/10.1016/j.hpb.2021.10.016.

Allemann, A., Staubli, S.M., Nebiker, C.A., 2024. Trypsin and trypsinogen activation peptide in the prediction of severity of acute pancreatitis. Life 14 (9). https://doi.org/10.3390/life14091055. Available from: www.mdpi.com/journal/life/.

Alter, D., Koch, D.D., 2021. A review of acute pancreatitis. JAMA, the Journal of the American Medical Association 325 (23), 2399–2402. https://doi.org/10.1001/jama.2021.6012.

Arvanitakis, M., Dumonceau, J.M., Albert, J., Badaoui, A., Bali, M.A., Barthet, M., Besselink, M., Deviere, J., Ferreira, A.O., Gyökeres, T., Hritz, I., Hucl, T., Milashka, M., Papanikolaou, I.S., Poley, J.W., Seewald, S., Vanbiervliet, G., Van Lienden, K., Van Santvoort, H., Voermans, R., Delhaye, M., Van Hooft, J., 2018. Endoscopic management of acute necrotizing pancreatitis: European Society of Gastrointestinal Endoscopy (ESGE) evidence-based multidisciplinary guidelines. Endoscopy 50 (5), 524–546. https://doi.org/10.1055/a-0588-5365.

Badalov, N., Baradarian, R., Iswara, K., Li, J., Steinberg, W., Tenner, S., 2007. Drug-induced acute pancreatitis: an evidence-based review. Clinical Gastroenterology and Hepatology 5 (6), 648. https://doi.org/10.1016/j.cgh.2006.11.023.

Baeza-Zapata, A.A., García-Compeán, D., Jaquez-Quintana, J.O., Scharrer-Cabello, S.I., Del Cueto-Aguilera, Á.N., Maldonado-Garza, H.J., 2021. Acute pancreatitis in elderly patients. Gastroenterology 161 (6), 1736–1740. https://doi.org/10.1053/j.gastro.2021.06.081.

Baillie, J., 1997. Rapid measurement of urinary trypsinogen-2 as a screening test for acute pancreatitis. Gastrointestinal Endoscopy 46. https://doi.org/10.1056/nejm199706193362504.

Bálint, E.R., Fűr, G., Kiss, L., Németh, D.I., Soós, A., Hegyi, P., Szakács, Z., Tinusz, B., Varjú, P., Vincze, Á., Erőss, B., Czimmer, J., Szepes, Z., Varga, G., Rakonczay, Z., 2020. Assessment of the course of acute pancreatitis in the light of aetiology: a systematic review and meta-analysis. Scientific Reports 10 (1). https://doi.org/10.1038/s41598-020-74943-8. Available from: www.nature.com/srep/index.html.

Balthazar, E.J., Robinson, D.L., Megibow, A.J., Ranson, J.H.C., 1990. Acute pancreatitis: value of CT in establishing prognosis. Radiology 174 (2), 331–336. https://doi.org/10.1148/radiology.174.2.2296641.

Balthazar, E.J., 2002. Acute pancreatitis: assessment of severity with clinical and CT evaluation. Radiology 223 (3), 603–613. https://doi.org/10.1148/radiol.2233010680.

Banks, P.A., Bollen, T.L., Dervenis, C., Gooszen, H.G., Johnson, C.D., Sarr, M.G., Tsiotos, G. G., Vege, S.S., Windsor, J.A., Horvath, K.D., Mortele, K.J., Gardner, T.B., Van Santvoort, H., Pelaez-Luna, M., Yadav, D., Stefanidis, G., Delakidis, S., Morgan, D.E., Thoeni, R.F.L., Acosta, J.M., Amann, S.T., Andren-Sandberg, A., Aranha, G.V., Asciutti, S., Barauskas, G., Baron, T.H., Bassi, C., Behrman, S., Behrns, K.E., Belliappa, V., Berzin, T.M., Besselink, M.G.H., Bhasin, D.K., Biankin, A., Bishop, M.

D., Bonini, C.J., Bradley, E.L., Buechler, M., Carter, R., Cavestro, G.M., Chari, S.T., Chavez-Rodriguez, J.J., Da Cunha, J.E., D'Agostino, H., De Campos, T., De-Madaria, E., Deprez, P.H., DiSario, J.A., Doria, C., Falconi, M., Fernandez-del Castillo, C., Freeny, P.C., Frey, C.F., Friess, H., Frossard, J.L., Fuchshuber, P., Gallagher, S.F., Garg, P.K., Ghattas, G., Glasgow, R., Gonzalez, J.A., Gress, T.M., Gumbs, A.A., Hallibruton, C., Helton, S., Hill, M.C., Hoyos, S., Imrie, C.W., Isenmann, R., Izbicki, J.R., Karagiannis, J.A., Klar, E., Kolokythas, O., Lau, J., Litvin, A.A., Longnecker, D.S., Lowenfels, A.B., Mackey, R., Mah'Moud, M., Malangoni, M., McFadden, D.W., Mishra, G., Moody, F.G., Morinville, V., Neoptolemos, J.P., Nordback, I., Pap, A., Papachristou, G.I., Parks, R., Pedrazolli, S., Pezzilli, R., Pitt, H.A., Prosanto, C., Ramesh, H., Ramirez, F.C., Raper, S.E., Rasheed, A., Reed, D.N., Romangnuolo, J., Rossaak, J., Sanabria, J., Schaefer, C., Schmidt, J., Schmidt, P.N., Serrablo, A., Senkowski, C.K., Sharma, M., Sigman, K.M., Singh, P., Steinberg, W., Steiner, J., Strasberg, S., Strum, W., Takada, T., Tanaka, M., Vaccaro, M., Villavicencio, R.L., Vrochides, D., Wagner, M., Warshaw, A.L., Wilcox, C.M., Wysocki, P., Zenilman, M.E., Zyromski, N.J., 2013. Classification of acute pancreatitis—2012: revision of the Atlanta classification and definitions by international consensus. Gut 62 (1), 102–111. https://doi.org/10.1136/gutjnl-2012-302779.

Baron, T.H., DiMaio, C.J., Wang, A.Y., Morgan, K.A., 2020. American gastroenterological association clinical practice update: management of pancreatic necrosis. Gastroenterology 158 (1), 67. https://doi.org/10.1053/j.gastro.2019.07.064.

Basak, F., Hasbahceci, M., Sisik, A., Acar, A., Tekesin, K., Bas, G., Alimoğlu, O., 2017. Can C-reactive protein levels increase the accuracy of the Ranson score in predicting the severity and prognosis of acute pancreatitis? A prospective cohort study. Turkish Journal of Gastroenterology 28 (3), 207–213. https://doi.org/10.5152/tjg.2017.16686.

Basnayake, C., Ratnam, D., 2015. Blood tests for acute pancreatitis. Australian Prescriber 38 (4), 128–130. https://doi.org/10.18773/austprescr.2015.043.

Berglund, L., Brunzell, J.D., Goldberg, A.C., Goldberg, I.J., Sacks, F., Murad, M.H., Stalenhoef, A.F.H., 2012. Evaluation and treatment of hypertriglyceridemia: an endocrine society clinical practice guideline. Journal of Clinical Endocrinology and Metabolism 97 (9), 2969–2989. https://doi.org/10.1210/jc.2011-3213.

Bhatia, M., Fei, L.W., Cao, Y., Hon, Y.L., Huang, J., Puneet, P., Chevali, L., 2005. Pathophysiology of acute pancreatitis. Pancreatology 5 (2–3), 132–144. https://doi.org/10.1159/000085265.

Block, S., Maier, W., Bittner, R., Büchler, M., Malfertheiner, P., Beger, H.G., 1986. Identification of pancreas necrosis in severe acute pancreatitis: imaging procedures versus clinical staging. Gut 27 (9), 1035–1042. https://doi.org/10.1136/gut.27.9.1035.

Boudghène, F., L'Herminé, C., Bigot, J.M., 1993. Arterial complications of pancreatitis: diagnostic and therapeutic aspects in 104 cases. Journal of Vascular and Interventional Radiology 4 (4), 551–558. https://doi.org/10.1016/S1051-0443(93)71920-X.

Bradley, E.L., 1993. A clinically based classification system for acute pancreatitis: summary of the international symposium on acute pancreatitis, Atlanta, Ga, September 11 through 13, 1992. Archives of Surgery 128 (5), 586–590. https://doi.org/10.1001/archsurg.1993.01420170122019.

Burnett, J.R., Hooper, A.J., Hegele, R.A., 1993. Familial Lipoprotein Lipase Deficiency.

Busireddy, K.K., AlObaidy, M., Ramalho, M., Kalubowila, J., Baodong, L., Santagostino, I., Semelka, R.C., 2014. Pancreatitis-imaging approach. World Journal of Gastrointestinal Pathophysiology 5 (3), 252. https://doi.org/10.4291/wjgp.v5.i3.252.

Buter, A., Imrie, C.W., Carter, C.R., Evans, S., McKay, C.J., 2002. Dynamic nature of early organ dysfunction determines outcome in acute pancreatitis. British Journal of Surgery 89 (3), 298–302. https://doi.org/10.1046/j.0007-1323.2001.02025.x.

Buxbaum, J.L., Freeman, M., Amateau, S.K., Chalhoub, J.M., Chowdhury, A., Coelho-Prabhu, N., Das, R., Desai, M., Elhanafi, S.E., Forbes, N., Fujii-Lau, L.L., Kohli, D.R., Kwon, R.S., Machicado, J.D., Marya, N.B., Pawa, S., Ruan, W.H., Sadik, J., Sheth, S. G., Thiruvengadam, N.R., Thosani, N.C., Zhou, S., Qumseya, B.J., 2023. American Society for Gastrointestinal Endoscopy guideline on post-ERCP pancreatitis prevention strategies: methodology and review of evidence. Gastrointestinal Endoscopy 97 (2), 163. https://doi.org/10.1016/j.gie.2022.09.011.

Cardoso, F.S., Ricardo, L.B., Oliveira, A.M., Canena, J.M., Horta, D.V., Papoila, A.L., Deus, J.R., 2013. C-reactive protein prognostic accuracy in acute pancreatitis: timing of measurement and cutoff points. European Journal of Gastroenterology and Hepatology 25 (7), 784–789. https://doi.org/10.1097/MEG.0b013e32835fd3f0.

Catalano, M.F., Sahai, A., Levy, M., Romagnuolo, J., Wiersema, M., Brugge, W., Freeman, M., Yamao, K., Canto, M., Hernandez, L.V., 2009. EUS-based criteria for the diagnosis of chronic pancreatitis: the Rosemont classification. Gastrointestinal Endoscopy 69 (7), 1251–1261. https://doi.org/10.1016/j.gie.2008.07.043.

Catena, F., De Simone, B., Coccolini, F., Di Saverio, S., Sartelli, M., Ansaloni, L., 2019. Bowel obstruction: a narrative review for all physicians. World Journal of Emergency Surgery 14 (1). https://doi.org/10.1186/s13017-019-0240-7.

Chen, L., Jiang, J., 2022. The diagnostic value of procalcitonin in patients with severe acute pancreatitis: a meta-analysis. Turkish Journal of Gastroenterology 33 (9), 722–730. https://doi.org/10.5152/tjg.2022.22098.

Chen, C.C., Wang, S.S., Lee, F.Y., Chang, F.Y., Lee, S.D., 1999. Proinflammatory cytokines in early assessment of the prognosis of acute pancreatitis. American Journal of Gastroenterology 94 (1), 213–218. https://doi.org/10.1016/s0002-9270(98)00590-5.

Chen, C., Cai, J., Wu, X., Hu, Q., Wu, X., Guo, X., Hu, S., Zhai, X., Qi, B., Zhang, W., Wang, N., Xu, F., Fang, Y., Li, C.k., Liang, C., Wang, L., Zhu, J., Deng, W., Tian, X., Ju, X., Zhang, G., Tang, J., Cheng, C., Yang, J., Shen, S., Pui, C.H., 2025. Risk factors and outcomes of asparaginase-associated pancreatitis in pediatric patients with ALL. Blood. https://doi.org/10.1182/blood.2025029785. Available from: https://www.sciencedirect.com/science/journal/00064971.

Coccolini, F., Kobayashi, L., Kluger, Y., Moore, E.E., Ansaloni, L., Biffl, W., Leppaniemi, A., Augustin, G., Reva, V., Wani, I., Kirkpatrick, A., Abu-Zidan, F., Cicuttin, E., Fraga, G.P., Ordonez, C., Pikoulis, E., Sibilla, M.G., Maier, R., Matsumura, Y., Masiakos, P.T., Khokha, V., Mefire, A.C., Ivatury, R., Favi, F., Manchev, V., Sartelli, M., Machado, F., Matsumoto, J., Chiarugi, M., Arvieux, C., Catena, F., Coimbra, R., Ben-Ishay, O., Tolonen, M., Bertelli, R., Horer, T., Ferrada, P., Di Carlo, I., Pereira, B.M., Parini, D., Montori, G., De Simone, B., Chiara, O., Hecker, A., Deangelis, N., Gomes, C.A., Galante, J., Bala, M., Mylonas, K.S., Pikoulis, A., Perfetti, P., Chirica, M., Bado, J., Inaba, K., Parry, N., Romeo, O., Stommel, M., Rajashekar, M., Tan, E., Salvetti, F., Sakakushev, B., 2019. Duodeno-pancreatic and extrahepatic biliary tree trauma: WSES-AAST guidelines. World Journal of Emergency Surgery 14 (1). https://doi.org/10.1186/s13017-019-0278-6. Available from: http://www.wjes.org/.

Cohn, J.A., Friedman, K.J., Noone, P.G., Knowles, M.R., Silverman, L.M., Jowell, P.S., 1998. Relation between mutations of the cystic fibrosis gene and idiopathic pancreatitis. New

England Journal of Medicine 339 (10), 653–658. https://doi.org/10.1056/NEJM199809033391002.

Comfort, M.W., Gambill, E.E., Baggenstoss, A.H., 1946. Chronic relapsing pancreatitis; a study of 29 cases without associated disease of the biliary or gastrointestinal tract. Gastroenterology 6, 376–408.

Copel, J., El-Sayed, Y., Heine, R.P., Wharton, K.R., 2017. Guidelines for diagnostic imaging during pregnancy and lactation. Obstetrics & Gynecology 130 (4), 933–934. https://doi.org/10.1097/AOG.0000000000002350.

Cosen-Binker, L.I., Gaisano, H.Y., 2007. Recent insights into the cellular mechanisms of acute pancreatitis. Canadian Journal of Gastroenterology 21 (1), 19–24. https://doi.org/10.1155/2007/930424.

da Costa, D.W., Dijksman, L.M., Bouwense, S.A., Schepers, N.J., Besselink, M.G., van Santvoort, H.C., Boerma, D., Gooszen, H.G., Dijkgraaf, M.G.W., 2016. Cost-effectiveness of same-admission versus interval cholecystectomy after mild gallstone pancreatitis in the PONCHO trial. British Journal of Surgery 103 (12), 1695–1703. https://doi.org/10.1002/bjs.10222. Available from: http://onlinelibrary.wiley.com/journal/10.1002/(ISSN)1365-2168.

de Oliveira, C., Khatua, B., Noel, P., Kostenko, S., Bag, A., Balakrishnan, B., Patel, K.S., Guerra, A.A., Martinez, M.N., Trivedi, S., McCullough, A., Lam-Himlin, D.M., Navina, S., Faigel, D.O., Fukami, N., Pannala, R., Phillips, A.E., Papachristou, G.I., Kershaw, E.E., Lowe, M.E., Singh, V.P., 2020. Pancreatic triglyceride lipase mediates lipotoxic systemic inflammation. Journal of Clinical Investigation 130 (4), 1931–1947. https://doi.org/10.1172/JCI132767.

de Pretis, N., Amodio, A., Frulloni, L., 2018. Hypertriglyceridemic pancreatitis: epidemiology, pathophysiology and clinical management. United European Gastroenterology Journal 6 (5), 649–655. https://doi.org/10.1177/2050640618755002.

Dermine, S., Masson, E., Girodon-Boulandet, E., Bienvenu, T., Férec, C., Lévy, P., Rebours, V., 2024. Diagnostic yield of repeat genetic testing in idiopathic chronic pancreatitis. Clinics and Research in Hepatology and Gastroenterology 48 (6), 102346. https://doi.org/10.1016/j.clinre.2024.102346.

Dos Santos Pereira, S.I., Alshatti, A., Nourhane, Al A., Rukh, S., Malik, N.Q., Ahmed, S., Kaur, R., Siddiqui, H.F., Ischemia, M., 2025. A comprehensive review of the role of radiology in diagnosis. Cureus 17 (8). https://doi.org/10.7759/cureus.90384.

Eachempati, S.R., Hydo, L.J., Barie, P.S., 2002. Severity scoring for prognostication in patients with severe acute pancreatitis. Archives of Surgery 137 (6). https://doi.org/10.1001/archsurg.137.6.730.

Elbanna, K.Y., Jang, H.J., Kim, T.K., 2020. Imaging diagnosis and staging of pancreatic ductal adenocarcinoma: a comprehensive review. Insights Into Imaging 11 (1). https://doi.org/10.1186/s13244-020-00861-y. Available from: http://www.springer.com/medicine/radiology/journal/13244.

Elman, R., 1937. The variations of blood amylase during acute transient disease of the pancreas. Annals of Surgery 105 (3), 379–384. https://doi.org/10.1097/00000658-193703000-00008.

Elmunzer, B.J., Scheiman, J.M., Lehman, G.A., Chak, A., Mosler, P., Higgins, P.D.R., Hayward, R.A., Romagnuolo, J., Elta, G.H., Sherman, S., Waljee, A.K., Repaka, A., Atkinson, M.R., Cote, G.A., Kwon, R.S., McHenry, L., Piraka, C.R., Wamsteker, E.J., Watkins, J.L., Korsnes, S.J., Schmidt, S.E., Turner, S.M., Nicholson, S., Fogel, E.L., 2012. A randomized trial of rectal indomethacin to prevent post-ERCP pancreatitis.

New England Journal of Medicine 366 (15), 1414–1422. https://doi.org/10.1056/NEJMoa1111103. Available from: http://www.nejm.org/doi/pdf/10.1056/NEJMoa1111103.

Fallat, R.W., Vester, J.W., Glueck, C.J., 1973. Suppression of amylase activity by hypertriglyceridemia. JAMA, the Journal of the American Medical Association 225 (11), 1331–1334. https://doi.org/10.1001/jama.1973.03220390031007.

Frossard, J.L., Steer, M.L., Pastor, C.M., 2008. Acute pancreatitis. The Lancet 371 (9607), 143–152. https://doi.org/10.1016/S0140-6736(08)60107-5.

Gan, S.I., Romagnuolo, J., 2004. Admission hematocrit: a simple, useful and early predictor of severe pancreatitis. Digestive Diseases and Sciences 49 (11–12), 1946–1952. https://doi.org/10.1007/s10620-004-9598-8.

Garg, R., Rustagi, T., 2018. Management of hypertriglyceridemia induced acute pancreatitis. BioMed Research International 2018. https://doi.org/10.1155/2018/4721357. Available from: http://www.hindawi.com/journals/biomed/.

Gaudet, D., Alexander, V.J., Baker, B.F., Brisson, D., Tremblay, K., Singleton, W., Geary, R. S., Hughes, S.G., Viney, N.J., Graham, M.J., Crooke, R.M., Witztum, J.L., Brunzell, J.D., Kastelein, J.J.P., 2015. Antisense inhibition of apolipoprotein C-III in patients with hypertriglyceridemia. New England Journal of Medicine 373 (5), 438–447. https://doi.org/10.1056/NEJMoa1400283. Available from: http://www.nejm.org/doi/pdf/10.1056/NEJMoa1400283.

Ge, P., Luo, Y., Zhang, G., Chen, H., 2024. The role of proteomics in acute pancreatitis: new and old knowledge. Expert Review of Proteomics 21 (1–3), 115–123. https://doi.org/10.1080/14789450.2024.2320810.

Goodwin, D., Rosenzweig, B., Zhang, J., Xu, L., Stewart, S., Thompson, K., Rouse, R., 2014. Evaluation of miR-216a and miR-217 as potential biomarkers of acute pancreatic injury in rats and mice. Biomarkers 19 (6), 517–529. https://doi.org/10.3109/1354750X.2014.944217.

Greenberg, J.A., Hsu, J., Bawazeer, M., Marshall, J., Friedrich, J.O., Nathens, A., Coburn, N., May, G.R., Pearsall, E., McLeod, R.S., 2016. Clinical practice guideline: management of acute pancreatitis. Canadian Journal of Surgery 59 (2), 128–140. https://doi.org/10.1503/cjs.015015.

Griffin, N., Charles-Edwards, G., Grant, L.A., 2012. Magnetic resonance cholangiopancreatography: the ABC of MRCP. Insights Into Imaging 3 (1), 11–21. https://doi.org/10.1007/s13244-011-0129-9.

Gubensek, J., 2023. The role of apheresis and insulin therapy in hypertriglyceridemic acute pancreatitis—a concise review. BMC Gastroenterology 23 (1). https://doi.org/10.1186/s12876-023-02957-3.

Habtezion, A., Gukovskaya, A.S., Pandol, S.J., 2019. Acute pancreatitis: a multifaceted set of organelle and cellular interactions. Gastroenterology 156 (7), 1941–1950. https://doi.org/10.1053/j.gastro.2018.11.082.

Hamesch, K., Hollenbach, M., Guilabert, L., Lahmer, T., Koch, A., 2025. Practical management of severe acute pancreatitis. European Journal of Internal Medicine 133, 1–13. https://doi.org/10.1016/j.ejim.2024.10.030.

Hao, X., Wang, Z., Niu, X., Li, L., Bi, Y., Chai, N., 2025. Biomarkers to predict the severity of acute pancreatitis. Frontiers of Medicine 12. https://doi.org/10.3389/fmed.2025.1619087.

Hedström, J., Korvuo, A., Kenkimäki, P., Tikanoja, S., Haapiainen, R., Kivilaakso, E., Stenman, U.H., 1996. Urinary trypsinogen-2 test strip for acute pancreatitis. Lancet 347 (9003), 729–731. https://doi.org/10.1016/S0140-6736(96)90078-1.

Hegyi, E., Sahin-Tóth, M., 2017. Genetic risk in chronic pancreatitis: the trypsin-dependent pathway. Digestive Diseases and Sciences 62 (7), 1692–1701. https://doi.org/10.1007/s10620-017-4601-3.

Huang, W., Booth, D.M., Cane, M.C., Chvanov, M., Javed, M.A., Elliott, V.L., Armstrong, J. A., Dingsdale, H., Cash, N., Li, Y., Greenhalf, W., Mukherjee, R., Kaphalia, B.S., Jaffar, M., Petersen, O.H., Tepikin, A.V., Sutton, R., Criddle, D.N., 2014. Fatty acid ethyl ester synthase inhibition ameliorates ethanol-induced Ca2+-dependent mitochondrial dysfunction and acute pancreatitis. Gut 63 (8), 1313–1324. https://doi.org/10.1136/gutjnl-2012-304058.

Hung, W.Y., 2014. Contemporary review of drug-induced pancreatitis: a different perspective. World Journal of Gastrointestinal Pathophysiology 5. https://doi.org/10.4291/wjgp.v5.i4.405.

Hyare, H., Desigan, S., Brookes, J.A., Guiney, M.J., Lees, W.R., 2007. Endovascular management of major arterial hemorrhage as a complication of inflammatory pancreatic disease. Journal of Vascular and Interventional Radiology 18 (5), 591–596. https://doi.org/10.1016/j.jvir.2007.02.035.

Ibrahim, H., Jumani, A., Elhassan, K., Ibrahim, M., AlNajjar, M., 2024. Acute pancreatitis with normal amylase and lipase: a diagnostic dilemma. Cureus 16 (6). https://doi.org/10.7759/cureus.62374.

Ishibashi, T., Zhao, H., Kawabe, K., Oono, T., Egashira, K., Suzuki, K., Nawata, H., Takayanagi, R., Ito, T., 2008. Blocking of monocyte chemoattractant protein-1 (MCP-1) activity attenuates the severity of acute pancreatitis in rats. Journal of Gastroenterology 43 (1), 79–85. https://doi.org/10.1007/s00535-007-2126-9.

Ismail, O.Z., Bhayana, V., 2017. Lipase or amylase for the diagnosis of acute pancreatitis? Clinical Biochemistry 50 (18), 1275–1280. https://doi.org/10.1016/j.clinbiochem.2017.07.003.

Johnson, A., Cluskey, B., Hooshvar, N., Tice, D., Devin, C., Kao, E., Nawabi, S., Jones, S., Zhang, L., Dola, C., 2015. Significantly elevated serum lipase in pregnancy with nausea and vomiting: acute pancreatitis or hyperemesis gravidarum? Case Reports in Obstetrics and Gynecology 2015, 1–4. https://doi.org/10.1155/2015/359239.

Kaurich, 2008. Drug-induced acute pancreatitis. Proceedings—Baylor University Medical Center 21, 77–81. https://doi.org/10.1080/08998280.2008.11928366.

Keim, V., Teich, N., Fiedler, F., Hartig, W., Thiele, G., Mössner, J., 1998. A comparison of lipase and amylase in the diagnosis of acute pancreatitis in patients with abdominal pain. Pancreas 16 (1), 45–49. https://doi.org/10.1097/00006676-199801000-00008.

Khan, Z., Vlodov, J., Horovitz, J., Jose, R.M., Iswara, K., Smotkin, J., Brown, A., Tenner, S., 2002. Urinary trypsinogen activation peptide is more accurate than hematocrit in determining severity in patients with acute pancreatitis: a prospective study. American Journal of Gastroenterology 97 (8), 1973–1977. https://doi.org/10.1111/j.1572-0241.2002.05953.x.

Kiss, S., Pintér, J., Molontay, R., Nagy, M., Farkas, N., Sipos, Z., Fehérvári, P., Pecze, L., Földi, M., Vincze, Á., Takács, T., Czakó, L., Izbéki, F., Halász, A., Boros, E., Hamvas, J., Varga, M., Mickevicius, A., Faluhelyi, N., Farkas, O., Váncsa, S., Nagy, R., Bunduc, S., Hegyi, P.J., Márta, K., Borka, K., Doros, A., Hosszúfalusi, N., Zubek, L., Erőss, B., Molnár, Z., Párniczky, A., Hegyi, P., Szentesi, A., Kiss, S., Sipos, Z., Fehérvári, P., Pecze, L., Földi, M., Vincze, Á., Takács, T., Czakó, L., Izbéki, F., Halász, A., Boros, E., Hamvas, J., Varga, M., Mickevicius, A., Faluhelyi, N., Váncsa, S., Nagy, R., Bunduc, S., Márta, K., Borka, K., Doros, A., Hosszúfalusi, N., Zubek, L., Molnár, Z., Párniczky, A., Szentesi, A., Bajor, J., Gódi, S., Sarlós, P.,

Czimmer, J., Szabó, I., Pár, G., Illés, A., Hágendorn, R., Németh, B.C., Kui, B., Illés, D., Gajdán, L., Dunás-Varga, V., Fejes, R., Papp, M., Vitális, Z., Novák, J., Török, I., Macarie, M., Ramírez-Maldonado, E., Sallinen, V., Galeev, S., Bod, B., Ince, A.T., Pécsi, D., Varjú, P., Juhász, M.F., Ocskay, K., Mikó, A., Szakács, Z., 2022. Early prediction of acute necrotizing pancreatitis by artificial intelligence: a prospective cohort-analysis of 2387 cases. Scientific Reports 12 (1). https://doi.org/10.1038/s41598-022-11517-w. Available from: www.nature.com/srep/index.html.

Kiss, L., Fűr, G., Pisipati, S., Rajalingamgari, P., Ewald, N., Singh, V., Rakonczay, Z., 2023. Mechanisms linking hypertriglyceridemia to acute pancreatitis. Acta Physiologica 237 (3). https://doi.org/10.1111/apha.13916. Available from: http://onlinelibrary.wiley.com/journal/10.1111/(ISSN)1748-1716.

Klöppel, G., Dreyer, T., Willemer, S., Kern, H.F., Adler, G., 1986. Human acute pancreatitis: its pathogenesis in the light of immunocytochemical and ultrastructural findings in acinar cells. Virchows Archiv A Pathological Anatomy and Histopathology 409 (6), 791–803. https://doi.org/10.1007/BF00710764.

Kong, N., Chang, P., Shulman, I.A., Haq, U., Amini, M., Nguyen, D., Khan, F., Narala, R., Sharma, N., Wang, D., Thompson, T., Sadik, J., Breze, C., Whitcomb, D.C., Buxbaum, J.L., 2025. Machine learning-guided fluid resuscitation for acute pancreatitis improves outcomes. Clinical and Translational Gastroenterology 16 (4), e00825. https://doi.org/10.14309/ctg.0000000000000825.

Kumar, A., Panda, A., Gamanagatti, S., 2016. Blunt pancreatic trauma: a persistent diagnostic conundrum? World Journal of Radiology 8 (2), 159. https://doi.org/10.4329/wjr.v8.i2.159.

Kusnierz-Cabala, B., Nowak, E., Sporek, M., Kowalik, A., Kuźniewski, M., Enguita, F.J., Stępień, E., 2015. Serum levels of unique MIR-551-5p and endothelial-specific MIR-126a-5p allow discrimination of patients in the early phase of acute pancreatitis. Pancreatology 15 (4), 344–351. https://doi.org/10.1016/j.pan.2015.05.475.

Lanas, A., Chan, F.K.L., 2017. Peptic ulcer disease. The Lancet 390 (10094), 613–624. https://doi.org/10.1016/S0140-6736(16)32404-7.

LaRusch, J., Michael Barmada, M., Solomon, S., Whitcomb, D.C., 2012. Whole exome sequencing identifies multiple, complex etiologies in an idiopathic hereditary pancreatitis kindred. Journal of the Pancreas 13 (3), 258–262.

Lee, D.W., Cho, C.M., 2022. Predicting severity of acute pancreatitis. Medicina 58 (6), 787. https://doi.org/10.3390/medicina58060787.

Lee, E.S., Lee, J.M., 2014. Imaging diagnosis of pancreatic cancer: a state-of-the-art review. World Journal of Gastroenterology 20 (24), 7864–7877. https://doi.org/10.3748/wjg.v20.i24.7864.

Leppäniemi, A., Tolonen, M., Tarasconi, A., Segovia-Lohse, H., Gamberini, E., Kirkpatrick, A.W., Ball, C.G., Parry, N., Sartelli, M., Wolbrink, D., Van Goor, H., Baiocchi, G., Ansaloni, L., Biffl, W., Coccolini, F., Di Saverio, S., Kluger, Y., Moore, E., Catena, F., 2019. WSES guidelines for the management of severe acute pancreatitis. World Journal of Emergency Surgery 14 (1). https://doi.org/10.1186/s13017-019-0247-0. Available from: http://www.wjes.org/.

Leser, H.-G., Gross, V., Scheibenbogen, C., Heinisch, A., Salm, R., Lausen, M., Rückauer, K., Andreesen, R., Farthmann, E.H., Schölmerich, J., 1991. Elevation of serum interleukin-6 concentration precedes acute-phase response and reflects severity in acute pancreatitis. Gastroenterology 101 (3), 782–785. https://doi.org/10.1016/0016-5085(91)90539-w.

Liang, H., Wang, M., Wen, Y., Du, F., Jiang, L., Geng, X., Tang, L., Yan, H., 2023. Predicting acute pancreatitis severity with enhanced computed tomography scans using convolutional neural networks. Scientific Reports 13 (1). https://doi.org/10.1038/s41598-023-44828-7. Available from: https://www.nature.com/srep/.

Librero-Jiménez, M., Valverde-López, F., Abellán-Alfocea, P., Fernández-Cano, M.C., Fernández-Fernández, E., Martínez-Cara, J.G., López-González, E., Jiménez-Rosales, R., Redondo-Cerezo, E., 2024. Usefulness of dynamic assessment of clinical and laboratory factors in severe acute pancreatitis. Journal of Clinical Medicine 13 (15). https://doi.org/10.3390/jcm13154412. Available from: https://www.mdpi.com/journal/jcm.

Lin, W., Huang, Y., Zhu, J., Sun, H., Su, N., Pan, J., Xu, J., Chen, L., 2024. Machine learning improves early prediction of organ failure in hyperlipidemia acute pancreatitis using clinical and abdominal CT features. Pancreatology 24 (3), 350–356. https://doi.org/10.1016/j.pan.2024.02.003.

López Gordo, S., Ramirez-Maldonado, E., Fernandez-Planas, M.T., Bombuy, E., Memba, R., Jorba, R., 2025. AI and machine learning for precision medicine in acute pancreatitis: a narrative review. Medicina 61 (4), 629. https://doi.org/10.3390/medicina61040629.

Lott, J.A., Patel, S.T., Sawhney, A.K., Kazmierczak, S.C., Love, J.E., 1986. Assays of serum lipase: analytical and clinical considerations. Clinical Chemistry 32 (7), 1290–1302. https://doi.org/10.1093/clinchem/32.7.1290.

Lowenfels, A.B., Maisonneuve, P., DiMagno, E.P., Elitsur, Y., Gates, L.K., Perrault, J., Whitcomb, D.C., 1997. Hereditary pancreatitis and the risk of pancreatic cancer. Journal of the National Cancer Institute 89 (6), 442–446. https://doi.org/10.1093/jnci/89.6.442.

Luo, J., Lan, L., Huang, S., Zeng, X., Xiang, Q., Li, M., Yang, S., Zhao, W., Zhou, X., 2023. Real-time prediction of organ failures in patients with acute pancreatitis using longitudinal irregular data. Journal of Biomedical Informatics 139. https://doi.org/10.1016/j.jbi.2023.104310. Available from: http://www.elsevier.com/inca/publications/store/6/2/2/8/5/7/index.htt.

Mądro, A., 2022. Pancreatitis in pregnancy—comprehensive review. International Journal of Environmental Research and Public Health 19. https://doi.org/10.3390/ijerph192316179.

Melnick, S., Nazir, S., Gish, D., Raj Aryal, M., 2016. Hypertriglyceridemic pancreatitis associated with confounding laboratory abnormalities. Journal of Community Hospital Internal Medicine Perspectives 6 (3), 31808. https://doi.org/10.3402/jchimp.v6.31808.

Mitra, B., Fitzgerald, M., Raoofi, M., Tan, G.A., Spencer, J.C., Atkin, C., 2014. Serum lipase for assessment of pancreatic trauma. European Journal of Trauma and Emergency Surgery 40 (3), 309–313. https://doi.org/10.1007/s00068-013-0341-z.

Moore, E.E., Cogbill, T.H., Malangoni, M.A., Jurkovich, G.J., Champion, H.R., Gennarelli, T. A., Mc aninch, J.W., Pachter, H.L., Shackford, S.R., Trafton, P.G., 1990. Organ injury scaling, II: pancreas, duodenum, small bowel, colon, and rectum. The Journal of Trauma, Injury, Infection, and Critical Care 30 (11), 1427–1429. https://doi.org/10.1097/00005373-199011000-00035.

Morgan, D.E., Baron, T.H., Smith, J.K., Robbin, M.L., Kenney, P.J., 1997. Pancreatic fluid collections prior to intervention: evaluation with MR imaging compared with CT and US. Radiology 203 (3), 773–778. https://doi.org/10.1148/radiology.203.3.9169703.

Moschetta, M., 2015. Pancreatic trauma: the role of computed tomography for guiding therapeutic approach. World Journal of Radiology 7 (11), 415. https://doi.org/10.4329/wjr.v7.i11.415.

Mossa-Basha, M., Zhu, C., Pandhi, T., Mendoza, S., Azadbakht, J., Safwat, A., Homen, D., Zamora, C., Gnanasekaran, D.K., Peng, R., Cen, S., Duddalwar, V., Alger, J.R., Wang, D.J.J., 2024. Deep learning denoising improves CT perfusion image quality in the setting of lower contrast dosing: a feasibility study. American Journal of Neuroradiology 45 (10), 1468–1474. https://doi.org/10.3174/ajnr.A8367.

Mounzer, R., Langmead, C.J., Wu, B.U., Evans, A.C., Bishehsari, F., Muddana, V., Singh, V. K., Slivka, A., Whitcomb, D.C., Yadav, D., Banks, P.A., Papachristou, G.I., 2012. Comparison of existing clinical scoring systems to predict persistent organ failure in patients with acute pancreatitis. Gastroenterology 142 (7), 1476–1482. https://doi.org/10.1053/j.gastro.2012.03.005.

Muthusamy, V.R., Chandrasekhara, V., Acosta, R.D., Bruining, D.H., Chathadi, K.V., Eloubeidi, M.A., Faulx, A.L., Fonkalsrud, L., Gurudu, S.R., Khashab, M.A., Kothari, S., Lightdale, J.R., Pasha, S.F., Saltzman, J.R., Shaukat, A., Wang, A., Yang, J., Cash, B.D., Dewitt, J.M., 2016. The role of endoscopy in the diagnosis and treatment of inflammatory pancreatic fluid collections. Gastrointestinal Endoscopy 83 (3), 481–488. https://doi.org/10.1016/j.gie.2015.11.027.

Neoptolemos, J.P., Kemppainen, E.A., Mayer, J.M., Fitzpatrick, J.M., Raraty, M.G.T., Slavin, J., Beger, H.G., Hietaranta, A.J., Puolakkainen, P.A., 2000. Early prediction of severity in acute pancreatitis by urinary trypsinogen activation peptide: a multicentre study. Lancet 355 (9219), 1955–1960. https://doi.org/10.1016/S0140-6736(00)02327-8.

Nitsche, C.J., Jamieson, N., Lerch, M.M., Mayerle, J.V., 2010. Drug induced pancreatitis . Best practice and research. Clinics in Gastroenterology 24 (2), 143–155. https://doi.org/10.1016/j.bpg.2010.02.002.

Norman, J.G., Franz, M.G., Fink, G.S., Messina, J., Fabri, P.J., Gower, W.R., Carey, L.C., 1995. Decreased mortality of severe acute pancreatitis after proximal cytokine blockade. Annals of Surgery 221 (6), 625–634. https://doi.org/10.1097/00000658-199506000-00002.

O'Connor, O.J., Buckley, J.M., Maher, M.M., 2011. Imaging of the complications of acute pancreatitis. American Journal of Roentgenology 197 (3), W375. https://doi.org/10.2214/AJR.10.4339Ireland.

Ocskay, K., Juhász, M.F., Farkas, N., Zádori, N., Szakó, L., Szakács, Z., Szentesi, A., Er 'Ss, B., Miklós, E.K., Zemplényi, A., Birkás, B., Csathó, Á., Hartung, I., Nagy, T., Czopf, L., Izbéki, F., Gajdán, L., Papp, M., Czakó, L., Illés, D., Marino, M.V., Mirabella, A., Małecka-Panas, E., Zatorski, H., Susak, Y., Opalchuk, K., Capurso, G., Apadula, L., Gheorghe, C., Saizu, I.A., Petersen, O.H., De-Madaria, E., Rosendahl, J., Párniczky, A., Hegyi, P., 2022. Re current acute pancreatitis prevention by the elimination of alcohol and ciga r ette smoking (REAPPEAR): protocol of a randomised controlled trial and a cohort study. BMJ Open 12 (1). https://doi.org/10.1136/bmjopen-2021-050821. Available from: http://bmjopen.bmj.com/content/early/by/section.

Opie, E.L., 1904. The relation of cholelithiasis to disease of the pancreas. Journal of the American Medical Association XLIII (16), 1102–1105. https://doi.org/10.1001/jama.1904.92500160001a.

Pandol, S.J., Saluja, A.K., Imrie, C.W., Banks, P.A., 2007. Acute pancreatitis: bench to the bedside. Gastroenterology 132 (3), 1127–1151. https://doi.org/10.1053/j.gastro.2007.01.055.

Petersen, O.H., Tepikin, A.V., 2008. Polarized calcium signaling in exocrine gland cells. Annual Review of Physiology 70, 273–299. https://doi.org/10.1146/annurev.physiol.70.113006.100618.

Peterson, 2004. Biomarkers for alcohol use and abuse—a summary. Alcohol Research & Health: The Journal of the National Institute on Alcohol Abuse and Alcoholism (2), 30.

Pieńkowska, J., Gwoździewicz, K., Skrobisz-Balandowska, K., Marek, I., Kostro, J., Szurowska, E., Studniarek, M., 2016. Perfusion-CT—can we predict acute pancreatitis outcome within the first 24 hours from the onset of symptoms? PLoS One 11 (1). https://doi.org/10.1371/journal.pone.0146965. Available from. http://www.plosone.org/article/fetchObject.action?uri=info:doi/10.1371/journal.pone.0146965&representation=PDF.

Porter, K.K., Zaheer, A., Kamel, I.R., Horowitz, J.M., Arif-Tiwari, H., Bartel, T.B., Bashir, M. R., Camacho, M.A., Cash, B.D., Chernyak, V., Goldstein, A., Grajo, J.R., Gupta, S., Hindman, N.M., Kamaya, A., McNamara, M.M., Carucci, L.R., 2019. ACR appropriateness Criteria® acute pancreatitis. Journal of the American College of Radiology 16 (11), S316. https://doi.org/10.1016/j.jacr.2019.05.017.

Qi, M., Lu, C., Dai, R., Zhang, J., Hu, H., Shan, X., 2024. Prediction of acute pancreatitis severity based on early CT radiomics. BMC Medical Imaging 24 (1). https://doi.org/10.1186/s12880-024-01509-9. Available from: https://bmcmedimaging.biomedcentral.com/.

Raghuwanshi, S., Gupta, R., Vyas, M.M., Sharma, R., 2016. CT evaluation of acute pancreatitis and its prognostic correlation with CT severity index. Journal of Clinical and Diagnostic Research 10 (6), TC06. https://doi.org/10.7860/JCDR/2016/19849.7934.

Rana, S.S., Vilmann, P., 2015. Endoscopic ultrasound features of chronic pancreatitis: a pictorial review. Endoscopic Ultrasound 4 (1), 10. https://doi.org/10.4103/2303-9027.151314.

Rao, S.A., Kunte, A.R., 2017. Interleukin-6: an early predictive marker for severity of acute pancreatitis. Indian Journal of Critical Care Medicine 21 (7), 424–428. https://doi.org/10.4103/ijccm.IJCCM_478_16.

Rau, B., Steinbach, G., Gansauge, F., Mayer, J.M., Grünert, A., Beger, H.G., 1997. The potential role of procalcitonin and interleukin 8 in the prediction of infected necrosis in acute pancreatitis. Gut 41 (6), 832–840. https://doi.org/10.1136/gut.41.6.832.

Rau, B.M., Kemppainen, E.A., Gumbs, A.A., Büchler, M.W., Wegscheider, K., Bassi, C., Puolakkainen, P.A., Beger, H.G., 2007. Early assessment of pancreatic infections and overall prognosis in severe acute pancreatitis by procalcitonin (PCT): a prospective international multicenter study. Annals of Surgery 245 (5), 745–754. https://doi.org/10.1097/01.sla.0000252443.22360.46.

Rifkind, L., Ranson, J.H., pancreatitis, A., 1976. Initial roentgenographic signs. New York State Journal of Medicine 76, 1968–1972.

Rosendahl, J., Landt, O., Bernadova, J., Kovacs, P., Teich, N., Bodeker, H., Keim, V., Ruffert, C., Mossner, J., Kage, A., Stumvoll, M., Groneberg, D., Kruger, R., Luck, W., Treiber, M., Becker, M., Witt, H., 2013. CFTR, SPINK1, CTRC and PRSS1 variants in chronic pancreatitis: is the role of mutated CFTR overestimated? Gut 62 (4), 582–592. https://doi.org/10.1136/gutjnl-2011-300645Germany.

Ryder, S.D., Beckingham, I.J., 2001. Acute hepatitis. BMJ 322 (7279), 151. https://doi.org/10.1136/bmj.322.7279.151.

Sato, Y., Inokuchi, R., Yamamoto, M., Horie, R., Asada, T., Matsubara, T., Doi, K., 2024. Impact of renal dysfunction on the diagnosis of acute pancreatitis using urinary trypsinogen-2: a retrospective study. Nephrology 29 (6), 338–343. https://doi.org/10.1111/nep.14276. Available from: http://onlinelibrary.wiley.com/journal/10.1111/(ISSN)1440-1797.

Schneider, B.J., Naidoo, J., Santomasso, B.D., Lacchetti, C., Adkins, S., Anadkat, M., Atkins, M.B., Brassil, K.J., Caterino, J.M., Chau, I., Davies, M.J., Ernstoff, M.S., Fecher, L., Ghosh, M., Jaiyesimi, I., Mammen, J.S., Naing, A., Nastoupil, L.J., Phillips, T., Porter, L.D., Reichner, C.A., Seigel, C., Song, J.M., Spira, A., Suarez-

Almazor, M., Swami, U., Thompson, J.A., Vikas, P., Wang, Y., Weber, J.S., Funchain, P., Bollin, K., 2021. Management of immune-related adverse events in patients treated with immune checkpoint inhibitor therapy: ASCO guideline update. Journal of Clinical Oncology 39 (36), 4073–4126. https://doi.org/10.1200/JCO.21.01440. Available from: https://ascopubs.org/doi/full/10.1200/JCO.21.01440.

Shah, S., Khosa, F., Rai, S., McLaughlin, P., Louis, L., Nicolaou, S., 2016. Imaging blunt pancreatic and duodenal trauma. Applied Radiology 45 (11), 22–28. https://doi.org/10.37549/ar2328.

Silva-Vaz, P., Abrantes, A.M., Castelo-Branco, M., Gouveia, A., Botelho, M.F., Tralhão, J.G., 2020. Multifactorial scores and biomarkers of prognosis of acute pancreatitis: applications to research and practice. International Journal of Molecular Sciences 21 (1). https://doi.org/10.3390/ijms21010338. Available from: https://www.mdpi.com/1422-0067/21/1/338/pdf.

Soleimani, N., Mohammadzadeh, S., Asadian, F., 2020. Lipemia interferences in biochemical tests, investigating the efficacy of different removal methods in comparison with ultracentrifugation as the gold standard. Journal of Analytical Methods in Chemistry 2020. https://doi.org/10.1155/2020/9857636. Available from: http://www.hindawi.com/journals/jamc/.

Sorribas, M., Carnaval, T., Peláez, N., Secanella, L., Salord, S., Sarret, S., Videla, S., Busquets, J., 2023. Home monitoring vs hospitalization for mild acute pancreatitis. A pilot randomized controlled clinical trials. Medicine (United States) 102 (20), E33853. https://doi.org/10.1097/MD.0000000000033853.

Spanier, B.W.M., Dijkgraaf, M.G.W., Bruno, M.J., 2008. Epidemiology, aetiology and outcome of acute and chronic pancreatitis: an update. Best Practice and Research Clinical Gastroenterology 22 (1), 45–63. https://doi.org/10.1016/j.bpg.2007.10.007.

Sun, H., Zuo, H.-D., Lin, Q., Yang, D.-D., Zhou, T., Tang, M.-Y., Wáng, Y.X.J., Zhang, X.-M., 2019. MR imaging for acute pancreatitis: the current status of clinical applications. Annals of Translational Medicine 7 (12). https://doi.org/10.21037/atm.2019.05.37, 269–269.

Takeda, K., 2007. Japanese guidelines for management of acute pancreatitis. Pancreas 35 (1), 88. https://doi.org/10.1097/01.mpa.0000278681.32942.d7.

Tan, Z., Li, G., Zheng, Y., Li, Q., Cai, W., Tu, J., Jin, S., 2025. Advances in the clinical application of machine learning in acute pancreatitis: a review. Frontiers of Medicine 11. https://doi.org/10.3389/fmed.2024.1487271.

Tarján, D., Szalai, E., Lipp, M., Verbói, M., Kói, T., Erőss, B., Teutsch, B., Faluhelyi, N., Hegyi, P., Mikó, A., 2024. Persistently high procalcitonin and C-reactive protein are good predictors of infection in acute necrotizing pancreatitis: a systematic review and meta-analysis. International Journal of Molecular Sciences 25 (2). https://doi.org/10.3390/ijms25021273. Available from: http://www.mdpi.com/journal/ijms.

Tempero, M.A., Malafa, M.P., Al-Hawary, M., Behrman, S.W., Benson, A.B., Cardin, D.B., Chiorean, E.G., Chung, V., Czito, B., Chiaro, M.D., Dillhoff, M., Donahue, T.R., Dotan, E., Ferrone, C.R., Fountzilas, C., Hardacre, J., Hawkins, W.G., Klute, K., Ko, A.H., Kunstman, J.W., LoConte, N., Lowy1, A.M., Moravek, C., Nakakura, E.K., Narang, A.K., Obando, J., Polanco, P.M., Reddy, S., Reyngold, M., Scaife, C., Shen, J., Vollmer, C., Wolff, R.A., Wolpin, B.M., Lynn, B., George, G.V., 2021. Pancreatic adenocarcinoma, version 2.2021. Journal of the National Comprehensive Cancer Network 19 (4), 439–457. https://doi.org/10.6004/jnccn.2021.0017.

Tenner, S., Dubner, H., Steinberg, W., 1994. Predicting gallstone pancreatitis with laboratory parameters: a meta-analysis. American Journal of Gastroenterology 89 (10), 1863–1866. https://doi.org/10.1111/j.1572-0241.1994.tb07710.x.

Tenner, S., Fernandez-del Castillo, C., Warshaw, A., Steinberg, W., Hermon-Taylor, J., Valenzuela, J.E., Hariri, M., Hughes, M., Banks, P.A., 1997. Urinary trypsinogen activation peptide (TAP) predicts severity in patients with acute pancreatitis. International Journal of Pancreatology 21 (2), 105–110.

Tenner, S., Baillie, J., Dewitt, J., Vege, S.S., 2013. American college of gastroenterology guideline: management of acute pancreatitis. American Journal of Gastroenterology 108 (9), 1400–1415. https://doi.org/10.1038/ajg.2013.218.

The etiology of acute hemorrhagic pancreatitis. JAMA, the Journal of the American Medical Association XXXVI (25), 1901, 1786. https://doi.org/10.1001/jama.1901.02470250040005.

Thygesen, K., Alpert, J.S., Jaffe, A.S., Chaitman, B.R., Bax, J.J., Morrow, D.A., White, H.D., Mickley, H., Crea, F., Van De Werf, F., Bucciarelli-Ducci, C., Katus, H.A., Pinto, F.J., Antman, E.M., Hamm, C.W., De Caterina, R., Januzzi, J.L., Apple, F.S., Garcia, M.A. A., Underwood, S.R., Canty, J.M., Lyon, A.R., Devereaux, P.J., Zamorano, J.L., Lindahl, B., Weintraub, W.S., Newby, L.K., Virmani, R., Vranckx, P., Cutlip, D., Gibbons, R.J., Smith, S.C., Atar, D., Luepker, R.V., Robertson, R.M., Bonow, R.O., Steg, P.G., O'Gara, P.T., Fox, K.A.A., Hasdai, D., Aboyans, V., Achenbach, S., Agewall, S., Alexander, T., Avezum, A., Barbato, E., Bassand, J.P., Bates, E., Bittl, J. A., Breithardt, G., Bueno, H., Bugiardini, R., Cohen, M.G., Dangas, G., De Lemos, J. A., Delgado, V., Filippatos, G., Fry, E., Granger, C.B., Halvorsen, S., Hlatky, M.A., Ibanez, B., James, S., Kastrati, A., Leclercq, C., Mahaffey, K.W., Mehta, L., Müller, C., Patrono, C., Piepoli, M.F., Piñeiro, D., Roffi, M., Rubboli, A., Sharma, S., Simpson, I. A., Tendera, M., Valgimigli, M., Van Der Wal, A.C., Windecker, S., 2019. Fourth universal definition of myocardial infarction (2018). European Heart Journal 40 (3), 237–269. https://doi.org/10.1093/eurheartj/ehy462.

Tran, Q.T., Tran, Van H., Sendler, M., Doller, J., Wiese, M., Bolsmann, R., Wilden, A., Glaubitz, J., Modenbach, J.M., Thiel, F.G., de Freitas Chama, L.L., Weiss, F.U., Lerch, M.M., Aghdassi, A.A., 2021. Role of bile acids and bile salts in acute pancreatitis. Pancreas 50 (1), 3–11. https://doi.org/10.1097/mpa.0000000000001706.

Tsuji, Y., Yamamoto, H., Yazumi, S., Watanabe, Y., Matsueda, K., Yamamoto, H., Chiba, T., 2007. Perfusion computerized tomography can predict pancreatic necrosis in early stages of severe acute pancreatitis. Clinical Gastroenterology and Hepatology 5 (12), 1484–1492. https://doi.org/10.1016/j.cgh.2007.07.014.

Tsuji, Y., Takahashi, N., Isoda, H., Koizumi, K., Koyasu, S., Sekimoto, M., Imanaka, Y., Yazumi, S., Asada, M., Nishikawa, Y., Yamamoto, H., Kikuchi, O., Yoshida, T., Inokuma, T., Katsushima, S., Esaka, N., Okano, A., Kawanami, C., Kakiuchi, N., Shiokawa, M., Kodama, Y., Moriyama, I., Kajitani, T., Kinoshita, Y., Chiba, T., 2017. Early diagnosis of pancreatic necrosis based on perfusion CT to predict the severity of acute pancreatitis. Journal of Gastroenterology 52 (10), 1130–1139. https://doi.org/10.1007/s00535-017-1330-5.

Uc, A., Husain, S.Z., 2019. Pancreatitis in children. Gastroenterology 156 (7), 1969–1978. https://doi.org/10.1053/j.gastro.2018.12.043.

Umans, D.S., Rangkuti, C.K., Sperna Weiland, C.J., Timmerhuis, H.C., Bouwense, S.A.W., Fockens, P., Besselink, M.G., Verdonk, R.C., Van Hooft, J.E., 2020. Endoscopic ultrasonography can detect a cause in the majority of patients with idiopathic acute pancreatitis: a systematic review and meta-analysis. Endoscopy 52 (11), 955–964. https://doi.org/10.1055/a-1183-3370.

Umans, D.S., Timmerhuis, H.C., Anten, M.P.G.F., Bhalla, A., Bijlsma, R.A., Boxhoorn, L., Brink, M.A., Bruno, M.J., Curvers, W.L., van Eijck, B.C., Erkelens, G.W., van Geenen, E.J.M., Hazen, W.L., Hoge, C.V., Hol, L., Inderson, A., Kager, L.M., Kuiken, S.D., Perk, L.E., Quispel, R., Römkens, T.E.H., Sperna Weiland, C.J., Thijssen, A.Y., Venneman, N.G., Verdonk, R.C., van Wanrooij, R.L.J., Witteman, B.J., Besselink, M.G., van Hooft, J.E., 2023. Prospective multicentre study of indications for surgery in patients with idiopathic acute pancreatitis following endoscopic ultrasonography (PICUS). British Journal of Surgery 110 (12), 1877–1882. https://doi.org/10.1093/bjs/znad318.

van Brunschot, S., van Grinsven, J., van Santvoort, H.C., Bakker, O.J., Besselink, M.G., Boermeester, M.A., Bollen, T.L., Bosscha, K., Bouwense, S.A., Bruno, M.J., Cappendijk, V.C., Consten, E.C., Dejong, C.H., van Eijck, C.H., Erkelens, W.G., van Goor, H., van Grevenstein, W.M.U., Haveman, J.W., Hofker, S.H., Jansen, J.M., Laméris, J.S., van Lienden, K.P., Meijssen, M.A., Mulder, C.J., Nieuwenhuijs, V.B., Poley, J.W., Quispel, R., de Ridder, R.J., Römkens, T.E., Scheepers, J.J., Schepers, N.J., Schwartz, M.P., Seerden, T., Spanier, B.W.M., Straathof, J.W.A., Strijker, M., Timmer, R., Venneman, N.G., Vleggaar, F.P., Voermans, R.P., Witteman, B.J., Gooszen, H.G., Dijkgraaf, M.G., Fockens, P., Manusama, E.R., Hadithi, M., Rosman, C., Schaapherder, A.F., Schoon, E.J., 2018. Endoscopic or surgical step-up approach for infected necrotising pancreatitis: a multicentre randomised trial. The Lancet 391 (10115), 51–58. https://doi.org/10.1016/S0140-6736(17)32404-2.

Van Santvoort, H.C., Besselink, M.G., Bakker, O.J., Hofker, H.S., Boermeester, M.A., Dejong, C.H., Van Goor, H., Schaapherder, A.F., Van Eijck, C.H., Bollen, T.L., Van Ramshorst, B., Nieuwenhuijs, V.B., Timmer, R., Laméris, J.S., Kruyt, P.M., Manusama, E.R., Van Der Harst, E., Van Der Schelling, G.P., Karsten, T., Hesselink, E. J., Van Laarhoven, C.J., Rosman, C., Bosscha, K., De Wit, R.J., Houdijk, A.P., Van Leeuwen, M.S., Buskens, E., Gooszen, H.G., 2010. A step-up approach or open necrosectomy for necrotizing pancreatitis. New England Journal of Medicine 362 (16), 1491–1502. https://doi.org/10.1056/NEJMoa0908821.

Wang, Z.H., Lv, J.H., Teng, Y., Michael, N., Zhao, Y.F., Xia, M., Wang, B., 2025. Phospholipase D2: a biomarker for stratifying disease severity in acute pancreatitis? World Journal of Gastroenterology 31 (11). https://doi.org/10.3748/wjg.v31.i11.104033. Available from: https://www.f6publishing.com/forms/main/DownLoadFile.aspx?Type=Digital&TypeId=22&id=10.3748%2fwjg.v31.i11.104033&FilePath=FAF578EC5F1F94E387BB2FCDEEB09FD9D1F5FD83D04EB17A74D5C7289D80D5A5A60485618A286BF7D7DDE1C9210192E607155A24802A408C.

Werge, M., Novovic, S., Schmidt, P.N., Gluud, L.L., 2016. Infection increases mortality in necrotizing pancreatitis: a systematic review and meta-analysis. Pancreatology 16 (5), 698–707. https://doi.org/10.1016/j.pan.2016.07.004.

Whitcomb, D.C., Gorry, M.C., Preston, R.A., Furey, W., Sossenheimer, M.J., Ulrich, C.D., Martin, S.P., Gates, L.K., Amann, S.T., Toskes, P.P., Liddle, R., McGrath, K., Uomo, G., Post, J.C., Ehrlich, G.D., 1996. Hereditary pancreatitis is caused by a mutation in the cationic trypsinogen gene. Nature Genetics 14 (2), 141–145. https://doi.org/10.1038/ng1096-141.

Wilson, C., Heads, A., Shenkin, A., Imrie, C.W., 1989. C-reactive protein, antiproteases and complement factors as objective markers of severity in acute pancreatitis. British Journal of Surgery 76 (2), 177–181. https://doi.org/10.1002/bjs.1800760224.

Witt, H., Luck, W., Hennies, H.C., Claßen, M., Kage, A., Laß, U., Landt, O., Becker, M., 2000. Mutations in the gene encoding the serine protease inhibitor, Kazal type 1 are associated with chronic pancreatitis. Nature Genetics 25 (2), 213–216. https://doi.org/10.1038/76088.

Wu, B.U., Johannes, R.S., Sun, X., Conwell, D.L., Banks, P.A., 2009. Early changes in blood urea nitrogen predict mortality in acute pancreatitis. Gastroenterology 137 (1), 129–135. https://doi.org/10.1053/j.gastro.2009.03.056.

Wu, B.U., Bakker, O.J., Papachristou, G.I., Besselink, M.G., Repas, K., van Santvoort, H.C., Muddana, V., Singh, V.K., Whitcomb, D.C., Gooszen, H.G., Banks, P.A., 2011. Blood urea nitrogen in the early assessment of acute pancreatitis: an international validation study. Archives of Internal Medicine 171 (7), 669–676. https://doi.org/10.1001/archinternmed.2011.126.

Xie, Q., Wu, J., Tang, Y., Dou, Y., Hao, S., Xu, F., Feng, X., Liang, Z., 2013. Whole-organ CT perfusion of the pancreas: impact of iterative reconstruction on image quality, perfusion parameters and radiation dose in 256-slice CT-preliminary findings. PLoS One 8 (11). https://doi.org/10.1371/journal.pone.0080468China. Available from. http://www.plosone.org/article/fetchObject.action?uri=info%3Adoi%2F10.1371%2Fjournal.pone.0080468&representation=PDF.

Yadav, D., Lowenfels, A.B., 2013. The epidemiology of pancreatitis and pancreatic cancer. Gastroenterology 144 (6), 1252–1261. https://doi.org/10.1053/j.gastro.2013.01.068.

Yao, Q., Duan, Y., Jin, C., Xiang, L., Wei, S., Shi, Y., Zhang, Y., Zhang, J., Liu, C., 2025. A nomogram for early prediction of infected pancreatic necrosis based on contrast-enhanced CT radiomics and inflammatory indicators. Journal of Inflammation Research 18, 13651–13663. https://doi.org/10.2147/jir.s538345.

Yasuda, H., Kataoka, K., Takeyama, Y., Takeda, K., Ito, T., Mayumi, T., Isaji, S., Mine, T., Kitagawa, M., Kiriyama, S., Sakagami, J., Masamune, A., Inui, K., Hirano, K., Akashi, R., Yokoe, M., Sogame, Y., Okazaki, K., Morioka, C., Kihara, Y., Kawa, S., Tanaka, M., Andoh, A., Kimura, W., Nishimori, I., Furuse, J., Yokota, I., Shimosegawa, T., 2019. Usefulness of urinary trypsinogen-2 and trypsinogen activation peptide in acute pancreatitis: a multicenter study in Japan. World Journal of Gastroenterology 25 (1), 107–117. https://doi.org/10.3748/wjg.v25.i1.107. Available from: https://www.f6publishing.com/forms/main/DownLoadFile.aspx?Type=Digital&TypeId=1&id=10.3748%2fwjg.v25.i1.107&FilePath=52858CEF481997A1E522C44A7167CD9096CB2C16945659F82B795BFE7162E75876CD963C838741F298FBF0EAB552BB9F60F58B70C99DD9D4.

Yokoe, M., Takada, T., Strasberg, S.M., Solomkin, J.S., Mayumi, T., Gomi, H., Pitt, H.A., Garden, O.J., Kiriyama, S., Hata, J., Gabata, T., Yoshida, M., Miura, F., Okamoto, K., Tsuyuguchi, T., Itoi, T., Yamashita, Y., Dervenis, C., Chan, A.C.W., Lau, W.Y., Supe, A.N., Belli, G., Hilvano, S.C., Liau, K.H., Kim, M.H., Kim, S.W., Ker, C.G., 2013. TG13 diagnostic criteria and severity grading of acute cholecystitis (with videos). Journal of Hepato-Biliary-Pancreatic Sciences 20 (1), 35–46. https://doi.org/10.1007/s00534-012-0568-9.

Zhang, Y., Yan, L., Han, W., 2018. Elevated level of miR-551b-5p is associated with inflammation and disease progression in patients with severe acute pancreatitis. Therapeutic

Apheresis and Dialysis 22 (6), 649–655. https://doi.org/10.1111/1744-9987.12720. Available from: http://onlinelibrary.wiley.com/journal/10.1111/(ISSN)1744-9987.

Zhang, T., Wang, G., Cao, Z., Huang, W., Xiao, H., Wei, H., Lu, J., Liu, R., Yin, C., 2022. Acute pancreatitis in pregnancy: a 10-year, multi-center, retrospective study in Beijing. BMC Pregnancy and Childbirth 22 (1). https://doi.org/10.1186/s12884-022-04742-8. Available from: http://www.biomedcentral.com/bmcpregnancychildbirth/.

Zhong, J., Hu, Y., Xing, Y., Ge, X., Ding, D., Zhang, H., Yao, W., 2022. A systematic review of radiomics in pancreatitis: applying the evidence level rating tool for promoting clinical transferability. Insights Into Imaging 13 (1). https://doi.org/10.1186/s13244-022-01279-4. Available from: http://www.springer.com/medicine/radiology/journal/13244.

Zhou, W., Dong, S., Chen, Z., Li, X., Jiang, W., 2022. New challenges for microRNAs in acute pancreatitis: progress and treatment. Journal of Translational Medicine 20 (1). https://doi.org/10.1186/s12967-022-03338-2. Available from: http://www.translational-medicine.com/home/.

Zilio, M.B., Eyff, T.F., Azeredo-Da-Silva, A.L.F., Bersch, V.P., Osvaldt, A.B., 2019. A systematic review and meta-analysis of the aetiology of acute pancreatitis. HPB 21 (3), 259–267. https://doi.org/10.1016/j.hpb.2018.08.003.

CHAPTER

Radiological assessment in acute pancreatitis

6

Masaya Kawaguchi
Gifu University, Department of Radiology, Gifu, Japan

6.1 Introduction

The two most common causes of acute pancreatitis are gallstones (30%–45%) and alcohol abuse (30%–35%). Other causes include drug-induced pancreatitis, pancreatic cancer, and idiopathic pancreatitis of unknown origin (Whitcomb, 2006; Spanier et al., 2008; Türkvatan et al., 2015a,b). Acute pancreatitis is a relatively common condition in daily clinical practice, and diagnosis is usually straightforward in typical cases. Imaging findings, however, vary depending on the severity of inflammation, the extent of spread to surrounding tissues, the presence of complications, and the underlying cause. Accurate interpretation requires careful evaluation, as both typical and atypical cases, presentations may be misinterpreted or overlooked by those unfamiliar with the spectrum of imaging features.

This chapter aims to review the causes of acute pancreatitis, illustrate typical and atypical cases, discuss differential diagnoses, and describe imaging features of associated complications.

6.2 Imaging diagnosis

6.2.1 Typical imaging findings of acute pancreatitis

According to the revised Atlanta classification, acute pancreatitis can be morphologically categorized into two types: interstitial edematous pancreatitis and necrotizing pancreatitis (Rocha et al., 2020; Banks et al., 2013). Interstitial edematous pancreatitis is characterized by diffuse (or occasionally localized) inflammatory enlargement of the pancreas due to edema, often accompanied by inflammatory changes in the peripancreatic fat and the presence of peripancreatic fluid. Clinical symptoms in these cases typically improve within the first week (Banks et al., 2013). Necrotizing pancreatitis, which occurs less frequently, is subclassified based on the extent of necrosis: pancreatic parenchyma, peripancreatic tissue, or both. Necrosis involving both the pancreatic parenchyma and peripancreatic tissue is the most common form, whereas necrosis confined solely to the pancreatic parenchyma is the least common (Rocha et al., 2020; Banks et al., 2013).

Advanced Insights into Acute Pancreatitis. https://doi.org/10.1016/B978-0-443-44736-5.00015-2

6.2.1.1 Interstitial edematous pancreatitis

Most patients with interstitial edematous pancreatitis present with diffuse or localized pancreatic enlargement due to inflammatory edema. Contrast-enhanced computed tomography (CT) typically demonstrates relatively homogeneous pancreatic enhancement and mild increased attenuation of the peripancreatic fat from inflammation (Fig. 6.1) (Banks et al., 2013; Fung et al., 2020). Small amounts of peripancreatic fluid may also be present, without evidence of peripancreatic necrosis.

6.2.1.2 Necrotizing pancreatitis

Necrotizing pancreatitis most commonly involves both the pancreatic parenchyma and peripancreatic tissues, less commonly involves only the peripancreatic tissue, and rarely affects the pancreatic parenchyma alone (Banks et al., 2013). During the first few days, contrast-enhanced CT may show patchy pancreatic perfusion with variable attenuation before areas of impaired enhancement become more clearly demarcated or confluent (Fig. 6.2). After the first week, unenhanced regions of pancreatic parenchyma should be considered necrotic. The extent of necrosis is best visualized as non-enhancing pancreatic tissue on contrast-enhanced CT (Banks et al., 2013; Fung et al., 2020).

In a normal pancreas, the strongest contrast enhancement, typically 100–150 HU, occurs during the pancreatic parenchymal phase (approximately 40 s after intravenous contrast injection), reflecting healthy tissue (Shyu et al., 2014; Sugawara et al., 2024). However, the portal venous phase (approximately 70–100 s post-injection) provides higher diagnostic accuracy for predicting pancreatic necrosis (Sugawara et al., 2024; Noda et al., 2018). Therefore, both the pancreatic parenchymal and portal venous phases should be assessed when evaluating the extent of pancreatic necrosis.

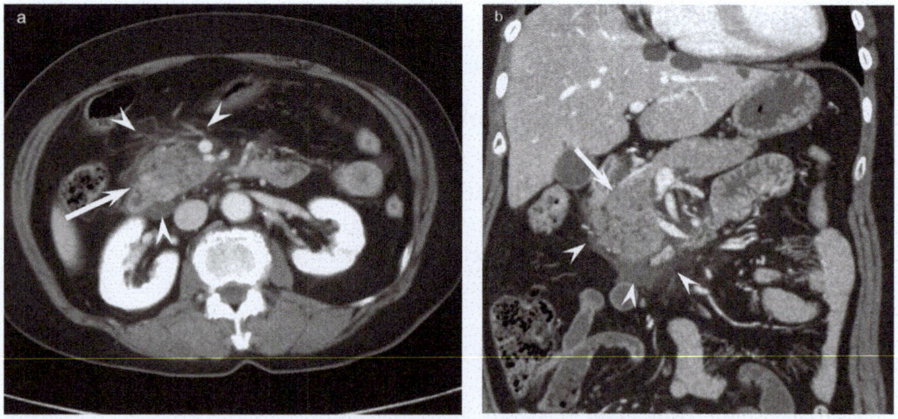

FIGURE 6.1 76-year-old female with interstitial edematous pancreatitis.

Axial (a) and coronal (b) contrast-enhanced CT images show swelling of the pancreatic head (arrow) with mildly increased peripancreatic fat density (arrowhead). Homogeneous contrast enhancement is observed in the pancreatic parenchyma.

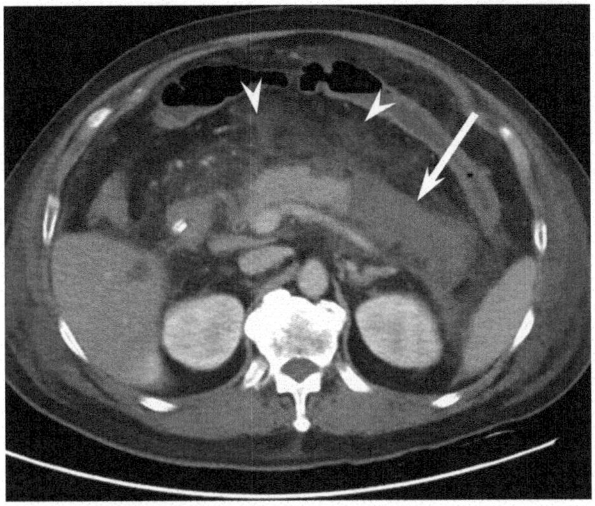

FIGURE 6.2 61-year-old male with necrotizing pancreatitis.

Delayed-phase contrast-enhanced CT demonstrates diffuse pancreatic enlargement with an unenhanced area (arrow). Markedly increased peripancreatic fat density is also noted (arrowhead).

6.2.2 Atypical imaging findings of acute pancreatitis

6.2.2.1 Groove pancreatitis

Groove pancreatitis is a rare form of pancreatitis that affects the "groove" between the pancreatic head, duodenum, and common bile duct. It is traditionally classified into two types: the pure type, which is confined to the groove itself (between the pancreatic head and duodenum), and the segmental type, which extends into the pancreatic head while centering on the pancreaticoduodenal groove (Raman et al., 2013; Shanbhogue et al., 2009). CT findings differ between the two types. In the pure type, imaging may show subtle fatty fibrosis and inflammatory changes within the groove, or more clearly defined soft tissue, which appears as a crescent-shaped, "sheet-like" structure best visualized on coronal multiplanar reconstruction images. On multiphase imaging, this soft tissue typically demonstrates gradual delayed enhancement due to significant fibrotic content. Thickening of the duodenal wall-—particularly evident in coronal sections—and small cysts within the thickened duodenal wall or the pancreaticoduodenal groove are common findings (Fig. 6.3) (Raman et al., 2013; Shanbhogue et al., 2009; Kwak et al., 2009). The segmental type can be more challenging to diagnose because the groove lesion may be obscured by tumor-like enlargement of the pancreatic head. It is frequently confused with pancreatic head tumors, and in some cases, distinguishing the two based on imaging alone is nearly impossible (Raman et al., 2013; Shanbhogue et al., 2009).

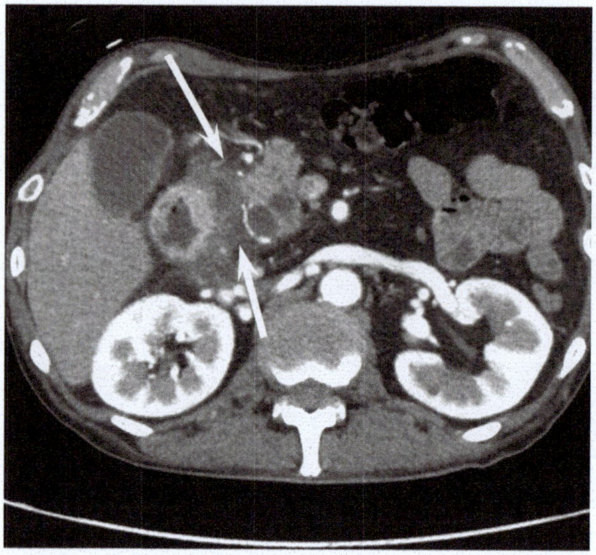

FIGURE 6.3 64-year-old male with groove pancreatitis.

Contrast-enhanced CT shows pancreatic head swelling with duodenal wall thickening. Fatty stranding and fluid are observed in the pancreaticoduodenal groove (arrow).

6.2.2.2 Ectopic pancreatitis

Ectopic pancreas is a congenital anomaly in which pancreatic tissue is anatomically separate from the main gland. The most common locations are in the upper gastrointestinal tract, including the stomach, duodenum, and proximal jejunum (Rezvani et al., 2017; Wei et al., 2011). Acute pancreatitis in ectopic pancreas is often clinically silent and may not be evident on imaging. Imaging findings are nonspecific and can include thickening of the involved bowel wall and surrounding inflammatory changes. On contrast-enhanced CT, the ectopic pancreatic mass appears hypoattenuating relative to the normally enhancing duodenal wall and pancreatic head. Pseudocysts may develop in the ectopic tissue, either secondary to pancreatitis or due to duct obstruction and accumulation of pancreatic secretions (Shanbhogue et al., 2009; Kung et al., 2010). Magnetic resonance imaging (MRI) is particularly useful for distinguishing ectopic pancreatic tissue from small tumors. Tumors typically show low signal intensity on T1-weighted images compared with normal pancreatic tissue (Rezvani et al., 2017; Blasbalg et al., 2007). In the early stages, masses may be edematous and show high signal intensity on T2-weighted images, which may later decrease due to fibrosis. On contrast-enhanced imaging, early phase enhancement is poor, but delayed-phase imaging demonstrates progressive heterogeneous enhancement (Rezvani et al., 2017; Blasbalg et al., 2007).

6.2.2.3 Focal acute pancreatitis

Focal acute pancreatitis can mimic pancreatic ductal adenocarcinoma and may manifest as either focal interstitial edematous pancreatitis or necrotizing pancreatitis. Imaging features include localized pancreatic enlargement, heterogeneous or decreased enhancement, peripancreatic inflammation, and fluid collections. The association between acute pancreatitis and pancreatic ductal adenocarcinoma is well established. Recent studies indicate that pancreatic ductal adenocarcinoma coexists in 6.8%–13.8% of patients presenting with acute pancreatitis, particularly in individuals over 40 years of age (Munigala et al., 2014; Kimura et al., 2015; Miller et al., 2023). This association is thought to result from ductal obstruction by a tumor, which can trigger clinically apparent acute pancreatitis while the underlying carcinoma remains undetected. Therefore, pancreatic ductal adenocarcinoma should be strongly suspected in cases of unexplained acute pancreatitis, especially in patients over 40, and further evaluation is recommended to identify potential underlying malignancy (Munigala et al., 2014; Kimura et al., 2015; Miller et al., 2023).

6.2.3 Differential diagnosis of acute pancreatitis

6.2.3.1 Pancreatic ductal carcinoma

Pancreatic ductal adenocarcinoma typically appears on cross-sectional imaging as a hypoenhancing, infiltrative mass relative to the surrounding pancreas on late arterial (pancreatic) and portal venous phase images. Delayed enhancement may be seen on later phases due to decreased vascularity and desmoplastic stroma (Miller et al., 2023; Kobi et al., 2020). Upstream pancreatic duct dilatation and parenchymal atrophy are common findings. Differentiating pancreatic ductal adenocarcinoma from pancreatitis is usually straightforward. However, when carcinoma coexists with inflammation, diagnosis can be challenging, as the tumor may be masked by pancreatitis (Fig. 6.4). The presence of dilated pancreatic and common bile ducts—the so-called "double duct" sign—suggests a pancreatic head tumor, even if the lesion itself is not clearly visualized (Ishigami et al., 2009). Tumors in the pancreatic tail often present late, frequently with metastases, and may appear infiltrative without pancreatic duct dilatation. Early recognition of liver metastases, lymph node involvement, and peritoneal dissemination is crucial for staging and management (Miller et al., 2023).

6.2.3.2 Autoimmune pancreatitis

Autoimmune pancreatitis represents a pancreatic manifestation of IgG4-related systemic disease. Although the clinical course usually distinguishes autoimmune pancreatitis from acute pancreatitis, their imaging findings can appear similar. Typical imaging features include sausage-shaped swelling of the focal or entire pancreas, the duct-penetrating sign, a capsule-like rim surrounding the affected areas, and homogeneous delayed enhancement or an enhanced duct sign after contrast administration (Takahashi et al., 2020; Ogawa et al., 2021). While both autoimmune and acute pancreatitis demonstrate pancreatic swelling, peripancreatic

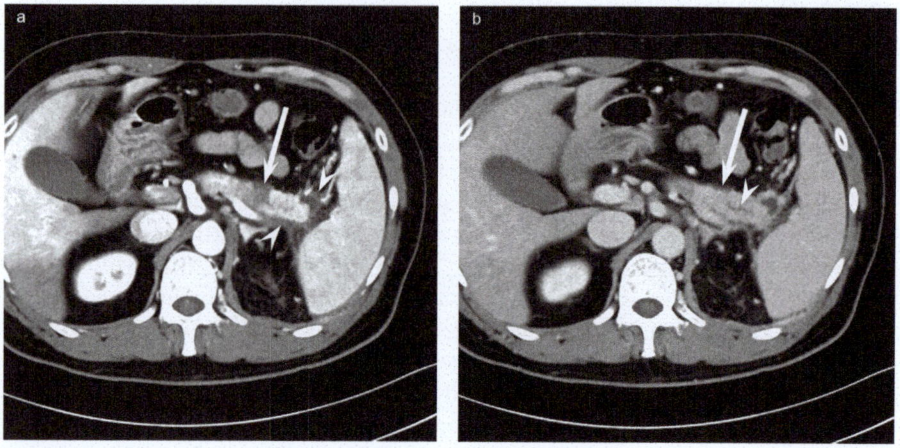

FIGURE 6.4 48-year-old male with pancreatic ductal carcinoma.

Early phase contrast-enhanced CT (a) shows a hypovascular mass in the pancreatic tail with pancreatic tail swelling (arrow) with increased peripancreatic fat density (arrowhead). Delayed phase contrast-enhanced CT (b) demonstrates delayed enhanced mass in the pancreatic tail (arrow) with upstream pancreatic duct dilatation (arrowhead).

fluid is typically absent in autoimmune pancreatitis. Furthermore, the capsule-like rim—a band-like hypodense area encircling the lesion—is not observed in acute pancreatitis, providing a key imaging distinction between the two conditions (Fig. 6.5) (Takahashi et al., 2020).

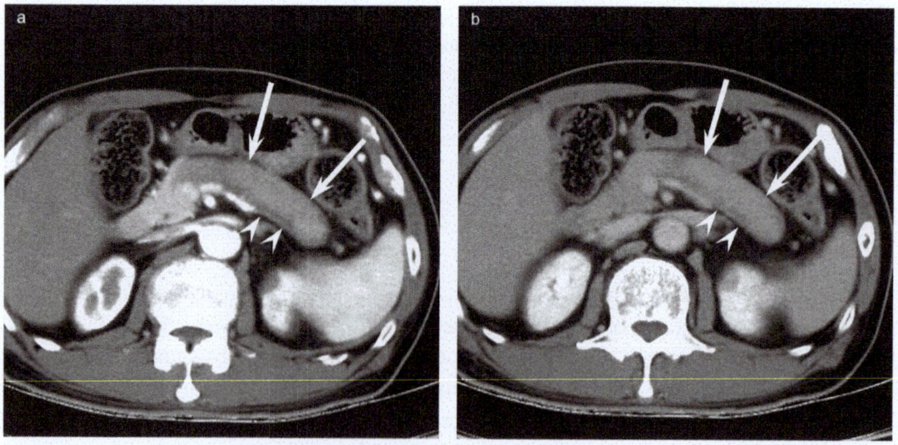

FIGURE 6.5 50-year-old female with autoimmune pancreatitis.

Contrast-enhanced CT (a) shows diffuse pancreatic enlargement with loss of the pancreatic cleft, producing a sausage-like appearance (arrow). Delayed-phase contrast-enhanced CT (b) showed a peripancreatic hypodense rim (capsule-like rim, arrowhead).

6.2.3.3 Pancreatic lymphoma

Although less common, pancreatic lymphoma can present with symptoms resembling acute pancreatitis (Rad et al., 2017). On CT, it typically appears as a large soft tissue mass with progressive contrast enhancement, absence of necrosis, lack of main pancreatic duct dilation, and vessel encasement without infiltration (Rad et al., 2017; Boninsegna et al., 2018). On MRI, pancreatic lymphoma demonstrates low signal intensity on T1-weighted images and marked restricted diffusion on diffusion-weighted imaging, reflecting high cellularity. Unlike acute pancreatitis, which often shows peripancreatic inflammation or fat stranding, pancreatic lymphoma generally presents as diffuse pancreatic enlargement with minimal or absent surrounding inflammatory changes. Fluid collections, pancreatic fat necrosis, and pancreatic duct rupture are not observed in lymphoma, providing key imaging distinctions from pancreatitis (Rad et al., 2017).

6.3 Imaging modality

The diagnosis of acute pancreatitis requires fulfillment of at least two of the following three criteria: (1) abdominal pain consistent with acute pancreatitis, typically acute-onset, persistent, severe epigastric pain often radiating to the back; (2) serum lipase or amylase levels elevated to three times or more above the upper limit of normal; and (3) characteristic imaging findings on contrast-enhanced CT, and less commonly MRI or transabdominal ultrasonography (Rocha et al., 2020; Banks et al., 2013). If the clinical diagnosis is established based on abdominal pain and elevated serum pancreatic enzymes, contrast-enhanced CT is generally not required at admission or in the emergency department. Cross-sectional imaging may be indicated in specific scenarios, such as confirmation of diagnosis in sedated or unconscious patients, clinical suspicion of duodenal perforation, or delayed presentation when serum lipase and amylase levels may have normalized (Rocha et al., 2020; Banks et al., 2013; IAP/APA evidence-based guidelines for the management of acute pancreatitis, 2013).

6.3.1 Ultrasonography

Ultrasound is often the first-line imaging modality for acute pancreatitis in many institutions, both to confirm thc diagnosis and to exclude other causes of acute abdomen. It is rapid, bedside-accessible, repeatable, and free of radiation. In addition to evaluating the pancreatic parenchyma, ultrasound allows assessment of the gallbladder, biliary tree, peripancreatic tissues, and regional vascular structures (Burrowes et al., 2020; Türkvatan et al., 2015a,b; Scaglione et al., 2008).

Pancreatic enlargement and decreased echogenicity due to interstitial edema may be observed in approximately 30% of cases. In early or mild acute pancreatitis, the pancreas may appear normal on ultrasonography. Focal areas of heterogeneous echogenicity—reflecting edema or hemorrhage—may be present within the parenchyma. Edema of the surrounding adipose tissue or fluid collections, particularly in the lesser sac or left anterior pararenal space, may obscure the pancreatic outline. Ultrasound is also useful for characterizing the contents of fluid collections and pseudocysts (Türkvatan et al., 2015a,b; Scaglione et al., 2008). Additionally, it is superior to other modalities for detecting common bile duct stones and is valuable in diagnosing cholelithiasis-associated pancreatitis (Burrowes et al., 2020).

In severely ill patients with acute pancreatitis, transabdominal ultrasound has limited utility for diagnosis and prognostication, particularly beyond 48–72 h after symptom onset. Complete visualization of the pancreas is often difficult due to concomitant ileus, and image quality may be further compromised in overweight or obese patients. Detection of choledocholithiasis is also frequently limited, as bowel gas can obscure the distal common bile duct and pancreatic head (Burrowes et al., 2020; Porter et al., 2019; Busireddy et al., 2014).

6.3.2 CT

Contrast-enhanced CT is the primary imaging modality for acute pancreatitis, used to evaluate disease severity, identify complications, guide treatment strategies, and monitor disease progression. Contrast-enhanced imaging is essential for evaluating pancreatic morphology, detecting and quantifying pancreatic necrosis, and assessing vascular complications such as pseudoaneurysms and splenic vein thrombosis.

CT imaging protocols vary by institution. A typical protocol for acute pancreatitis includes a single-phase scan during the pancreatic parenchymal phase (approximately 40 s after intravenous contrast administration) or the portal venous phase (60–80 s), covering the region from above the diaphragm to the entire abdomen (Fung et al., 2020; Avanesov et al., 2016; Ding et al., 2019). Images are generally reconstructed in axial slices at 3–5 mm intervals. Thin-slice reconstructions (0.75–1.5 mm) allow for advanced post-processing, including 3D and multiplanar reconstructions (Fung et al., 2020).

In cases with suspected arterial complications, such as pseudoaneurysms or active bleeding, dual-phase CT—including both arterial and venous phases—is preferred for lesion evaluation (Bae, 2010). The arterial phase is typically acquired 25–30 s after the start of intravenous contrast administration. However, performing early dual-phase CT more than 72 h after symptom onset has not been shown to provide additional benefit over single-phase CT in assessing the severity of pancreatic and peripancreatic changes (Fung et al., 2020; Avanesov et al., 2016; Ding et al., 2019).

6.3.2.1 Dual energy CT

Dual-energy CT is a computed tomography technique that utilizes two different X-ray photon energy spectra, enabling the analysis of materials exhibiting different attenuation characteristics at different energies. The role of dual-energy CT in acute pancreatitis is not yet fully established, but it has several potential applications for evaluating complications such as ischemia, necrosis, pseudocyst formation, and vascular events, including thrombosis and pseudoaneurysms (Marin et al., 2014; George et al., 2017; Murray et al., 2019).

On conventional single-energy contrast-enhanced CT, the inflamed pancreas often demonstrates low attenuation or reduced enhancement, making it challenging to distinguish between edema, ischemia, or necrosis. Dual-energy CT, particularly low-keV virtual monoenergetic imaging, enhances the contrast between normal and pathological tissues, improving the detection of areas with low or absent enhancement, such as pancreatic ischemia or necrosis (Murray et al., 2019; Silva et al., 2011; Aran et al., 2014). Necrotizing pancreatitis can be more accurately identified using iodine overlay maps, which highlight regions lacking iodine uptake. Additionally, dual-energy CT can differentiate complex pancreatic and peripancreatic fluid collections from necrotic debris, hematomas, or residual parenchyma with retained enhancement (George et al., 2017; Murray et al., 2019).

Approximately 40%–70% of acute pancreatitis cases are associated with gallstones (Walling and Freelove, 2017). Dual-energy CT with virtual monoenergetic imaging has been shown to improve the detection of noncalcified gallstones in the gallbladder (Murray et al., 2019; Uyeda et al., 2017). Similarly, this technique enhances the visibility of isodense cholesterol stones in the common bile duct. These stones demonstrate higher attenuation at high keV values and lower attenuation at low keV values compared with the surrounding bile, facilitating their identification (Murray et al., 2019; Uyeda et al., 2017; Li et al., 2015).

6.3.2.2 Deep learning

Recent studies have shown that deep learning (DL) can be valuable for the clinical diagnosis and severity assessment of acute pancreatitis using CT images (López Gordo et al., 2025; Zhang et al., 2024; Chen et al., 2023; Liang et al., 2023).

The pancreas exhibits considerable variability in shape, size, and location among individuals, and anatomical variations further complicate imaging evaluation. Its deep abdominal position makes clear visualization difficult, particularly inflamed tissue (Deng et al., 2023). To address these challenges, radiologists are increasingly applying segmentation technologies to facilitate structural analysis of the pancreas, improving diagnostic accuracy and precision (Pan et al., 2024). DL models can detect the presence of acute pancreatitis and assess its severity. For example, Chi Zhang et al. described two key modules: (1) an Acute Pancreatitis Classifier, designed to detect pancreatitis on CT images, and (2) a Lesion Segmentation Module, used to identify affected pancreatic areas (López Gordo et al., 2025; Zhang et al., 2024). Other studies have shown that combining abdominal unenhanced CT images with clinical markers in DL models can predict severity in the early stages of acute pancreatitis (Chen et al., 2023). These AI-based approaches have the potential to enhance diagnostic efficiency, guide management, and improve patient outcomes globally (López Gordo et al., 2025).

6.3.3 MRI

While CT is the most widely used imaging modality for acute pancreatitis due to its accessibility, speed, and lower cost, MRI offers advantages in specific situations because of its superior soft tissue contrast and detailed evaluation of the bile and pancreatic ducts, such as in cases of bile duct rupture (Brizi et al., 2021). Contrast-enhanced CT is generally performed 5–7 days after hospital admission in the early phase of acute pancreatitis. In the late phase of moderate to severe disease, local complications—such as infection within the necrotic areas—are fully developed, and MRI is particularly useful for evaluating complications involving the bile and pancreatic ducts (Brizi et al., 2021).

Normal pancreatic parenchyma is best visualized on fat-suppressed T1-weighted images, where it typically appears hyperintense due to pancreatic acinar protein content (Noda et al., 2016; Ditkofsky et al., 2014). On T2-weighted images, the pancreas generally appears hypointense. In interstitial edematous pancreatitis, the pancreas is enlarged, shows increased signal intensity on T2-weighted and diffusion-weighted images, and demonstrates high signal intensity in the peripancreatic fat on fat-suppressed T2-weighted images. The pancreatic parenchyma may also exhibit high signal intensity on T1-weighted images. In necrotizing pancreatitis, necrotic areas are hypointense on T1-weighted images, are hyperintense on T2-weighted images, and show no enhancement on contrast-enhanced T1-weighted images (Brizi et al., 2021).

On MRI, acute peripancreatic fluid collections (APFCs) appear homogeneously hypointense on T1-weighted images and hyperintense on T2-weighted images, without visible cystic walls or solid components. Pancreatic pseudocysts, in contrast, have a thin, smooth wall and show homogeneous signal intensity in the liquid component on both T1-and T2-weighted images, with no internal debris or solid component. Rupture of the pancreatic duct results in continuous leakage of pancreatic fluid, leading to progressive enlargement of the cyst. Accurate visualization of the location and extent of the duct rupture is crucial for determining surgical or interventional management (Sandrasegaran et al., 2007). For this reason, T2-weighted imaging, magnetic resonance cholangiopancreatography (MRCP), and multiplanar reconstruction are typically performed. MRI combined with MRCP is particularly useful in treatment planning for pancreatic pseudocysts, as it allows evaluation of internal contents and demonstrates the relationship of the cyst to adjacent structures, such as the stomach and duodenum (Brizi et al., 2021).

Acute necrotic collections (ANCs) lack a capsule and exhibit heterogeneous signal intensity on both T1-and T2-weighted images. Although MRI is highly sensitive for distinguishing components within peripancreatic collections, it may be unable to differentiate small necrotic tissue fragments from inflammatory fluid on fat-suppressed T2-weighted images, limiting accurate characterization of ANCs (Dhaka et al., 2015). The mature stage of ANCs is walled-off necrosis (WON), which appears on MRI as encapsulated collections containing non-liquid material, including flocculent and band-like tissue fragments. These areas show no enhancement on contrast-enhanced images (Brizi et al., 2021).

Disconnected pancreatic ductal syndrome (DPDS) represents severe necrotizing pancreatitis with pancreatic duct disruption. In such cases, combined MRI and MRCP provide a noninvasive method to evaluate changes in the pancreas and peripancreatic region, as well as the proximal and distal ends of the disrupted main pancreatic duct and associated fistulas. MRI and MRCP are particularly advantageous in evaluating DPDS (Xiao et al., 2010). Pancreatic duct disruption should be suspected when there is at least a 2 cm necrotic area around the pancreas and the main pancreatic duct in the upstream pancreatic tissue is seen to traverse through the WON area, either within the intrapancreatic or extrapancreatic tissue, on MRCP.

6.4 Complications of acute pancreatitis

6.4.1 Acute peripancreatic fluid collections

APFCs occur within the first 4 weeks of interstitial edematous pancreatitis in the absence of peripancreatic necrosis and without features of pseudocysts (Banks et al., 2013; Fung et al., 2020). On contrast-enhanced CT, APFCs appear as homogeneous fluid-density collections without a distinct wall or internal solid components (Fig. 6.6). Most acute fluid collections remain sterile and resolve spontaneously without intervention (Banks et al., 2013; Fung et al., 2020; Brizi et al., 2021) (Table 6.1).

6.4.2 Pancreatic pseudocysts

If an APFC does not resolve within 4 weeks, it may become more organized and develop a non-epithelialized, enhancing wall on contrast-enhanced CT, forming a pseudocyst (Banks et al., 2013; Fung et al., 2020). Pseudocysts have well-defined boundaries, contain homogeneous fluid (<20 HU) without debris, and are surrounded by a capsule of fibrous or granulation tissue (Fig. 6.7). Over time, the walls may thicken, become irregular, and even develop calcifications. Pseudocysts can occur anywhere from the mediastinum to the pelvis, although they are most commonly found in the lesser sac (Fung et al., 2020; Türkvatan et al., 2015a,b; Brizi et al., 2021) (Table 6.1).

Distinguishing pseudocysts from pancreatic cystic tumors can be challenging. Approximately 30% of pancreatic cystic tumors are initially misdiagnosed as pseudocysts. Clinical history is essential: in patients without a prior episode of acute pancreatitis, or if imaging shows septations or wall nodularity, a pancreatic cystic tumor should be considered. Solid nodules at the periphery or intra-cavitary enhancement are not observed in pseudocysts (Türkvatan et al., 2015a,b; Warshaw and Rutledge, 1987).

6.4.3 Acute necrotic collections

ANCs develop within the first 4 weeks after the onset of necrotizing pancreatitis and represent poorly organized collections of necrotic tissue. Necrosis may involve the pancreatic parenchyma, the peripancreatic tissue, or both (Banks et al., 2013; Fung et al., 2020). On contrast-enhanced CT, ANCs appear heterogeneous and lack a

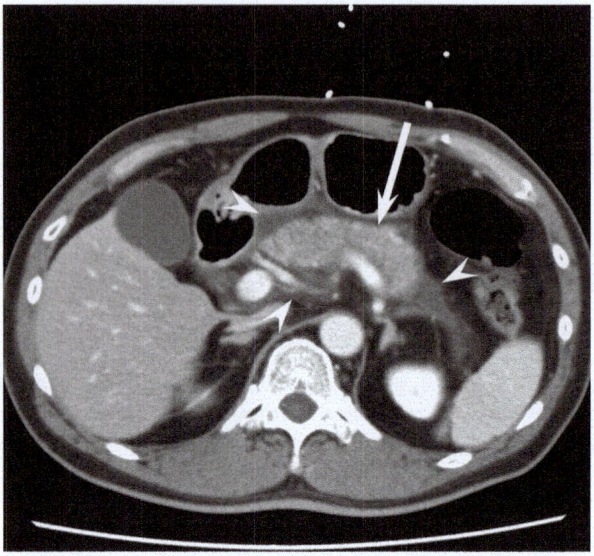

FIGURE 6.6 53-year-old male with acute peripancreatic fluid collections.

Contrast-enhanced CT demonstrates diffuse pancreatic enlargement with homogeneous enhancement (arrow). A homogeneous fluid-density collection without a distinct wall is noted in the peripancreatic region (arrowhead).

Table 6.1 Pancreatic and peripancreatic collection in interstitial edematous pancreatitis.

Collection	Time after onset	Location	Imaging features on contrast-enhanced CT	Imaging features on MRI
APFC	≤4	Extra-pancreatic	Homogeneous collection with fluid density Confined by normal peripancreatic fascial planes No definable wall Adjacent to pancreas (no intrapancreatic extension)	Homogeneously hypointense on T1WIs and hyperintense on T2WI No cystic walls and no solid components
Pseudocyst	≥4	Extra-pancreatic	Well defined wall Usually round or oval Homogeneous fluid density No non-liquid component	Thin smooth wall Homogeneous intensity on T1WIs and T2WIs of the liquid content No solid component or debris in the fluid

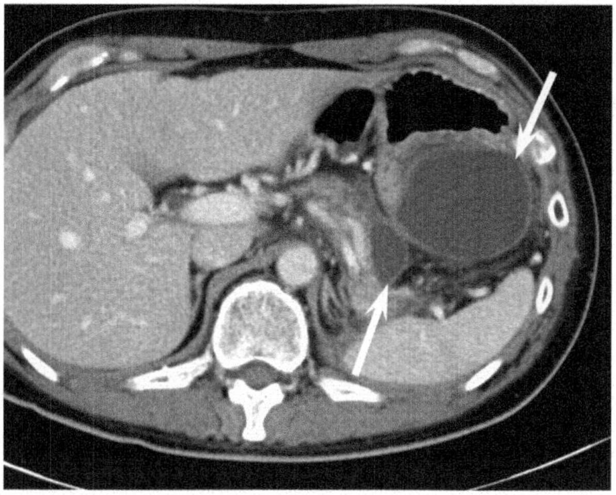

FIGURE 6.7 36-year-old male with pancreatic pseudocysts.

Contrast-enhanced CT shows a homogeneous fluid-density mass in the peripancreatic region and posterior to the stomach (arrow). The collection is surrounded by a smooth, enhancing capsule.

well-defined enclosing wall. Even if the collection appears homogeneous, it is classified as an ANC if it is associated with known pancreatic parenchymal necrosis (Brizi et al., 2021) (Table 6.2).

6.4.4 Walled off necrosis

After approximately 4 weeks of necrotizing pancreatitis, ANCs evolve into WON. Similar to ANCs, WON may be located within or outside the pancreas. Although WON can resemble a pseudocyst, it can be differentiated on contrast-enhanced CT by the presence of internal solid components (Fig. 6.8). When WON develops in a pancreas that enhances normally on contrast-enhanced CT, additional imaging with T2-weighted MRI or ultrasonography may be necessary to identify debris within the fluid collection and distinguish WON from a pseudocyst (Banks et al., 2013; Fung et al., 2020; Brizi et al., 2021) (Table 6.2).

6.4.5 Disconnected pancreatic duct syndrome

DPDS is an important but often overlooked complication of acute necrotizing pancreatitis, resulting from necrosis of the main pancreatic duct. This necrosis creates a disconnection between the viable upstream pancreatic parenchyma and the duodenum (Verma and Rana, 2020).

Contrast-enhanced CT can be useful in raising suspicion for DPDS. A necrotic area of at least 2 cm, along with viable pancreatic tissue upstream of the necrosis that enhances intravenous contrast, is highly suggestive of DPDS (Tann et al.,

Table 6.2 Pancreatic and peripancreatic collection in necrotizing pancreatitis.

Collection	Time after onset	Location	Imaging features on contrast-enhanced CT	Imaging features on MRI
ANC	≤4	Intra- and/or extra-pancreatic	Heterogeneous and nonliquid density of varying degrees in different locations No definable wall encapsulating the collection	Mixed signals on T1WI and T2WI Necrotic tissue fragments in the collection
WON	≥4	Intra- and/or extra-pancreatic	Encapsulated collection of pancreatic and/or peripancreatic necrosis Heterogeneous Well defined wall	Encapsulated effusion contains non-liquid substances, flocculent, with banded tissue fragments, free and floating

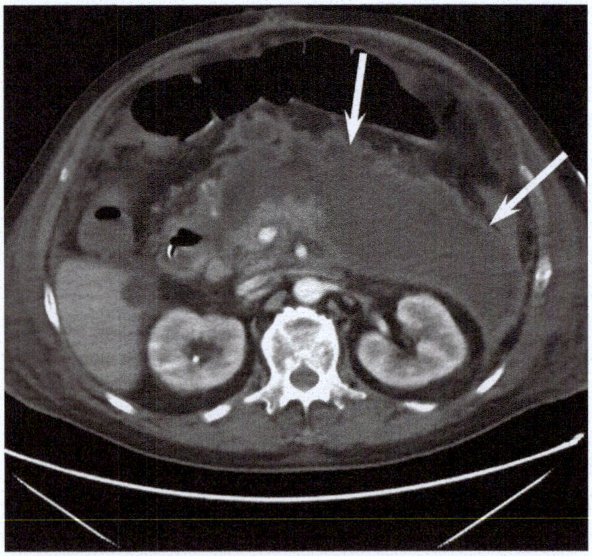

FIGURE 6.8 61-year-old male with walled-off necrosis.

Contrast-enhanced CT demonstrates a heterogeneous soft-tissue density area without contrast enhancement, enclosed by a well-defined wall (arrow).

2003). The upstream main pancreatic duct may be dilated or non-dilated. However, while CT can suggest the diagnosis, definitive confirmation requires ductal imaging using MRCP or endoscopic retrograde cholangiopancreatography (ERCP) (Verma and Rana, 2020). MRI provides a more detailed assessment of pancreatic duct anatomy than CT (Verma and Rana, 2020; Tann et al., 2003). On MRCP, sudden disruption of the pancreatic duct at a right angle to the ductal contents may indicate DPDS. In a retrospective study, MRCP detected pancreatic duct rupture in 21 of 23 cases (91.3%) confirmed by ERCP (Drake et al., 2012). However, mechanical compression or localized narrowing of the pancreatic duct caused by fluid collections can mimic rupture, potentially leading to false-positive findings. Compared with ERCP, MRCP offers the advantage of noninvasively visualizing the upstream segment of the ruptured pancreatic duct (Verma and Rana, 2020).

6.4.6 Pseudoaneurysm and hemorrhage

The release of pancreatic enzymes in acute pancreatitis can erode local blood vessels, leading to pseudoaneurysm formation or spontaneous arterial hemorrhage. Although rare, spontaneous arterial hemorrhage is a critical complication (Fung et al., 2020; Türkvatan et al., 2015a,b; Brizi et al., 2021). The most commonly affected arteries are the splenic artery (40%), gastroduodenal artery (30%), and pancreaticoduodenal artery (20%) (Fig. 6.9) (Kim et al., 2007; Araki et al., 2009). Bleeding may extend into the gastrointestinal tract or the abdominal cavity.

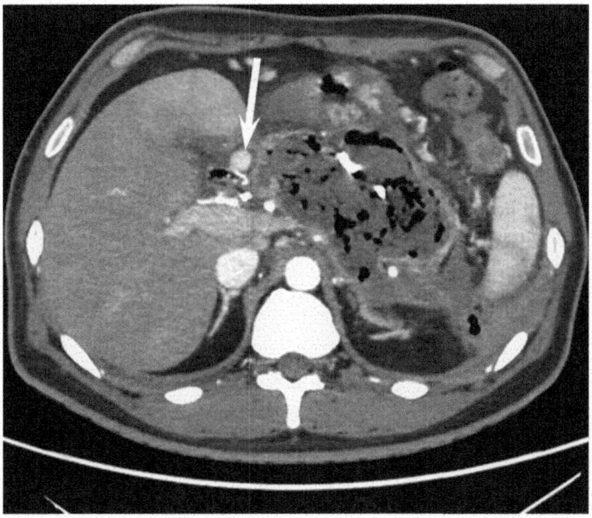

FIGURE 6.9 39-year-old male with necrotizing pancreatitis undergoing drainage of walled-off necrosis.

A pseudoaneurysm of the gastroduodenal artery (arrow) is visible without evidence of active bleeding.

Pseudoaneurysms can rupture into the abdominal cavity, retroperitoneal space, pseudocysts, and, rarely, pancreatic duct. On unenhanced CT, acute hemorrhage appears as heterogeneous high-attenuation material with a hazy margin in the retroperitoneum. Contrast-enhanced CT demonstrates an unenhanced hematoma if no active bleeding is present. Active bleeding is indicated by contrast extravasation, which can be seen expanding from the arterial to the delayed phase.

6.4.7 Thrombosis

Splanchnic venous thrombosis is a relatively common vascular complication of acute pancreatitis, occurring in approximately 27.1%–30.6% of cases (Uddin Ahmed et al., 2018; Nawacki et al., 2021). Local inflammation, reduced venous blood flow, and compression of veins by adjacent pancreatic fluid collections contribute to thrombus formation. The splenic vein is the most frequently affected site, followed by the portal vein and superior mesenteric vein (Fung et al., 2020; Uddin Ahmed et al., 2018; Nawacki et al., 2021). Splanchnic venous thrombosis is typically evaluated using contrast-enhanced CT during the portal or delayed phase (Fig. 6.10). The presence of necrotizing pancreatitis is significantly associated with an increased risk of thrombosis (Uddin Ahmed et al., 2018). Additionally, splanchnic venous thrombosis correlates with more severe acute pancreatitis,

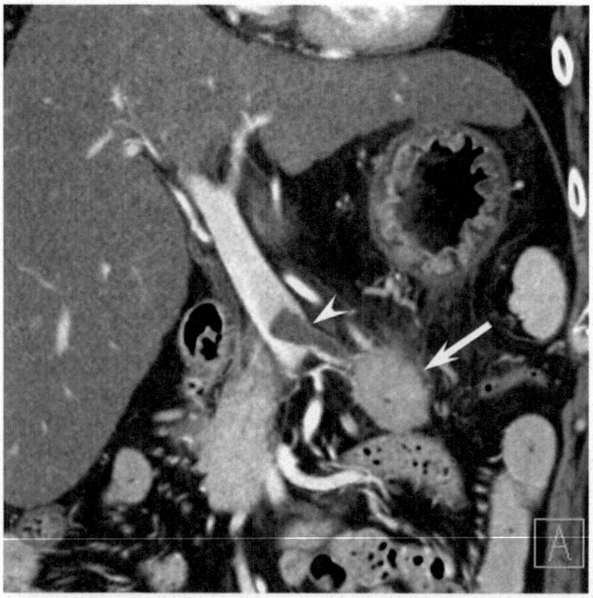

FIGURE 6.10 53-year-old male with acute pancreatitis and portal vein thrombosis.

Coronal contrast-enhanced CT shows pancreatic parenchyma swelling with increased peripancreatic fat density (arrow). A contrast defect extending from the portal vein to the splenic vein is consistent with thrombosis (arrowhead).

prolonged hospitalization, and other local complications (Nawacki et al., 2021). Early detection, prevention, and timely recanalization of affected veins are essential for optimal patient outcomes.

6.5 Conclusion

Imaging, particularly contrast-enhanced CT, plays a crucial role in the evaluation of acute pancreatitis. During the initial week, it contributes to assessing the severity of acute pancreatitis and determining treatment strategies. In the later phase, after 4 weeks, it becomes a useful tool for detecting complications. Radiologists, internists, and surgeons must be familiar with the imaging findings of acute pancreatitis and evaluate it by the appropriate imaging modality at the proper time. Suitable imaging diagnosis leads to improved patient prognosis.

References

Araki, K.I., Shimura, T., Watanabe, A., Kobayashi, T., Suzuki, H., Suehiro, T., Kuwano, H., 2009. Gastric bleeding from a penetrating pancreatic pseudocyst with pseudoaneurysm of the splenic artery. Hepato-Gastroenterology 56 (94–95), 1411–1413.

Aran, S., Besheli, L.D., Karcaaltincaba, M., Gupta, R., Flores, E.J., Abujudeh, H.H., 2014. Applications of dual-energy CT in emergency radiology. American Journal of Roentgenology 202 (4), W314. https://doi.org/10.2214/ajr.13.11682.

Avanesov, M., Weinrich, J.M., Kraus, T., Derlin, T., Adam, G., Yamamura, J., Karul, M., 2016. MDCT of acute pancreatitis: intraindividual comparison of single-phase versus dual-phase MDCT for initial assessment of acute pancreatitis using different CT scoring systems. European Journal of Radiology 85 (11), 2014–2022. https://doi.org/10.1016/j.ejrad.2016.09.013.

Bae, K.T., 2010. Intravenous contrast medium administration and scan timing at CT: considerations and approaches. Radiology 256 (1), 32–61. https://doi.org/10.1148/radiol.10090908UnitedStates.

Banks, P.A., Bollen, T.L., Dervenis, C., Gooszen, H.G., Johnson, C.D., Sarr, M.G., Tsiotos, G. G., Vege, S.S., Windsor, J.A., Horvath, K.D., Mortele, K.J., Gardner, T.B., Van Santvoort, H., Pelaez-Luna, M., Yadav, D., Stefanidis, G., Delakidis, S., Morgan, D.E., Thoeni, R.F.L., Acosta, J.M., Amann, S.T., Andren-Sandberg, A., Aranha, G.V., Asciutti, S., Barauskas, G., Baron, T.H., Bassi, C., Behrman, S., Behrns, K.E., Belliappa, V., Berzin, T.M., Besselink, M.G.H., Bhasin, D.K., Biankin, A., Bishop, M.D., Bonini, C.J., Bradley, E.L., Buechler, M., Carter, R., Cavestro, G.M., Chari, S.T., Chavez-Rodriguez, J.J., Da Cunha, J.E., D'Agostino, H., De Campos, T., De-Madaria, E., Deprez, P.H., DiSario, J.A., Doria, C., Falconi, M., Fernandez-del Castillo, C., Freeny, P. C., Frey, C.F., Friess, H., Frossard, J.L., Fuchshuber, P., Gallagher, S.F., Garg, P.K., Ghattas, G., Glasgow, R., Gonzalez, J.A., Gress, T.M., Gumbs, A.A., Hallibruton, C., Helton, S., Hill, M.C., Hoyos, S., Imrie, C.W., Isenmann, R., Izbicki, J.R., Karagiannis, J. A., Klar, E., Kolokythas, O., Lau, J., Litvin, A.A., Longnecker, D.S., Lowenfels, A.B., Mackey, R., Mah'Moud, M., Malangoni, M., McFadden, D.W., Mishra, G., Moody, F.G.,

Morinville, V., Neoptolemos, J.P., Nordback, I., Pap, A., Papachristou, G.I., Parks, R., Pedrazolli, S., Pezzilli, R., Pitt, H.A., Prosanto, C., Ramesh, H., Ramirez, F.C., Raper, S. E., Rasheed, A., Reed, D.N., Romangnuolo, J., Rossaak, J., Sanabria, J., Schaefer, C., Schmidt, J., Schmidt, P.N., Serrablo, A., Senkowski, C.K., Sharma, M., Sigman, K.M., Singh, P., Steinberg, W., Steiner, J., Strasberg, S., Strum, W., Takada, T., Tanaka, M., Vaccaro, M., Villavicencio, R.L., Vrochides, D., Wagner, M., Warshaw, A.L., Wilcox, C. M., Wysocki, P., Zenilman, M.E., Zyromski, N.J., 2013. Classification of acute pancreatitis—2012: revision of the Atlanta classification and definitions by international consensus. Gut 62 (1), 102–111. https://doi.org/10.1136/gutjnl-2012-302779UnitedStates.

Blasbalg, R., Baroni, R.H., Costa, D.N., Machado, M.C.C., 2007. MRI features of groove pancreatitis. American Journal of Roentgenology 189 (1), 73–80. https://doi.org/10.2214/AJR.06.1244.

Boninsegna, E., Zamboni, G.A., Facchinelli, D., Triantopoulou, C., Gourtsoyianni, S., Ambrosetti, M.C., Veneri, D., Ambrosetti, A., Pozzi Mucelli, R., 2018. CT imaging of primary pancreatic lymphoma: experience from three referral centres for pancreatic diseases. Insights Into Imaging 9 (1), 17–24. https://doi.org/10.1007/s13244-017-0585-y.

Brizi, M.G., Perillo, F., Cannone, F., Tuzza, L., Manfredi, R., 2021. The role of imaging in acute pancreatitis. La Radiologia Medica 126 (8), 1017–1029. https://doi.org/10.1007/s11547-021-01359-3.

Burrowes, D.P., Choi, H.H., Rodgers, S.K., Fetzer, D.T., Kamaya, A., 2020. Utility of ultrasound in acute pancreatitis. Abdominal Radiology 45 (5), 1253–1264. https://doi.org/10.1007/s00261-019-02364-x.

Busireddy, K.K., AlObaidy, M., Ramalho, M., Kalubowila, J., Baodong, L., Santagostino, I., Semelka, R.C., 2014. Pancreatitis-imaging approach. World Journal of Gastrointestinal Pathophysiology 5 (3), 252. https://doi.org/10.4291/wjgp.v5.i3.252.

Chen, Z., Wang, Y., Zhang, H., Yin, H., Hu, C., Huang, Z., Tan, Q., Song, B., Deng, L., Xia, Q., 2023. Deep learning models for severity prediction of acute pancreatitis in the early phase from abdominal nonenhanced computed tomography images. Pancreas 52 (1), e45. https://doi.org/10.1097/mpa.0000000000002216.

Deng, Y., Lan, L., You, L., Chen, K., Peng, L., Zhao, W., Song, B., Wang, Y., Ji, Z., Zhou, X., 2023. Automated CT pancreas segmentation for acute pancreatitis patients by combining a novel object detection approach and U-Net. Biomedical Signal Processing and Control 81. https://doi.org/10.1016/j.bspc.2022.104430.

Dhaka, N., Samanta, J., Kochhar, S., Kalra, N., Appasani, S., Manrai, M., Kochhar, R., 2015. Pancreatic fluid collections: what is the ideal imaging technique? World Journal of Gastroenterology 21 (48), 13403–13410. https://doi.org/10.3748/wjg.v21.i48.13403. Available from: http://www.wjgnet.com/esps/DownLoadFile.aspx?Type=Digital&SubType=1&DOI=10.3748%2fwjg.v21.i48.1340&FilePath=Pub%5c10.3748%5cv21%5ci48%5cWJG-21-13403.pdf.

Ding, L., Yu, C., Deng, F., He, W.H., Xia, L., Zhou, M., Lan, G.L., Huang, X., Lei, Y.P., Zhou, X.J., Zhu, Y., Lu, N.H., 2019. New risk factors for infected pancreatic necrosis secondary to severe acute pancreatitis: the role of initial contrast-enhanced computed tomography. Digestive Diseases and Sciences 64 (2), 553–560. https://doi.org/10.1007/s10620-018-5359-y.

Ditkofsky, N.G., Singh, A., Avery, L., Novelline, R.A., 2014. The role of emergency MRI in the setting of acute abdominal pain. Emergency Radiology 21 (6), 615–624. https://doi.org/10.1007/s10140-014-1232-2.

Drake, L.M., Anis, M., Lawrence, C., 2012. Accuracy of magnetic resonance cholangiopancreatography in identifying pancreatic duct disruption. Journal of Clinical Gastroenterology 46 (8), 696–699. https://doi.org/10.1097/mcg.0b013e31825003b3.

Fung, C., Svystun, O., Fouladi, D.F., Kawamoto, S., 2020. CT imaging, classification, and complications of acute pancreatitis. Abdominal Radiology 45 (5), 1243–1252. https://doi.org/10.1007/s00261-019-02236-4.

George, E., Wortman, J.R., Fulwadhva, U.P., Uyeda, J.W., Sodickson, A.D., 2017. Dual energy CT applications in pancreatic pathologies. British Journal of Radiology 90 (1080), 20170411. https://doi.org/10.1259/bjr.20170411.

IAP/APA evidence-based guidelines for the management of acute pancreatitis. Pancreatology 13, 2013.

Ishigami, K., Yoshimitsu, K., Irie, H., Tajima, T., Asayama, Y., Nishie, A., Hirakawa, M., Ushijima, Y., Okamoto, D., Nagata, S., Nishihara, Y., Yamaguchi, K., Taketomi, A., Honda, H., 2009. Diagnostic value of the delayed phase image for iso-attenuating pancreatic carcinomas in the pancreatic parenchymal phase on multidetector computed tomography. European Journal of Radiology 69 (1), 139–146. https://doi.org/10.1016/j.ejrad.2007.09.012.

Kim, J.H., Kim, J.S., Kim, C.D., Lee, H.S., Cho, Y.J., Lee, J.S., Choi, D.W., Han, W.S., Kim, Y.H., Yeon, J.E., Park, J.J., Byun, K.S., Bak, Y.T., 2007. Clinical features of pseudoaneurysms complicating pancreatitis: single center experience and review of Korean literature. The Korean Journal of Gastroenterology 50 (2), 108–115.

Kimura, Y., Kikuyama, M., Kodama, Y., 2015. Acute pancreatitis as a possible indicator of pancreatic cancer: the importance of mass detection. Internal Medicine 54 (17), 2109–2114. https://doi.org/10.2169/internalmedicine.54.4068.

Kobi, M., Veillette, G., Narurkar, R., Sadowsky, D., Paroder, V., Shilagani, C., Gilet, A., Flusberg, M., 2020. Imaging and management of pancreatic cancer. Seminars in Ultrasound, CT and MRI 41 (2), 139–151. https://doi.org/10.1053/j.sult.2019.12.003.

Kung, J.W., Brown, A., Kruskal, J.B., Goldsmith, J.D., Pedrosa, I., 2010. Heterotopic pancreas: typical and atypical imaging findings. Clinical Radiology 65 (5), 403–407. https://doi.org/10.1016/j.crad.2010.01.005.

Kwak, S.W., Kim, S., Lee, J.W., Lee, N.K., Kim, C.W., Yi, M.S., Kim, G.H., Kang, D.H., 2009. Evaluation of unusual causes of pancreatitis: role of cross-sectional imaging. European Journal of Radiology 71 (2), 296–312. https://doi.org/10.1016/j.ejrad.2008.04.006.

Li, H., He, D., Lao, Q., Chen, X., Liu, M., Yin, B., Zhao, K., Wang, R., Chen, L., 2015. Clinical value of spectral CT in diagnosis of negative gallstones and common bile duct stones. Abdominal Imaging 40 (6), 1587–1594. https://doi.org/10.1007/s00261-015-0387-y.

Liang, H., Wang, M., Wen, Y., Du, F., Jiang, L., Geng, X., Tang, L., Yan, H., 2023. Predicting acute pancreatitis severity with enhanced computed tomography scans using convolutional neural networks. Scientific Reports 13 (1). https://doi.org/10.1038/s41598-023-44828-7.

López Gordo, S., Ramirez-Maldonado, E., Fernandez-Planas, M.T., Bombuy, E., Memba, R., Jorba, R., 2025. AI and machine learning for precision medicine in acute pancreatitis: a narrative review. Medicina 61 (4), 629. https://doi.org/10.3390/medicina61040629.

Marin, D., Boll, D.T., Mileto, A., Nelson, R.C., 2014. State of the art: dual-energy CT of the abdomen. Radiology 271 (2), 327–342. https://doi.org/10.1148/radiol.14131480. Available from: http://pubs.rsna.org/doi/pdf/10.1148/radiol.14131480.

Miller, F.H., Vendrami, C.L., Hammond, N.A., Mittal, P.K., Nikolaidis, P., Jawahar, A., 2023. Pancreatic cancer and its mimics. RadioGraphics 43 (11). https://doi.org/10.1148/rg.230054. Available from: https://pubs.rsna.org/doi/10.1148/rg.230054.

Munigala, S., Kanwal, F., Xian, H., Scherrer, J.F., Agarwal, B., 2014. Increased risk of pancreatic adenocarcinoma after acute pancreatitis. Clinical Gastroenterology and Hepatology 12 (7), 1143. https://doi.org/10.1016/j.cgh.2013.12.033.

Murray, N., Darras, K.E., Walstra, F.E., Mohammed, M.F., McLaughlin, P.D., Nicolaou, S., 2019. Dual-energy CT in evaluation of the acute abdomen. RadioGraphics 39 (1), 264–286. https://doi.org/10.1148/rg.2019180087. Available from: https://pubs.rsna.org/doi/pdf/10.1148/rg.2019180087.

Nawacki, Ł., Matykiewicz, J., Stochmal, E., Głuszek, S., 2021. Splanchnic vein thrombosis in acute pancreatitis and its consequences. Clinical and Applied Thrombosis/Hemostasis 27. https://doi.org/10.1177/10760296211010260.

Noda, Y., Goshima, S., Tanaka, K., Osada, S., Tomita, H., Hara, A., Horikawa, Y., Takeda, J., Kajita, K., Watanabe, H., Kawada, H., Kawai, N., Kanematsu, M., Bae, K.T., 2016. Findings in pancreatic MRI associated with pancreatic fibrosis and HbA1c values. Journal of Magnetic Resonance Imaging 43 (3), 680–687. https://doi.org/10.1002/jmri.25019.

Noda, Y., Goshima, S., Fujimoto, K., Kawada, H., Kawai, N., Tanahashi, Y., Matsuo, M., 2018. Utility of the portal venous phase for diagnosing pancreatic necrosis in acute pancreatitis using the CT severity index. Abdominal Radiology 43 (11), 3035–3042. https://doi.org/10.1007/s00261-018-1579-z.

Ogawa, H., Takehara, Y., Naganawa, S., 2021. Imaging diagnosis of autoimmune pancreatitis: computed tomography and magnetic resonance imaging. Journal of Medical Ultrasonics 48 (4), 565–571. https://doi.org/10.1007/s10396-021-01145-8.

Pan, X., Jiao, K., Li, X., Feng, L., Tian, Y., Wu, L., Zhang, P., Wang, K., Chen, S., Yang, B., Wen, C., 2024. Artificial intelligence-based tools with automated segmentation and measurement on CT images to assist accurate and fast diagnosis in acute pancreatitis. British Journal of Radiology 97 (1159), 1268–1277. https://doi.org/10.1093/bjr/tqae091.

Porter, K.K., Zaheer, A., Kamel, I.R., Horowitz, J.M., Arif-Tiwari, H., Bartel, T.B., Bashir, M. R., Camacho, M.A., Cash, B.D., Chernyak, V., Goldstein, A., Grajo, J.R., Gupta, S., Hindman, N.M., Kamaya, A., McNamara, M.M., Carucci, L.R., 2019. ACR appropriateness Criteria® acute pancreatitis. Journal of the American College of Radiology 16 (11), S316. https://doi.org/10.1016/j.jacr.2019.05.017.

Rad, N., Khafaf, A., Mohammad Alizadeh, A.H., 2017. Primary pancreatic lymphoma: what we need to know. Journal of Gastrointestinal Oncology 8 (4), 749–757. https://doi.org/10.21037/jgo.2017.06.03.

Raman, S.P., Salaria, S.N., Hruban, R.H., Fishman, E.K., 2013. Groove pancreatitis: spectrum of imaging findings and radiology-pathology correlation. American Journal of Roentgenology 201 (1), W29. https://doi.org/10.2214/AJR.12.9956UnitedStates. Available from: http://www.ajronline.org/doi/pdf/10.2214/AJR.12.9956.

Rezvani, M., Menias, C., Sandrasegaran, K., Olpin, J.D., Elsayes, K.M., Shaaban, A.M., 2017. Heterotopic pancreas: histopathologic features, imaging findings, and complications. RadioGraphics 37 (2), 484–499. https://doi.org/10.1148/rg.2017160091. Available from: http://pubs.rsna.org/doi/pdf/10.1148/rg.2017160091.

Rocha, A.P.C., Schawkat, K., Mortele, K.J., 2020. Imaging guidelines for acute pancreatitis: when and when not to image. Abdominal Radiology 45 (5), 1338–1349. https://doi.org/10.1007/s00261-019-02319-2.

Sandrasegaran, K., Tann, M., Jennings, S.G., Maglinte, D.D., Peter, S.D., Sherman, S., Howard, T.J., 2007. Disconnection of the pancreatic duct: an important but overlooked complication of severe acute pancreatitis. RadioGraphics 27 (5), 1389–1400. https://doi.org/10.1148/rg.275065163UnitedStates.

Scaglione, M., Casciani, E., Pinto, A., Andreoli, C., De Vargas, M., Gualdi, G.F., 2008. Imaging assessment of acute pancreatitis: a review. Seminars in Ultrasound, CT and MRI 29 (5), 322–340. https://doi.org/10.1053/j.sult.2008.06.009.

Shanbhogue, A.K.P., Fasih, N., Surabhi, V.R., Doherty, G.P., Shanbhogue, D.K.P., Sethi, S.K., 2009. A clinical and radio-logic review of uncommon types and causes of pancreatitis. RadioGraphics 29 (4), 1003–1026. https://doi.org/10.1148/rg.294085748UnitedStates.

Shyu, J.Y., Sainani, N.I., Anik Sahni, V., Chick, J.F., Chauhan, N.R., Conwell, D.L., Clancy, T. E., Banks, P.A., Silverman, S.G., 2014. Necrotizing pancreatitis: diagnosis, imaging, and intervention. RadioGraphics 34 (5), 1218–1240. https://doi.org/10.1148/rg.345130012. Available from: http://pubs.rsna.org/doi/pdf/10.1148/rg.345130012.

Silva, A.C., Morse, B.G., Hara, A.K., Paden, R.G., Hongo, N., Pavlicek, W., 2011. Dual-energy (spectral) CT: applications in abdominal imaging. RadioGraphics 31 (4), 1031–1046. https://doi.org/10.1148/rg.314105159.

Spanier, B.W.M., Dijkgraaf, M.G.W., Bruno, M.J., 2008. Epidemiology, aetiology and outcome of acute and chronic pancreatitis: an update. Best Practice and Research Clinical Gastroenterology 22 (1), 45–63. https://doi.org/10.1016/j.bpg.2007.10.007.

Sugawara, D., Matsuda, M., Shirayama, K., Satoh, K., Asano, T., Nakae, H., Iijima, K., Mori, N., 2024. Guidelines for acute pancreatitis 2021 in Japan: reassurance of severity assessment using repeated dynamic contrast-enhanced CT including pancreatic parenchymal and portal venous phases. Japanese Journal of Radiology 42 (5), 553–554. https://doi.org/10.1007/s11604-024-01532-4.

Takahashi, M., Fujinaga, Y., Notohara, K., Koyama, T., Inoue, D., Irie, H., Gabata, T., Kadoya, M., Kawa, S., Okazaki, K., 2020. Diagnostic imaging guide for autoimmune pancreatitis. Japanese Journal of Radiology 38 (7), 591–612. https://doi.org/10.1007/s11604-020-00971-z.

Tann, M., Maglinte, D., Howard, T.J., Sherman, S., Fogel, E., Madura, J.A., Lehman, G.A., 2003. Disconnected pancreatic duct syndrome: imaging findings and therapeutic implications in 26 surgically corrected patients. Journal of Computer Assisted Tomography 27 (4), 577–582. https://doi.org/10.1097/00004728-200307000-00023.

Türkvatan, A., Erden, A., Türkoğlu, M.A., Seçil, M., Yener, Ö., 2015a. Imaging of acute pancreatitis and its complications. Part 1: acute pancreatitis. Diagnostic and Interventional Imaging 96 (2), 151–160. https://doi.org/10.1016/j.diii.2013.12.017.

Türkvatan, A., Erden, A., Türkoğlu, M.A., Seçil, M., Yüce, G., 2015b. Imaging of acute pancreatitis and its complications. Part 2: complications of acute pancreatitis. Diagnostic and Interventional Imaging 96 (2), 161–169. https://doi.org/10.1016/j.diii.2013.12.018.

Uddin Ahmed, S., Rana, S.S., Ahluwalia, J., Varma, N., Sharma, R., Gupta, R., Kang, M., 2018. Role of thrombophilia in splanchnic venous thrombosis in acute pancreatitis. Annals of Gastroenterology 31 (3), 371–378. https://doi.org/10.20524/aog.2018.0242.

Uyeda, J.W., Richardson, I.J., Sodickson, A.D., 2017. Making the invisible visible: improving conspicuity of noncalcified gallstones using dual-energy CT. Abdominal Radiology 42 (12), 2933–2939. https://doi.org/10.1007/s00261-017-1229-x.

Verma, S., Rana, S.S., 2020. Disconnected pancreatic duct syndrome: updated review on clinical implications and management. Pancreatology 20 (6), 1035–1044. https://doi.org/10.1016/j.pan.2020.07.402.

Walling, A., Freelove, R., 2017. Pancreatitis and pancreatic cancer. Primary Care: Clinics in Office Practice 44 (4), 609–620. https://doi.org/10.1016/j.pop.2017.07.004.

Warshaw, A.L., Rutledge, P.L., 1987. Cystic tumors mistaken for pancreatic pseudocysts. Annals of Surgery 205 (4), 393–398. https://doi.org/10.1097/00000658-198704000-00009.

Wei, R., Wang, Q.B., Chen, Q.H., Liu, J.S., Zhang, B., 2011. Upper gastrointestinal tract heterotopic pancreas: findings from CT and endoscopic imaging with histopathologic correlation. Clinical Imaging 35 (5), 353–359. https://doi.org/10.1016/j.clinimag.2010.10.001.

Whitcomb, D.C., 2006. Acute pancreatitis. New England Journal of Medicine 354 (20), 2142–2150. https://doi.org/10.1056/NEJMcp054958UnitedStates.

Xiao, B., Zhang, X.M., Tang, W., Zeng, N.L., Zhai, Z.H., 2010. Magnetic resonance imaging for local complications of acute pancreatitis: a pictorial review. World Journal of Gastroenterology 16 (22), 2735–2742. https://doi.org/10.3748/wjg.v16.i22.2735.

Zhang, C., Peng, J., Wang, L., Wang, Y., Chen, W., Sun, M.W., Jiang, H., 2024. A deep learning-powered diagnostic model for acute pancreatitis. BMC Medical Imaging 24 (1). https://doi.org/10.1186/s12880-024-01339-9. Available from: https://bmcmedimaging.biomedcentral.com/.

CHAPTER

Severity assessment of acute pancreatitis

7

Takahito Miyake
Advanced Critical Care Center, Gifu University Hospital, Gifu, Japan

7.1 Introduction

Acute pancreatitis (AP) is a disease in which inflammation occur at the pancreas. They are not always limited to the injuries of one organ; they could induce the systematic disorders and varying clinical courses (Boxhoorn et al., 2020). Although most patients present with mild AP and overall mortality has been reported as 3%–5% (Vincent and Shashirekha, 2024; Li et al., 2021), approximately 20% of patients develop moderate or severe AP, with necrotic pancreas and organ failures, and mortality could be up to 20%–40% (Leppaniemi et al., 2019; Schepers et al., 2019; Gurusamy et al., 2016). Patients with severe AP often need multiple interventions and intensive care approaches, including surgical approaches (Heckler et al., 2021). Based on these backgrounds, the classification of the severity of AP should include both local and systemic assessments.

7.2 Overview of history of severity assessments of acute pancreatitis

In 1974, Ranson et al. presented the first prognostic scoring system developed for AP (Ranson and Pasternack, 1977), at a time when surgical laparotomy was commonly employed to manage severe AP (Metri et al., 2024). Although the Ranson scoring system has been widely used, some criticism has been raised about its need of 48 h requirement of final score, or its low predictive power (Ong and Shelat, 2021). The first definition of severe acute pancreatitis (SAP) was proposed by the Atlanta classification in 1992. The classification was established at the 1992 Atlanta symposium, which produced a clinically based classification system (Bradley, 1993). In the symposium, definitions of AP, its severity, organ failure, and local complications were proposed (Bollen et al., 2008). Although it has been widely used for decades, the knowledge in pathophysiology and technical radiological findings has been developed, the classification faced some shortcomings (Bollen et al., 2008). In 2012, the renewed classification of Atlanta criteria was published by the Acute Pancreatitis Classification working group, incorporating responses from the members of 11 national and international pancreatic societies (Banks et al., 2013),

Advanced Insights into Acute Pancreatitis. https://doi.org/10.1016/B978-0-443-44736-5.00006-1

generated by an iterative, web-based consultation process. The classification offered a comprehensive classification of AP, severity, and peripancreatic collections, and it differentiated acute necrotic collections and walled-off necrosis. The severity of AP had been defined as mild, moderately severe, or severe (Table 7.1). The classification demonstrated the redefined radiological classification, introducing new concepts (Maldonado et al., 2021). They also presented the definition of types of AP: Interstitial edematous pancreatitis, necrotizing pancreatitis, and infected pancreatic necrosis. Three organ failures: respiratory, cardiovascular, and renal, were presented as complications of AP. They introduced the scoring systems of the complications of AP, for example, the modified scoring (Table 7.2) system or the SOFA scoring system. In 2012, another classification called "Determinant-based classification" of AP was presented (Dellinger et al., 2012) based on the survey of 249 pancreatologists from 49 different countries (Metri et al., 2024). It is based on the actual local and systemic determinants of severity, rather than a description of events that are correlated with severity. The local determinant relates to whether there is (peri) pancreatic necrosis or not, and if present, whether it is sterile or infected. And it indicated the systemic determinant categories relating to whether presentation of organ failure occurs or not, and whether they are transient or not. Based on these assessments, the disease would be categorized as "mild," "moderate", "severe", "critical" (Table 7.3). The key difference between these classifications is that the determinant-based classification accords greater importance to the presence of infected necrosis as a determinant of severity (Metri et al., 2024). The revised Atlanta classification offers details of pathophysiology and timelines of AP. It is more widely used as the global standard. After some studies assessed the better classifications (Chen et al., 2015; Nawaz et al., 2013; Choi et al., 2014), as the manuscript reported from 1655 patients cohort of 23 multicenter study, the revised Atlanta classification and the determinant-based classification were equally superior to the Atlanta classification for describing the clinical course of AP (Sternby et al., 2019). It also indicated that persistent organ failure was the most significant determinant of severity.

Table 7.1 Grades of AP severity revised Atlanta classification.

Grades of severity	Organ failure	Local/Systemic complications
Mild	No	No
Moderate	Resolves within 48h (transient organ failure)	Without persistent organ failure
Severe	Persistent for more than 48h	Single/Multiple Organ failure

Modified from the revised Atlanta classification. Banks, P.A., Bollen, T.L., Dervenis, C., Gooszen, H.G., Johnson, C.D., Sarr, M.G., Tsiotos, G.G., Vege, S.S., 2013. Acute Pancreatitis Classification Working G: Classification of acute pancreatitis–2012: revision of the Atlanta classification and definitions by international consensus. Gut 62 (1), 102–111.

Table 7.2 Modified Marshall scoring system for organ dysfunction.

	Score				
Organ system	**0**	**1**	**2**	**3**	**4**
Respiratory (PaO_2/FiO_2)	>400	301–400	201–300	101–200	≤101
Renal[a]					
(Serum creatinine, μmol/L)	≤134	134–169	170–310	311–439	>439
(Serum creatinine, mg/dL)	<1.4	1.4–1.8	1.9–3.6	3.6–4.9	>4.9
Cardiovascular (systolic blood pressure, mmHg)[b]	>90	<90, fluid responsive	<90, not fluid responsive	<90, pH < 7.3	<90, pH < 7.2
For nonventilated patients, the FiO_2 can be estimated flow below:					
Supplemental oxygen (L/min)	**FiO_2 (%)**				
Room air	21				
2	25				
4	30				
6–8	40				
9–10	50				

A score of 2 or more in any system defines the presence of organ failure.

[a] *A score for patients with preexisting chronic renal failure depends on the extent of further deterioration of baseline renal function. No formal correction exists for a baseline serum Creatinine ≥134 μmol/L or ≥1.4 mg/dL.*

[b] *Off inotropic support.*

Banks, P.A., Bollen, T.L., Dervenis, C., Gooszen, H.G., Johnson, C.D., Sarr, M.G., Tsiotos, G.G., Vege, S.S., 2013. Acute Pancreatitis Classification Working G: Classification of acute pancreatitis–2012: revision of the Atlanta classification and definitions by international consensus. Gut 62 (1), 102–111.

Table 7.3 Determinant-based classification of acute pancreatitis severity.

	Mild AP	**Moderate AP**	**Severe AP**	**Critical AP**
(Peri)pancreatic necrosis	No	Sterile	Infected	Infected
	AND	AND/OR	OR	AND
Organ failure	No	Transient	Persistent	Persistent

Modified from Determinant-Based Classification. Dellinger, E.P., Forsmark, C.E., Layer, P., Levy, P., Maravi-Poma, E., Petrov, M.S., Shimosegawa, T., Siriwardena, A.K., Uomo, G., Whitcomb, D.C., et al., 2012. Determinant-based classification of acute pancreatitis severity: an international multidisciplinary consultation. Ann Surg 256 (6), 875–880.

There have been some reports of biomarkers to predict severity of AP (Pongprasobchai et al., 2010; Staubli et al., 2019; Stirling et al., 2017; Chen et al., 2017; Rau et al., 2007; Modrau et al., 2005; Lu et al., 2020; Nawaz et al., 2015; Jain et al., 2018; Tan et al., 2020). In the systematic review and meta-analysis article reported in 2020 (van den Berg et al., 2020), for the primary outcome admission, summary sensitivities and specificities were 87% and 88%, respectively, for IL-6 at a threshold of >50 pg/mL, concluding the superiority of IL-6 for the early prediction of SAP.

There have been other severity assessments of AP from other groups. For example, the Pancreatitis Activity Scoring System (PASS) was reported in 2017 (Wu et al., 2017) (Table 7.5). It was designed to provide an objective tool for the measurement of disease activity in patients with AP. In the prospective cohort study of consecutive patients in single center, For the prediction of moderately severe/severe pancreatitis, the PASS score (AUC = 0.71) was comparable to the more established Ranson's (AUC = 0.63), Glasgow (AUC = 0.72), Panc3 (AUC = 0.57), and HAPS (AUC = 0.54) scoring systems (Buxbaum et al., 2018). And they reported that PASS score>140 was associated with moderately severe and severe pancreatitis (OR 3.5[95% CI2.0, 6.3]), ICU admission (OR 4.9[95% CI2.5, 9.4]), local complications (OR 3.0[1.6, 5.7]), and development of SIRS (OR 2.9[1.8, 4.5]) (Table 7.4).

Recently, some guidelines have been published by several international organizations (Leppaniemi et al., 2019; Yokoe et al., 2015; IAEIJWGEa and Group IAEIJW, 2025; Hirota et al., 2006). Through a panel of experts meeting during the World Society of Emergency Surgery (WSES) in 2018, they presented the WSES guidelines for the management of SAP (Leppaniemi et al., 2019). Their statements were formulated and graded according to the Grading of Recommendations Assessments. In the guideline, SAP has been described as a disease associated with persistent organ failure and high mortality. And "Revised Atlanta Classification and Determinant-based Classification of Acute Pancreatitis Severity are similar in establishing the diagnosis and severity of acute pancreatitis", in which they are graded as 1C, strong recommendations. In the guideline, patients with organ failure should be urgently transferred to an ICU. In the guideline of the International Association of Pancreatology Revised Guidelines on Acute Pancreatitis 2025 (IAEIJWGEa and Group IAEIJW, 2025), systemic inflammatory response syndrome (SIRS) at admission and persistent SIRS at 48

Table 7.4 Systemic inflammatory response syndrome (SIRS).

Signs of systemic inflammatory response syndrome
SIRS- defined by presence of two or more criteria:
✓ Heart rate >90 beats/min
✓ Core temperature <36°C or >38°C
✓ White blood count <4000 or >12,000/mm^3
✓ Respirations >20/min or PCO2 <32 mmHg

Banks, P.A., Bollen, T.L., Dervenis, C., Gooszen, H.G., Johnson, C.D., Sarr, M.G., Tsiotos, G.G., Vege, S.S., 2013. Acute Pancreatitis Classification Working G: Classification of acute pancreatitis–2012: revision of the Atlanta classification and definitions by international consensus. Gut 62 (1), 102–111.

Table 7.5 Sequential organ failure assessment (SOFA).

	SOFA score				
	0	**1**	**2**	**3**	**4**
Respiration					
PaO_2/FiO_2 (torr)	>400	≤400	≤300	≤200 With respiratory support	≤100 With respiratory support
Coagulation					
Platelets ($\times 10^3/mm^3$)	>150	≤150	≤100	≤50	≤20
Liver					
Bilirubin (mg/dL) (µmol/L)	<1.2 <20	1.2–1.9 20–32	2.0–5.9 33–101	6.0–11.9 102–204	>12.0 >204
Cardiovascular					
Hypotension	No hypotension	MAP <70 mmHg	Dopamin ≤5 or dobutamine (any dose)[a]	Dopamin >5 or epi ≤0.1 or norepi ≤0.1[a]	Dopamin >15 or epi >0.1 or norepi >0.1[a]
Central Nervous system					
Glasgow Coma Score	15	13–14	10–12	6–9	<6
Renal					
Creatinine (mg/dL) (µmol/L) or urine output	<1.2 <110	1.2–1.9 110–170	2.0–3.4 171–299	3.5–4.9 300–440 or < 500 mL/day	>5.0 >440 or < 200 mL/day

epi, *epinephrine;* norepi, *norepinephrine.*

To convert torr to kPa, multiply the value by 0.1333.

[a] *Adrenergic agents administered for at least 1h (doses given are in µg/kg/min).*

Vincent, J.L., de Mendonca, A., Cantraine, F., Moreno, R., Takala, J., Suter, P.M., Sprung, C.L., Colardyn, F., Blecher, S., 1998. Use of the SOFA score to assess the incidence of organ dysfunction/failure in intensive care units: results of a multicenter, prospective study. Working group on "sepsis-related problems" of the European Society of Intensive Care Medicine. Crit Care Med 26 (11), 1793–1800.

h from the onset of abdominal pain, either alone or with some serum cytokines had been described as the parameter to predict AP severity.

Just as the importance of local findings, systemic conditions should be focused on in the patients' assessments. As described earlier, in the revised Atlanta classification, they recommended the Modified Marshall scoring or SOFA scoring system. From the report from 2019 of systematic research which had aimed to investigate whether the computed tomography severity index (CTSI) can predict the outcomes of AP better than other scoring systems such as acute physiologic assessment and chronic health evaluation (APACHE II) (Knaus et al., 1985)or bedside index for severity in acute pancreatitis (BISAP) score or not, APACHE II is the most accurate predictor of mortality which is highest predictive value(area under the curve 0.91, CI; 0.95–0.99), and CTSI is a good predictor of both mortality and AP severity (Miko et al., 2019). In another prospective multicenter study in 2022, which aimed to assess the performance of the PASS in multicenter studies of AP patients and to validate the PASS (Paragomi et al., 2021). In the study, the PASS can quantify AP activity, which means it can propose an AP activity.

For the assessment of the outcome of AP, persistent (>48h) organ failure is the most important clinical determinant of the outcome of AP in terms of mortality and should be used for predicting outcomes (IAEIJWGEa and Group IAEIJW, 2025). In the literature, persistent organ failure is associated with high mortality (Schepers et al., 2019; Padhan et al., 2018). From these discussions, the importance of organ failure assessments is indicated.

7.3 Major assessment systems

As indicated above, there have been many scores that have been addressed as AP severities. Although each scoring would have different characteristics, the revised Atlanta classification could be standard for AP assessment, and it is crucial to assess systemic conditions, including single or multiple organ failures.

7.3.1 Revised Atlanta classification

As indicated above, the revised Atlanta classification was presented in 2012 as the renewed version of the previous classification (Banks et al., 2013) (Table 7.1). The diagnosis and the onset of AP were clearly defined as described in Tables 7.1 and 7.2, and as "the time of onset of pain", not the time of hospital arrival. They presented the definition of types of AP: Interstitial edematous pancreatitis, Necrotizing pancreatitis, Infected pancreatic necrosis based on the radiological findings. They described the natural history and clinical consequences of different local complications. These are named as "acute peripancreatic fluid collection", "pancreatic pseudocyst", "acute necrotic collection", and "walled-off necrosis". These should be suspected when there is persistence or recurrence of abdominal pain, secondary increases in serum pancreatic enzyme activity, increasing organ dysfunction, and/or development of clinical signs of sepsis.

They also indicated the two peaks of mortality: early and late, and these phases are usually overlapping. The early phase usually lasts for the first week, with systemic disturbances resulting from the host response to local pancreatic injury. In that phase, pancreatic inflammation would cause the systemic inflammatory response syndrome (SIRS) and organ failure. If the organ failure recovers in 48 h, it can be described as "transient" organ failure; if it is longer, "persistent organ failure" (Fig. 7.1).

7.3.2 Determinant-based classification

The classification is principally based on the factors that are causally associated with the severity of AP (Dellinger et al., 2012), actual local and systemic determinants of severity, rather than description of events that are correlated with severity. The local determinant relates to whether there is (peri)pancreatic necrosis or not, and if present, whether it is sterile or infected (Table 7.3).

7.3.3 Systemic inflammatory response syndrome

First described at the conference of the American College of Chest Physicians-Society of Critical Care Medicine Consensus (American College of Chest Physicians/Society of Critical Care Medicine, 1992), and then it was introduced as the patients' conditions of sepsis or its endothelial inflammations (Bone, 1992) (Table 7.4). And now it is widely used as a complex and multifactorial life-threatening reaction triggered by trauma or infection (Marti-Chillon et al., 2025; Singer et al., 2016). This score helps assess the patient's status of "now," from basic vital signs and laboratory data.

In the International Association of Pancreatology Revised Guidelines on Acute Pancreatitis 2025 or other guidelines (Boxhoorn et al., 2020; Rana et al., 2019; Working Group IAPAPAAPG, 2013), they recommend using SIRS criteria at admission and 48 h to predict the severity of AP, preferably along with CRP or IL-6 (IAEIJWGEa and Group IAEIJW, 2025).

7.3.4 Sequential organ failure assessment

The SOFA score was presented in 1994 by a group of critical care physicians (Vincent et al., 1996, 1998) to describe the critical patients' organ failure as a simple way to be used widely (Table 7.5). The SOFA score is composed of scores from six organ systems, graded from 0 to 4 according to the degree of dysfunction/failure.

Many studies have described the usefulness of SOFA (Arts et al., 2005; Tallgren et al., 2009; Peres et al., 2002). It can be used as a routine assessment of organ failure in the settings of the ICU.

7.3.5 Acute physiology and chronic health evaluation II

The APACHE II score was first described by Knaus et al. (1985) (Table 7.6). In a meta-analysis of 30 studies that compared Ranson's, APACHE II, CTSI, modified CTSI, BISAP score, and CRP, APACHE II had the highest predictive value for mortality (area under the curve (AUC) 0.91, 95% CI 0.88-0.93) (Miko et al., 2019).

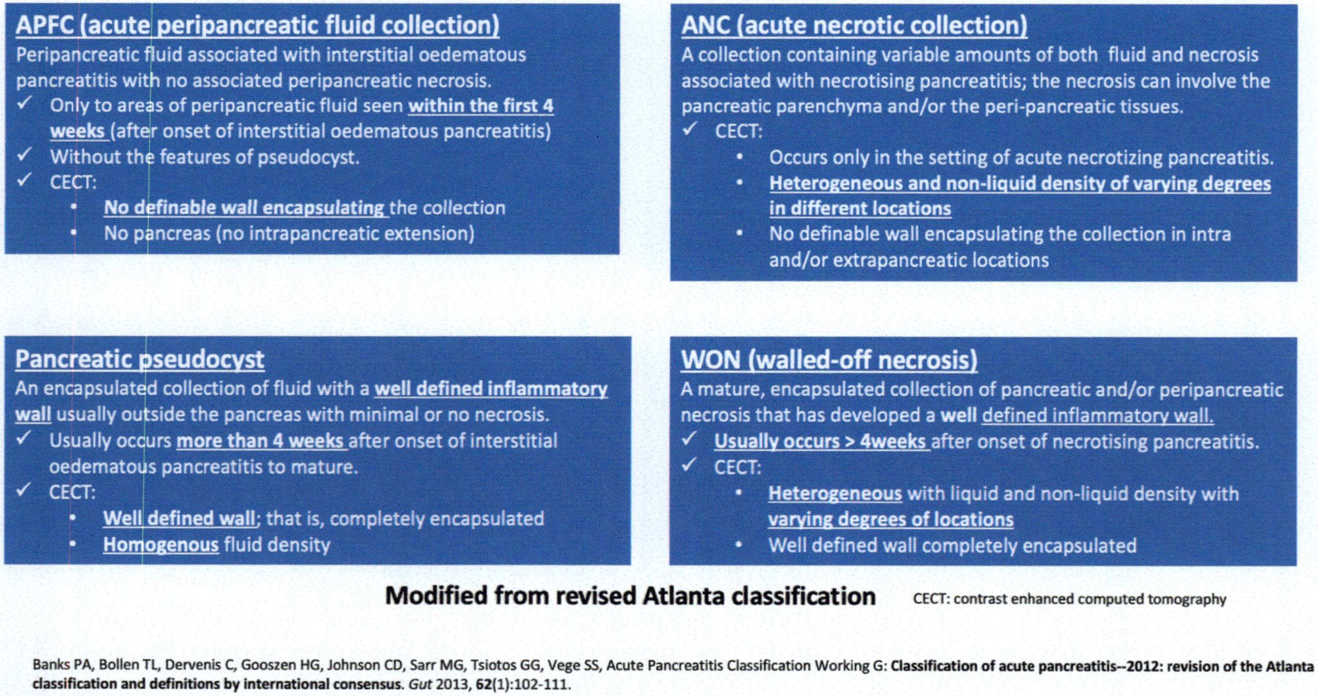

FIGURE 7.1

The morphologic features of local complications of AP from the revised Atlanta classification.

Table 7.6 Acute physiologic assessment and chronic health evaluation (APACHE II) score.

Physiologic variable[a]	+4	+3	+2	+1	0	+1	+2	+3	+4
1. Temperature (°C)	≥41	39–40.9	–	38.5–38.9	36–38.4	34–35.9	<29.9	30–31.9	≤29.9
2 Mean arterial pressure (mmHg)	≥160	130–159	110–129	–	70–109	–	50–69	–	≤49
3 Heart rate	≥180	140–179	110–139	–	70–109	–	55–69	40–54	≤39
4 Respiratory rate	≥50	35–49	–	25–34	12–24	10–11	6–9	–	≤5
5 Oxygenation									
a) $FiO_2 \geq 0.5$ Use A–aDO_2/PaO_2	≥500	350–499	200–349		<200	$PaO_2 \geq 70$	$PaO_2 < 70$	–	–
b) $FiO_2 < 0.5$ Use PaO_2(mmHg)	–	–	–	–	>70	61–70	–	55–60	<55
Arterial pH	≥7.7	7.6–7.69	–	7.5–7.59	7.33–7.49	–	7.25–7.32	7.15–7.24	<7.15
Serum Na^+ (mMol/L)	≥180	160–179	155–159	150–154	130–149	–	120–129	111–119	≤110
Serum K^+ (mMol/L)	≥7.0	6.0–6.9	–	5.5–5.9	3.5–5.4	3.0–3.4	2.5–2.9	–	<2.5
Serum creatinine (mg/dL): Double point score for acute renal failure	≥3.5	2–3.4	1.5–1.9	–	0.6–1.4	–	<0.6	–	–
Hematocrit (%)	≥60	–	50–59.9	46–49.9	30–45.9	–	20–29.9	–	<20
WBC (1000s)	≥40	–	20–39.9	15–19.9	3–14.9	–	1–2.9	–	<1
Serum HCO_3 (venous-mMol/L) Use only if no ABGs [b]	≥52	41–51.9	–	31–40.9	22–31.9	–	18–21.9	15–17.9	<15
GCS	Score = 15 – actual GCS								

[a] *Choose worst value in the past 24h*

[b] ***Optional value:** use only if no ABGs*

Increasing score is associated with increasing risk of hospital death.

APACHE II Score = Acute physiology score + Age points + Chronic health points.

Minimum score = 0, maximum score 71.

***Acute physiology score**: Calculated from table above*

***Age points, Chronic health points:** Calculated from table below*

Organ sufficiency (e.g., hepatic, cardiovascular, renal, pulmonary) or immune-compromised state must have preceded current admission.

0–4 = –4% death rate 10–14 = –15% death rate 20–24 = –40% death rate 30–34 = –75% death rate.

5–9 = –8% death rate 15–19 = –25% death rate 25–29 = –55% death rate 34 = –85% death rate

7.3.6 Pancreatitis Activity Scoring System

PASS was first described in 2017 (Wu et al., 2017) (Tables 7.7 and 7.8), and is introduced as the dynamic measure to quantify AP activity (Paragomi et al., 2021). Unlike other severity scores, which classify AP based on the outcomes, PASS can address the progressive nature of the disease by its dynamic manifestations (Dellinger et al., 2012).

7.3.7 Modified Marshall score

First described by Marshall et al. (1995), the modified for simplicity, universal applicability across international usefulness (Working Party of the British Society of G et al., 2005) (Table 7.2). They are introduced at revised Atlanta classification (Banks

Table 7.7 Pancreatitis activity scoring system (PASS).

Parameter	Weight
Organ failure (Modified Marshall Score >2)	× 100 per system
Systemic inflammatory response syndrome	× 25 per criterion
Intravenous morphine equivalents (mg)	× 5
Abdominal pain (0–10 scale)	× 5
Toleration solid died (no = 1, yes = 0)	× 40

Scoring system of Pancreatitis Activity Scoring System (PASS) (Wu et al., 2017).
Admission PASS score >140 was associated with
✓ moderately severe and severe pancreatitis (OR 3.5 [95% CI 2.0, 6.3]).
✓ ICU admission (OR 4.9 [2.5, 9.4]).
✓ local complications (3.0 [1.6, 5.7]).
✓ Development of SIRS (OR 2.9 [1.8, 4.5]).
✓ Prolongation of hospitalization by a mean of 1.5 (1.3–1.7) days.
✓ Discharge PASS score >60 was associated with early readmission (OR 5.0 [2.4, 10.7]) (Buxbaum et al., 2018).

Table 7.8 Definition of acute pancreatitis from the revised Atlanta classification.

Features	Notes
Abdominal pain	Consistent with acute pancreatitis (acute onset of a persistent, severe, epigastric pain often radiating to the back)
Serum lipase activity (or amylase activity)	At least three times greater than the upper limit of normal
Contrast-enhanced computed tomography	Characteristic findings of acute pancreatitis

The diagnosis of acute pancreatitis requires two of the above three features.
Banks, P.A., Bollen, T.L., Dervenis, C., Gooszen, H.G., Johnson, C.D., Sarr, M.G., Tsiotos, G.G., Vege, S.S., 2013. Acute Pancreatitis Classification Working G: Classification of acute pancreatitis–2012: revision of the Atlanta classification and definitions by international consensus. Gut 62 (1), 102–111.

et al., 2013). In the manuscript, this score system is preferred to the SOFA scoring system, which is for patients managed in a critical care unit and which considers the use of inotropic and respiratory support.

7.3.8 Other parameters

Although many markers have been presented as predicting markers of severity of AP (Pongprasobchai et al., 2010; Staubli et al., 2019; Stirling et al., 2017; Chen et al., 2017; Rau et al., 2007; Modrau et al., 2005; Lu et al., 2020; Nawaz et al., 2015; Jain et al., 2018; Tan et al., 2020), IL-6 or CRP could be the better markers.

7.4 Clinical instructions

What can we do to assess the patients of AP and their severity.

1. Assess the general conditions and make a diagnosis.

   Based on the patients' symptoms or physical examinations, additional laboratory tests and radiological examinations should be added.
   Once the diagnosis is confirmed, the additional findings to assess the severity of AP, multiple or single organ failures, respiratory problems, and hemodynamic status should be identified.
   It is valuable to assess via scoring systems like SOFA or Marshall, not only the local findings of AP, but also systemic findings. What specific data should be used depends on which scoring system is chosen.

2. Additional radiological findings

   Radiological findings, including contrast-enhanced computed tomography, are a useful way to
   Assess the local findings of AP, edematous pancreatitis/necrotizing or not/infectious or not.
   <Once the assessments have been completed, the treatment strategy, including transferring to another high-volume center or not, should be decided.>

3. Tertiary survey

   Daily assessments by laboratory data or scoring system (SOFA, Marshall etc.) are valuable to check the treatment response. PASS might be a better way for continuous checkups.

7.5 Future insights of severity assessments

Although the revised Atlanta classification might be the gold standard of this field, many other scorings or laboratory tests have also been reported. AP presents complications not only in local findings, but also in systemic inflammatory damage. Each scoring or laboratory tests have several advantages.

Multiple approaches to advancing the assessment of the AP would be developed in the future.

References

American College of Chest Physicians/Society of Critical Care Medicine Consensus Conference: definitions for sepsis and organ failure and guidelines for the use of innovative therapies in sepsis. Critical Care Medicine 20 (6), 1992, 864–874.

Arts, D.G., de Keizer, N.F., Vroom, M.B., de Jonge, E., 2005. Reliability and accuracy of Sequential Organ Failure Assessment (SOFA) scoring. Critical Care Medicine 33 (9), 1988–1993.

Banks, P.A., Bollen, T.L., Dervenis, C., Gooszen, H.G., Johnson, C.D., Sarr, M.G., Tsiotos, G. G., Vege, S.S., Acute Pancreatitis Classification Working, G., 2013. Classification of acute pancreatitis–2012: revision of the Atlanta classification and definitions by international consensus. Gut 62 (1), 102–111.

Bollen, T.L., van Santvoort, H.C., Besselink, M.G., van Leeuwen, M.S., Horvath, K.D., Freeny, P.C., Gooszen, H.G., 2008. Dutch Acute pancreatitis study G: the Atlanta classification of acute pancreatitis revisited. British Journal of Surgery 95 (1), 6–21.

Bone, R.C., 1992. Toward an epidemiology and natural history of SIRS (systemic inflammatory response syndrome). JAMA 268 (24), 3452–3455.

Boxhoorn, L., Voermans, R.P., Bouwense, S.A., Bruno, M.J., Verdonk, R.C., Boermeester, M. A., van Santvoort, H.C., Besselink, M.G., 2020. Acute pancreatitis. Lancet 396 (10252), 726–734.

Bradley, E.L., 1993. 3rd: A clinically based classification system for acute pancreatitis. Summary of the International Symposium on Acute Pancreatitis, Atlanta, Ga, September 11 through 13, 1992. Archives of Surgery 128 (5), 586–590.

Buxbaum, J., Quezada, M., Chong, B., Gupta, N., Yu, C.Y., Lane, C., Da, B., Leung, K., Shulman, I., Pandol, S., et al., 2018. The Pancreatitis Activity Scoring System predicts clinical outcomes in acute pancreatitis: findings from a prospective cohort study. American Journal of Gastroenterology 113 (5), 755–764.

Chen, Y., Ke, L., Tong, Z., Li, W., Li, J., 2015. Association between severity and the determinant-based classification, Atlanta 2012 and Atlanta 1992, in acute pancreatitis: a clinical retrospective study. Medicine 94 (13), e638.

Chen, H.Z., Ji, L., Li, L., Wang, G., Bai, X.W., Cheng, C.D., Sun, B., 2017. Early prediction of infected pancreatic necrosis secondary to necrotizing pancreatitis. Medicine 96 (30), e7487.

Choi, J.H., Kim, M.H., Oh, D., Paik, W.H., Park, D.H., Lee, S.S., Seo, D.W., Lee, S.K., 2014. Clinical relevance of the revised Atlanta classification focusing on severity stratification system. Pancreatology 14 (5), 324–329.

Dellinger, E.P., Forsmark, C.E., Layer, P., Levy, P., Maravi-Poma, E., Petrov, M.S., Shimosegawa, T., Siriwardena, A.K., Uomo, G., Whitcomb, D.C., et al., 2012. Determinant-based classification of acute pancreatitis severity: an international multidisciplinary consultation. Annals of Surgery 256 (6), 875–880.

Gurusamy, K.S., Belgaumkar, A.P., Haswell, A., Pereira, S.P., Davidson, B.R., 2016. Interventions for necrotising pancreatitis. Cochrane Database of Systematic Reviews 4 (4), CD011383.

Heckler, M., Hackert, T., Hu, K., Halloran, C.M., Buchler, M.W., Neoptolemos, J.P., 2021. Severe acute pancreatitis: surgical indications and treatment. Langenbeck's Archives of Surgery 406 (3), 521–535.

Hirota, M., Takada, T., Kawarada, Y., Hirata, K., Mayumi, T., Yoshida, M., Sekimoto, M., Kimura, Y., Takeda, K., Isaji, S., et al., 2006. JPN Guidelines for the management of acute pancreatitis: severity assessment of acute pancreatitis. Journal of Hepato-Biliary-Pancreatic Surgery 13 (1), 33–41.

IAEIJWGEa, Group IAEIJW, 2025. International Association of Pancreatology Revised Guidelines on Acute Pancreatitis 2025: Supported and Endorsed by the American Pancreatic Association, European Pancreatic Club, Indian Pancreas Club, and Japan Pancreas Society. Pancreatology 25 (6), 770–814.

Jain, S., Midha, S., Mahapatra, S.J., Gupta, S., Sharma, M.K., Nayak, B., Jacob, T.G., Shalimar, Garg, P.K., 2018. Interleukin-6 significantly improves predictive value of systemic inflammatory response syndrome for predicting severe acute pancreatitis. Pancreatology 18 (5), 500–506.

Knaus, W.A., Draper, E.A., Wagner, D.P., Zimmerman, J.E., 1985. APACHE II: a severity of disease classification system. Critical Care Medicine 13 (10), 818–829.

Leppaniemi, A., Tolonen, M., Tarasconi, A., Segovia-Lohse, H., Gamberini, E., Kirkpatrick, A.W., Ball, C.G., Parry, N., Sartelli, M., Wolbrink, D., et al., 2019. WSES guidelines for the management of severe acute pancreatitis. World Journal of Emergency Surgery 14, 27.

Li, C.L., Jiang, M., Pan, C.Q., Li, J., Xu, L.G., 2021. The global, regional, and national burden of acute pancreatitis in 204 countries and territories, 1990-2019. BMC Gastroenterology 21 (1), 332.

Lu, Z., Zhang, G., Guo, F., Li, M., Ding, Y., Zheng, H., Wang, D., 2020. Elevated triglycerides on admission positively correlate with the severity of hypertriglyceridaemic pancreatitis. International Journal of Clinical Practice 74 (3), e13458.

Maldonado, I., Shetty, A., Estay, M.C., Sina, E., Rojas, A., Narra, V., Varela, C., 2021. Acute pancreatitis imaging in MDCT: state of the art of usual and unusual local complications. 2012 Atlanta Classification revisited. Current Problems in Diagnostic Radiology 50 (2), 186–199.

Marshall, J.C., Cook, D.J., Christou, N.V., Bernard, G.R., Sprung, C.L., Sibbald, W.J., 1995. Multiple organ dysfunction score: a reliable descriptor of a complex clinical outcome. Critical Care Medicine 23 (10), 1638–1652.

Marti-Chillon, G.J., Muntion, S., Flores-Montero, J., Gonzalez-Robledo, J., Corchete, L., Pozo, J., Osugui, L., Preciado, S., Gomez-Redondo, M., Blanco, J.F., et al., 2025. Immune cell dynamics, cytokines, and extracellular vesicles in systemic inflammatory response syndrome (SIRS): a multiparametric analysis. Journal of Translational Medicine 23 (1), 1331.

Metri, A., Bush, N., Singh, V.K., 2024. Predicting the severity of acute pancreatitis: current approaches and future directions. Surgery Open Science 19, 109–117.

Miko, A., Vigh, E., Matrai, P., Soos, A., Garami, A., Balasko, M., Czako, L., Mosdosi, B., Sarlos, P., Eross, B., et al., 2019. Computed tomography severity index vs. Other indices in the prediction of severity and mortality in acute pancreatitis: a predictive accuracy meta-analysis. Frontiers in Physiology 10, 1002.

Modrau, I.S., Floyd, A.K., Thorlacius-Ussing, O., 2005. The clinical value of procalcitonin in early assessment of acute pancreatitis. American Journal of Gastroenterology 100 (7), 1593–1597.

Nawaz, H., Mounzer, R., Yadav, D., Yabes, J.G., Slivka, A., Whitcomb, D.C., Papachristou, G. I., 2013. Revised Atlanta and determinant-based classification: application in a prospective cohort of acute pancreatitis patients. American Journal of Gastroenterology 108 (12), 1911–1917.

Nawaz, H., Koutroumpakis, E., Easler, J., Slivka, A., Whitcomb, D.C., Singh, V.P., Yadav, D., Papachristou, G.I., 2015. Elevated serum triglycerides are independently associated with persistent organ failure in acute pancreatitis. American Journal of Gastroenterology 110 (10), 1497–1503.

Ong, Y., Shelat, V.G., 2021. Ranson score to stratify severity in Acute Pancreatitis remains valid - old is gold. Expert Review of Gastroenterology & Hepatology 15 (8), 865–877.

Padhan, R.K., Jain, S., Agarwal, S., Harikrishnan, S., Vadiraja, P., Behera, S., Jain, S.K., Dhingra, R., Dash, N.R., Sahni, P., et al., 2018. Primary and secondary organ failures cause mortality differentially in acute pancreatitis and should be distinguished. Pancreas 47 (3), 302–307.

Paragomi, P., Tuft, M., Pothoulakis, I., Singh, V.K., Stevens, T., Nawaz, H., Easler, J.J., Thakkar, S., Cote, G.A., Lee, P.J., et al., 2021. Dynamic changes in the pancreatitis activity scoring system during hospital course in a multicenter, prospective cohort. Journal of Gastroenterology and Hepatology 36 (9), 2416–2423.

Peres, B.D., Melot, C., Lopes, F.F., Nguyen Ba, V., Vincent, J.L., 2002. The multiple organ dysfunction score (MODS) versus the sequential organ failure assessment (SOFA) score in outcome prediction. Intensive Care Medicine 28 (11), 1619–1624.

Pongprasobchai, S., Jianjaroonwong, V., Charatcharoenwitthaya, P., Komoltri, C., Tanwandee, T., Leelakusolvong, S., Pausawasdi, N., Srikureja, W., Chainuvati, S., Prachayakul, V., et al., 2010. Erythrocyte sedimentation rate and C-reactive protein for the prediction of severity of acute pancreatitis. Pancreas 39 (8), 1226–1230.

Rana, S.S., Sharma, R.K., Gupta, R., 2019. Endoscopic management of pancreatic ascites due to duct disruption following acute necrotizing pancreatitis. JGH Open 3 (2), 111–116.

Ranson, J.H., Pasternack, B.S., 1977. Statistical methods for quantifying the severity of clinical acute pancreatitis. Journal of Surgical Research 22 (2), 79–91.

Rau, B.M., Kemppainen, E.A., Gumbs, A.A., Buchler, M.W., Wegscheider, K., Bassi, C., Puolakkainen, P.A., Beger, H.G., 2007. Early assessment of pancreatic infections and overall prognosis in severe acute pancreatitis by procalcitonin (PCT): a prospective international multicenter study. Annals of Surgery 245 (5), 745–754.

Schepers, N.J., Bakker, O.J., Besselink, M.G., Ahmed Ali, U., Bollen, T.L., Gooszen, H.G., van Santvoort, H.C., Bruno, M.J., 2019. Dutch Pancreatitis Study G: impact of characteristics of organ failure and infected necrosis on mortality in necrotising pancreatitis. Gut 68 (6), 1044–1051.

Singer, M., Deutschman, C.S., Seymour, C.W., Shankar-Hari, M., Annane, D., Bauer, M., Bellomo, R., Bernard, G.R., Chiche, J.D., Coopersmith, C.M., et al., 2016. The third international consensus definitions for sepsis and Septic Shock (Sepsis-3). JAMA 315 (8), 801–810.

Staubli, S.M., Schafer, J., Rosenthal, R., Zeindler, J., Oertli, D., Nebiker, C.A., 2019. The role of CRP and Pentraxin 3 in the prediction of systemic inflammatory response syndrome and death in acute pancreatitis. Scientific Reports 9 (1), 18340.

Sternby, H., Bolado, F., Canaval-Zuleta, H.J., Marra-Lopez, C., Hernando-Alonso, A.I., Del-Val-Antonana, A., Garcia-Rayado, G., Rivera-Irigoin, R., Grau-Garcia, F.J., Oms, L., et al., 2019. Determinants of severity in acute pancreatitis: a nation-wide multicenter prospective cohort study. Annals of Surgery 270 (2), 348–355.

Stirling, A.D., Moran, N.R., Kelly, M.E., Ridgway, P.F., Conlon, K.C., 2017. The predictive value of C-reactive protein (CRP) in acute pancreatitis - is interval change in CRP an additional indicator of severity? HPB (Oxford) 19 (10), 874–880.

Tallgren, M., Backlund, M., Hynninen, M., 2009. Accuracy of sequential organ failure assessment (SOFA) scoring in clinical practice. Acta Anaesthesiologica Scandinavica 53 (1), 39–45.

Tan, J.W., Zhang, X.Q., Geng, C.M., Peng, L.L., 2020. Development of the national early warning score-calcium model for predicting adverse outcomes in patients with acute pancreatitis. Journal of Emergency Nursing 46 (2), 171–179.

van den Berg, F.F., de Bruijn, A.C., van Santvoort, H.C., Issa, Y., Boermeester, M.A., 2020. Early laboratory biomarkers for severity in acute pancreatitis; A systematic review and meta-analysis. Pancreatology 20 (7), 1302–1311.

Vincent, A., Shashirekha, C.A., 2024. Predicting severity of acute pancreatitis-evaluation of neutrophil-to-lymphocyte count ratio as emerging biomarker: a retrospective analytical study. Cureus 16 (11), e74881.

Vincent, J.L., Moreno, R., Takala, J., Willatts, S., De Mendonca, A., Bruining, H., Reinhart, C. K., Suter, P.M., Thijs, L.G., 1996. The SOFA (Sepsis-related organ failure assessment) score to describe organ dysfunction/failure. On behalf of the working group on sepsis-related problems of the European society of intensive care medicine. Intensive Care Medicine 22 (7), 707–710.

Vincent, J.L., de Mendonca, A., Cantraine, F., Moreno, R., Takala, J., Suter, P.M., Sprung, C. L., Colardyn, F., Blecher, S., 1998. Use of the SOFA score to assess the incidence of organ dysfunction/failure in intensive care units: results of a multicenter, prospective study. Working group on "sepsis-related problems" of the European Society of Intensive Care Medicine. Critical Care Medicine 26 (11), 1793–1800.

Working Group IAPAPAAPG, 2013. IAP/APA evidence-based guidelines for the management of acute pancreatitis. Pancreatology 13 (4 Suppl. 2), e1–e15.

Working Party of the British Society of G, Association of Surgeons of Great B, Ireland, 2005. Pancreatic society of Great B, Ireland, association of Upper GISoGB, Ireland: UK guidelines for the management of acute pancreatitis. Gut 54 (Suppl. 3), iii1–iii9.

Wu, B.U., Batech, M., Quezada, M., Lew, D., Fujikawa, K., Kung, J., Jamil, L.H., Chen, W., Afghani, E., Reicher, S., et al., 2017. Dynamic measurement of disease activity in acute pancreatitis: the pancreatitis activity scoring system. American Journal of Gastroenterology 112 (7), 1144–1152.

Yokoe, M., Takada, T., Mayumi, T., Yoshida, M., Isaji, S., Wada, K., Itoi, T., Sata, N., Gabata, T., Igarashi, H., et al., 2015. Japanese guidelines for the management of acute pancreatitis: Japanese Guidelines 2015. Journal of Hepato-Biliary-Pancreatic Sciences 22 (6), 405–432.

CHAPTER

Treatment of pancreatitis 1: Fluid resuscitation strategies

8

Kodai Suzuki

Gifu University, Graduate School of Medicine, Department of Infection Control, and the Advanced Critical Care Center at Gifu University Hospital, Gifu, Japan

8.1 Chapter 1: Introduction

8.1.1 Background and epidemiology of acute pancreatitis

The prognosis of acute pancreatitis (AP), an acute inflammatory disease of the pancreas caused by various factors (Pacella et al., 2025), depends on the presence or absence of multiple-organ failure and secondary infection due to necrosis of the pancreas or peripancreatic tissues (Boxhoorn et al., 2020). Globally, the incidence of AP is 33.74 cases per 100,000 population per year, with no difference in incidence or mortality between men and women (Xiao et al., 2016; Tanaka et al., 2025). However, considerable differences in etiology—such as gallstones, alcohol consumption, idiopathic causes, and metabolic disorders—are observed between men and women and across regions (Xiao et al., 2016; Tanaka et al., 2025). In women, gallstones are the most common cause, whereas alcohol-related AP remains a major factor in men (Xiao et al., 2016; Tanaka et al., 2025). Most patients experience mild illness; however, approximately 20% develop organ failure in the heart, lungs, kidneys, and other organs due to systemic inflammation caused by pancreatitis, leading to multiple-organ failure (Garg and Singh, 2019). Early detection and appropriate treatment are essential to improve patient outcomes (Tenner et al., 2024).

8.1.2 Pathophysiological rationale for fluid resuscitation

The pathophysiology of AP involves complex interactions among factors that contribute to its onset and progression, including abnormal activation of trypsin within the pancreas, production of inflammatory mediators, genetic predisposition, intracellular stress responses, microcirculatory dysfunction, and endothelial glycocalyx injury (Tenner et al., 2024; Mayerle et al., 2019; Turunen et al., 2024; Li et al., 2025). This cascade leads to increased endothelial permeability, fluid loss into the third space, and a reduction in intravascular volume (Tenner et al., 2024). A decrease in intravascular blood volume contributes to pancreatic ischemia,

Advanced Insights into Acute Pancreatitis. https://doi.org/10.1016/B978-0-443-44736-5.00007-3

exacerbating acinar cell injury and prolonging inflammation (Foitzik et al., 1995). In summary, fluid resuscitation aims to prevent progression to multiple-organ failure by restoring intravascular volume and improving pancreatic microcirculation (Tenner et al., 2024). This rationale is supported by experimental models demonstrating that early volume resuscitation reduces pancreatic necrosis and systemic complications (Aggarwal et al., 2014).

8.1.3 Impact of early fluid therapy on outcomes

Multiple observational studies and randomized controlled trials have demonstrated that appropriate fluid resuscitation within 24 h of onset leads to a reduced incidence of organ failure, shorter hospital stays, and lower mortality rates (Mao et al., 2009; Li et al., 2023; Guo et al., 2023). The WATERFALL trial compared aggressive versus moderate fluid resuscitation in AP and found that overly aggressive hydration increased the risk of fluid overload. Although no difference was observed in tumor assessment scores, patients receiving aggressive hydration had a higher risk of worsening disease, more severe symptoms, longer hospital stays, and a higher incidence of necrotizing pancreatitis (de-Madaria et al., 2022). Therefore, current infusion therapy emphasizes the need for a treatment approach that evaluates fluid balance over time. Early fluid administration is effective in preventing hypoperfusion-related complications; however, excessive fluid administration may cause pulmonary edema or abdominal compartment syndrome, potentially worsening outcomes (Mao et al., 2009).

8.1.4 Scope and limitations of this review

This review integrates evidence from international guidelines, randomized controlled trials, observational studies, and metaanalyses focusing on fluid resuscitation strategies for adult patients with AP. Limitations include heterogeneity in study designs, variability in definitions of "early" and "appropriate" resuscitation, and differences in patient populations across regions.

8.2 Chapter 2: Pathophysiology and hemodynamic changes in acute pancreatitis

8.2.1 Pathophysiology of AP

In AP, localized pancreatic microcirculatory impairment plays a central role in disease progression (Cuthbertson and Christophi, 2006). Digestive enzymes such as trypsin, which are activated by damage to pancreatic acinar cells, directly injure pancreatic vascular endothelial cells, promoting microthrombus formation and blood flow stasis (Tenner et al., 2024; Mayerle et al., 2019). AP is characterized by the abnormal activation of pancreatic enzymes, resulting in autolytic inflammation within the pancreas and surrounding tissues (Mayerle et al., 2019). In severe cases, this localized inflammation can spread, potentially progressing to systemic inflammatory response syndrome (SIRS) and multiple-organ failure (Garg and Singh, 2019).

In the early stages of AP, extensive release of inflammatory cytokines (including TNF-α, IL-1β, and IL-6) from damaged pancreatic acinar cells promotes the systemic spread of localized inflammation (Norman, 1998). This inflammatory response increases endothelial permeability, leading to fluid loss into the third space (Tenner et al., 2024). Inflammatory mediators induce vasodilation via nitric oxide production and prostaglandin pathways, while simultaneously increasing endothelial permeability (Zuckerbraun et al., 2005).

AP causes intense systemic inflammation, leading to widespread destruction of the vascular endothelial glycocalyx, a gel-like glycoprotein layer that lines the luminal surface of blood vessels (Li et al., 2025). The vascular endothelial glycocalyx regulates vascular permeability and separates the vascular endothelium from the blood (Suzuki et al., 2023). Destruction of the vascular endothelial glycocalyx directly increases vascular permeability (Li et al., 2025).

Consequently, plasma protein leakage occurs, leading to a substantial accumulation of exudate around the pancreas and within the abdominal cavity (Sinha et al., 2016). This rapidly reduces effective circulating blood volume, leading to hypotension and subsequent organ hypoperfusion (Foitzik et al., 1995). In the pathophysiology of pancreatitis, inflammation increases the oxygen demand of pancreatic tissue, whereas microcirculatory impairment limits oxygen supply. This oxygen deficit shifts cellular metabolism toward an anaerobic state, resulting in decreased ATP production and cellular damage (Luo et al., 2025).

8.2.2 Microcirculatory disturbances and pancreatic ischemia

In AP, local inflammation within the pancreas causes vasoconstriction, endothelial damage, and capillary obstruction by leukocytes, thereby affecting microcirculation within the pancreas (Aggarwal et al., 2014; Keck et al., 2005; Wu et al., 2024; Klar et al., 1994). These alterations impair oxygen supply to the pancreas and surrounding tissues, thereby promoting ischemia (Pień et al., 2016).

Neutrophil extracellular traps (NETs) also contribute to microcirculatory dysfunction in pancreatitis. NETs are reticular structures composed of DNA, histones, and granular enzymes released by neutrophils (Brinkmann et al., 2004; Rada, 2019). Their physiological role is to capture and inactivate pathogens; however, they are also immunological mechanisms deeply involved in the onset, progression, and severity of AP. NET-derived histones, although useful in infection defense, possess strong cytotoxicity, directly damaging pancreatic and vascular endothelial cells and worsening microcirculatory dysfunction. Excessive NET formation causes not only local necrosis but also distant organ damage such as acute respiratory distress syndrome (ARDS) and acute kidney injury (AKI) (Hu et al., 2020).

8.2.3 Systemic inflammatory response and capillary leak

Inflammation in AP is characterized by the release of inflammatory cytokines (such as TNF-α, IL-1β, and IL-6) and activation of the complement pathway (Norman, 1998). These mediators increase capillary permeability, causing plasma proteins and fluid to leak into the interstitial space. This results in decreased blood volume, blood concentration, and reduced organ perfusion (Aggarwal et al., 2014; Pień et al., 2016). This systemic inflammatory response-associated capillary leak extends beyond the local pancreas, contributing to pulmonary and renal dysfunction (Komara et al., 2020). The decrease in circulating blood volume leads to reduced cardiac output and worsened tissue hypoxia. In particular, renal hypoperfusion predisposes patients to AKI, which is a poor prognostic indicator. Since increased capillary permeability peaks within the first 24–48 h, this timeframe is critical for fluid resuscitation (Sinha et al., 2016). The challenge lies in determining the optimal fluid volume to restore effective circulating volume without exacerbating interstitial edema in vital organs such as the lungs.

8.2.4 Hemodynamic monitoring parameters relevant to fluid therapy

Key parameters guiding fluid resuscitation in AP include mean arterial pressure (MAP), heart rate, urine output, hematocrit, blood urea nitrogen (BUN), C-reactive protein (CRP), and serum lactate levels (Tenner et al., 2024; Brown et al., 2000; Mentula et al., 2014). Short-term increases in hematocrit, BUN, CRP, and serum creatinine predict worsening prognosis; however, there remains insufficient evidence to recommend targeting these parameters in fluid resuscitation (Wilms et al., 2014). Central venous pressure (CVP), although traditionally used to assess intravascular volume status, has shown limited correlation with actual vascular volume (Marik et al., 2011). However, dynamic indicators such as changes in Pleth Variability Index and stroke volume induced by passive leg raising (PLR), when measured using hemodynamic monitoring, are useful for assessing fluid responsiveness, particularly in mechanically ventilated patients (Cannesson et al., 2008; Biais et al., 2009).

8.2.5 Fluid therapy from a pathological perspective

Intravenous fluid therapy compensates for fluid loss associated with increased vascular permeability caused by inflammatory cytokines (Garg and Singh, 2019; Tenner et al., 2024; Aggarwal et al., 2014). It improves pancreatic microcirculation by correcting intravascular volume depletion and tissue hypoperfusion (Tenner et al., 2024). This represents the most critical initial intervention for maintaining oxygen delivery to the pancreas and other systemic organs by suppressing pancreatic ischemic injury (Aggarwal et al., 2014). Furthermore, intravascular volume depletion and tissue hypoperfusion themselves cause damage to the vascular endothelial glycocalyx (Oiwa et al., 2023). Therefore, inadequate fluid administration can further increase vascular permeability and contribute to the development of multiple organ failure, thereby creating a vicious cycle. Lactate levels serve as an indicator of tissue hypoperfusion and are also used to assess fluid responsiveness to infusion therapy (Aggarwal et al., 2014). Reducing lactate levels and improving tissue oxygenation through appropriate fluid administration constitute physiologically sound therapeutic goals (Aggarwal et al., 2014). Although adequate fluid administration does not directly prevent increased intraabdominal pressure, it mitigates secondary damage by maintaining organ perfusion. Experimental studies have shown that early fluid resuscitation improves microvascular hypoperfusion, reduces leukocyte adhesion, and suppresses necrosis (Juvonen et al., 1999).

Adequate fluid administration also contributes to the prevention of AKI by maintaining cardiac output and ensuring adequate renal blood flow. However, excessive fluid administration may further increase intraabdominal pressure (Mentula et al., 2014). Excessive fluid administration can reportedly exacerbate damage to the vascular endothelial glycocalyx (Alphonsus and Rodseth, 2014). Furthermore, the presence of progressive inflammatory pulmonary edema indicates that excessive fluid administration may worsen pulmonary edema, underscoring the importance of appropriate fluid management (Li et al., 2023).

8.3 Chapter 3: Evidence base for fluid resuscitation strategies

8.3.1 Summary of international guidelines (Japan, the USA, IAP/APA) (Tenner et al., 2024; Takada et al., 2022; International Association of Pancreatology Revised Guidelines on Acute Pancreatitis 2025, 2025)

International guidelines on fluid therapy in AP (including those from Japan, the USA, and IAP/APA) share common principles emphasizing early, adequate, and goal-directed resuscitation. However, they differ in recommended fluid type, infusion rate, monitoring parameters, and duration of aggressive hydration.

8.3.1.1 Similarities

1. Early initiation

All guidelines recommend starting fluid resuscitation as early as possible after diagnosis, ideally within the first 12–24 h, to prevent progression to severe pancreatitis and organ failure.

2. Goal-directed approach

Fluid therapy should be adjusted to the individual patient condition, with adjustments based on clinical and laboratory parameters (such as heart rate, MAP, urine output, hematocrit, and BUN) rather than fixed volumes alone.

3. Preferred fluid type

Balanced fluids are recommended as the first-line treatment, with lactated Ringer's solution generally favored over normal saline owing to a lower risk of hyperchloremic metabolic acidosis.

4. Avoidance of overhydration

All guidelines caution against excessive fluid administration, which may increase the risk of abdominal compartment syndrome, pulmonary edema, and worsening outcomes.

8.3.1.2 Differences

1. Initial rate and volume

- Japan (2021): Recommends aggressive early fluid therapy, with an initial administration of 1 L within the first 4 h and at least 4.3 L within 24 h.

USA (ACG): Suggests a moderately aggressive fluid replacement strategy. The initial resuscitation rate should be administered at $\leq$1.5 mL/kg/h. In patients with hypovolemia, a bolus dose of 10 mL/kg is recommended.

- IAP/APA: Recommends a moderate infusion rate of 1.5 mL/kg/h. If hypovolemia or hypotension occurs at the time of presentation, an infusion bolus is recommended. Additional fluid may be administered based on hematocrit levels and clinical signs of hypovolemia.

8.3.1.3 Monitoring parameters

Japan: Hematocrit, BUN, CVP, heart rate, blood pressure, and urine output. Since individual indicators have low reliability, a combination of parameters is recommended.

USA: Focuses primarily on lowering BUN levels.

- IAP/APA: MAP 65–85 mmHg, urine output $\geq$0.5 mL/kg/h, BUN <20 mg/dL (or blood urea <40 mg/dL), and hematocrit <44% are reasonable targets reflecting adequate fluid status. Therefore, invasive monitoring is recommended for patients admitted to the intensive care unit (ICU).

8.3.1.4 Duration of aggressive hydration

- Japan: Recommends active fluid administration of at least 1 L within the first 4 h after admission and at least 4.3 L within 24 h.

USA: Emphasizes achieving adequate resuscitation within the first 24 h. Fluid status should be reassessed frequently within 6 h and again between 24 and 48 h after presentation.

- IAP/APA: No explicit timeframe.

8.3.1.5 Special considerations

Japan: Strong caution in elderly patients and those with cardiac/renal dysfunction; recommends early transition to the maintenance phase.

- USA: Caution in patients with concomitant cardiovascular or renal disease. Monitoring for fluid overload is necessary.

- IAP/APA: High-risk patients—including those with a Charlson Comorbidity Index ≥ 3, morbid obesity, elderly patients (≥ 65 years), and those with organ dysfunction requiring intensive monitoring or treatment—should be evaluated for admission to an advanced care unit or ICU to prevent rapid clinical deterioration.

8.3.2 Key randomized controlled trials (including the WATERFALL trial) (de-Madaria et al., 2022)

The WATERFALL trial is a prospective, randomized, controlled study designed to examine the efficacy and safety of aggressive fluid therapy in AP. It addressed whether the traditional "early massive fluid infusion" approach is truly effective. Early observational studies of AP recommended early and sufficient fluid infusion because the inflammatory response and increased vascular permeability following the onset of pancreatitis can lead to progressive intravascular dehydration, which may contribute to pancreatic necrosis. However, excessive fluid infusion poses a clinical challenge, as it increases the risk of fluid overload, potentially leading to increased intraabdominal pressure and worsening disease severity.

In the WATERFALL trial, patients with AP were randomized into the aggressive (20 mL/kg bolus followed by 3 mL/kg/h) and moderate (10 mL/kg bolus only if hypovolemic, followed by 1.5 mL/kg/h) resuscitation groups. The primary endpoint was the development of moderate or severe AP (based on the revised Atlanta classification) during hospitalization. Secondary endpoints included the occurrence of organ failure, local complications, length of hospital stay, and mortality.

The results showed that, although no significant difference was observed between the groups, the aggressive group experienced a higher risk of severe disease, more severe symptoms, a longer length of hospital stay, and a higher incidence of necrotizing pancreatitis. This clearly supports the recommendation of individualized fluid management based on the hemodynamics of patients and clinical indicators, rather than routine "aggressive massive fluid resuscitation in AP."

The WATERFALL trial has prompted a reconsideration of initial fluid resuscitation strategies in AP and has influenced current guidelines. Although early fluid infusion at the onset of symptoms is important, ensuring safety requires avoiding excessive volume or overly rapid infusion and adjusting therapy based on frequent reassessment. This clinically confirms that the management of AP is based on "rapid and appropriate" fluid infusion, and avoiding secondary complications caused by excessive fluid infusion is essential.

8.3.3 Controversies and unresolved questions

Key points of discussion include the optimal infusion rate and volume, choice of crystalloid solution, and role of invasive hemodynamic monitoring. The required fluid volume varies depending on disease severity, preexisting conditions, age, and changes over time. Therefore, balancing the prevention of organ hypoperfusion with the risk of fluid overload remains an unresolved issue.

8.4 Chapter 4: Fluid type selection

8.4.1 Crystalloids: Lactated Ringer's solution versus normal saline

In the management of AP, early and adequate fluid resuscitation is a cornerstone of therapy to prevent progression to severe disease and organ failure. Among crystalloid solutions, LR and NS are the most commonly used. Evidence from randomized controlled trials and observational studies suggests that LR may be superior to NS in reducing systemic inflammation and improving clinical outcomes (Tenner et al., 2024; Takada et al., 2022; International Association of Pancreatology Revised Guidelines on Acute Pancreatitis 2025, 2025).

8.4.2 Advantages of the LR solution from a basic medical perspective

- Acid–base effects:

NS has a high chloride content (154 mEq/L), which can lead to hyperchloremic metabolic acidosis. This acidosis may exacerbate AKI and systemic inflammation (Wu et al., 2011). LR contains lower chloride (109 mEq/L) and includes lactate, which is metabolized to bicarbonate, helping to maintain physiological pH (Wu et al., 2011; Semler and Rice, 2016). Hyperchloremic acidosis activates the NF-κB pathway and induces inflammation, whereas LR stabilizes pH, thereby exerting an antiinflammatory effect (Wu et al., 2011).

- Inflammatory modulation:

1. Suppression of proinflammatory cytokines

Hyperchloremic acidosis activates the NF-κB pathway, increasing the production of IL-1β, IL-6, and TNF-α (Norman, 1998; Kellum et al., 2006; Liu et al., 2017). LR, with its low chloride content, produces HCO_3^- through lactate metabolism, which stabilizes pH, suppresses NF-κB activation, and prevents increases in blood levels of these cytokines (Wu et al., 2011; Yang et al., 2020; de-Madaria et al., 2018).

2. Promotion of antiinflammatory cytokines

Stabilization of pH exerts antiinflammatory effects, promotes IL-10 production via the STAT3 pathway, and induces a shift from proinflammatory macrophages (M1 type) to reparative macrophages (M2 type) (Yang et al., 2020; de-Madaria et al., 2018; Xia et al., 2023). IL-10 contributes to the suppression of neutrophil activation, reduction of inflammatory cytokine production, and promotion of tissue repair (Steen et al., 2020).

- Mechanisms of action at the immune cell level

1. Modulation of neutrophil function

An acidotic environment promotes neutrophil degranulation and NET formation, exacerbating local pancreatic tissue damage (Hu et al., 2020). pH stabilization by LR inhibits neutrophil overactivation and reduces NET formation (Wu et al., 2011; Mauerö et al., 2016).

2. Changes in macrophage polarization

LR promotes differentiation into M2 macrophages, supporting antiinflammatory cytokine production and enhancing tissue repair (de-Madaria et al., 2018).

8.4.3 Advantages of the LR solution from a clinical perspective

Multiple clinical studies have demonstrated that, in initial fluid therapy for AP, LR solution is superior to NS in suppressing inflammatory responses, reducing organ dysfunction, and inhibiting disease progression. This is primarily due to the avoidance of metabolic acidosis through differences in acid–base balance and electrolyte composition, as well as the reduction of SIRS. Consequently, LR is associated with shorter hospital stays, lower ICU admission rates, and reduced incidence of organ dysfunction (Tenner et al., 2024; Takada et al., 2022; International Association of Pancreatology Revised Guidelines on Acute Pancreatitis 2025, 2025).
- Efficacy comparison based on clinical research

1. Suppression of SIRS

LR has a lower chloride concentration than NS, making it less likely to cause metabolic hyperchloremic acidosis (Lee et al., 2021). In AP, pancreatic enzyme overactivity and inflammatory cytokine production induce SIRS, while acidosis further enhances inflammatory cytokine production and alters neutrophil function (Hu et al., 2020; Kellum et al., 2006). LR solution stabilizes pH and reduces the incidence of SIRS by suppressing the overproduction of inflammatory mediators (de-Madaria et al., 2018). A randomized controlled trial reported a significant reduction in the incidence of SIRS persisting for 24 h in patients receiving LR solution compared with NS (Wu et al., 2011).

2. Prevention of organ damage

AP is associated with multiple-organ failure, particularly affecting the respiratory, renal, and circulatory systems (Garg and Singh, 2019). LR solution is advantageous for maintaining intravascular volume and improving microcirculation owing to its composition, thereby reducing ischemia and reperfusion injury around the pancreas (Lee et al., 2021). This reduces the incidence of organ dysfunction (Lee et al., 2021). With NS, a decrease in renal blood flow due to metabolic hyperchloremic acidosis has been reported, resulting in a relatively higher risk of renal impairment (Lee et al., 2021).

3. Reduction in severity and mortality risk

Multiple prospective studies have demonstrated that administration of LR solution reduces the rate of disease progression and mortality risk by suppressing early inflammation and stabilizing hemodynamics. The quality of early fluid administration is directly associated with prognosis, particularly in moderate-to-severe cases, making the use of LR solution recommended (Tenner et al., 2024; Takada et al., 2022; International Association of Pancreatology Revised Guidelines on Acute Pancreatitis 2025, 2025; Lee et al., 2021).

4. Reduced ICU admission rate

Patients administered LR solution experience a reduced rate of ICU admission due to suppressed inflammatory responses and prevention of organ dysfunction (Tenner et al., 2024; Takada et al., 2022; International Association of Pancreatology Revised Guidelines on Acute Pancreatitis 2025, 2025; Lee et al., 2021). This also contributes to more efficient utilization of ICU resources.

5. Reduced length of hospital stay

Early suppression of inflammation and prevention of complications shortens the average length of hospital stay for patients receiving LR solution. For example, one prospective study demonstrated that the median length of stay in the LR solution group was 1.1 days shorter than that in the NS group. This also contributes to more efficient utilization of healthcare resources (Tenner et al., 2024; Takada et al., 2022; International Association of Pancreatology Revised Guidelines on Acute Pancreatitis 2025, 2025; Lee et al., 2021).

8.4.4 Colloids: Current evidence and limitations

Colloid solutions, such as hydroxyethyl starch (HES) and albumin, have been evaluated in severe illnesses; however, their roles in AP remain unclear. Colloids have a large molecular weight and tend to remain in blood vessels for a long time; therefore, a small amount may maintain plasma colloid osmotic pressure and improve circulating blood volume. HES decreases the risk of intraabdominal hypertension in early severe AP and reduces the duration of mechanical ventilation (Du et al., 2011). However, few studies have reported improved survival rates in critically ill patients with colloid use (Lewis et al., 2018). When microvascular permeability increases under pathological conditions, colloids lose their ability to remain within the blood vessels. Furthermore, HES increases the risk of AKI (Lewis et al., 2018). Colloids are more expensive and less safe than crystalloids; therefore, their use is likely limited.

8.4.5 Risks associated with inappropriate fluid choice

Balanced solutions, such as the LR solution, may suppress progression to SIRS, reduce ICU admission rates, and shorten hospital stays compared to NS (Lee et al., 2021).

Conversely, inappropriate fluid selection may lead to metabolic abnormalities, worsening organ dysfunction, and increased mortality (Tenner et al., 2024). NS-induced acidosis, for example, may induce renal dysfunction (Lee et al., 2021; Yunos et al., 2012). Accumulating reports suggest that initial fluid administration likely affects subsequent prognosis, indicating the necessity of selecting balanced fluids.

8.5 Chapter 5: Fluid resuscitation protocol—general adult cases

Effective fluid resuscitation in AP requires a structured, timeline-based approach that balances the benefits of early fluid therapy with the risks of fluid overload.

8.5.1 Initial assessment and triage

Upon arrival, patients must be rapidly assessed for disease severity using valid scoring systems (such as SIRS criteria, BISAP, and APACHE) and hemodynamics (Deshpande and Litake, 2025). Initial triage should identify patients at risk of rapid deterioration and prompt early intensive monitoring (International Association of Pancreatology Revised Guidelines on Acute Pancreatitis 2025, 2025).

8.5.2 Timeline-based protocol (Tenner et al., 2024; Takada et al., 2022; International Association of Pancreatology Revised Guidelines on Acute Pancreatitis 2025, 2025)

Strategies for Fluid Resuscitation in Acute Pancreatitis (Summary)

Fluid management in acute pancreatitis should be tailored to the stage of the illness, with the goal of ensuring **early resuscitation while preventing subsequent over-resuscitation**.

1. Initial Phase (0–24 h): Resuscitation and Correction of Organ Hypoperfusion

Type of Fluid	Isotonic crystalloids, especially Lactated Ringer's (LR) solution, are recommended (preferred over normal saline).
Initial Rate	Moderately aggressive hydration (usually 1.5 mL/kg/h). Bolus of 10 mL/kg for hypoperfusion.
Monitoring Targets	MAP ≧ 65 mmHg, Urine Output ≧ 0.5 mL/kg, Hematocrit 35–44%, BUN ≦20.
Re-evaluation	Within 6 h, then frequently (every 2–3 h) up to 24–48 h.
Fluid Rate Adjustment	To avoid fluid overload, a positive fluid balance of >3–4 L/day should be avoided.
Goal	Correction of SIRS and organ hypoperfusion, prevention of pancreatic necrosis and severe disease.

2. Early Stabilization Phase (24–48 h): Slowing Down and Monitoring after Circulatory Stability

Fluid Rate Adjustment	Slow down the rate after circulatory stabilization.
Monitoring	Body weight measurement, blood gas/acid-base balance, chest auscultation/oxygen saturation for pulmonary edema assessment.
Goal	Maintain organ perfusion while preventing excessive circulatory load and abdominal compartment syndrome.

3. Maintenance Phase (After 48 h): Metabolic Needs and Early Oral Intake

Type of Fluid/Nutrition	Early oral intake is allowed (within 24–48 h for mild cases).
Fluid Rate Adjustment	Individualize based on patient losses and comorbidities.
Monitoring	Na, K, Ca, Mg electrolytes, signs of ascites/third-space fluid, renal function.
Goal	Replenish metabolic requirements and prevent fluid overload.

4. Special Management (Severe Acute Pancreatitis): Intensive Care and Risk Management

Invasive Monitoring	CVP or echocardiography and Intensive Care Unit (ICU) management are recommended.
Risk Avoidance	Fluid overload exacerbates abdominal compartment syndrome and respiratory failure.
Comorbidity Management	Extremely careful adjustment is needed in patients with heart failure or renal impairment to prevent circulatory overload.

- 0–24 h:

LR solution is selected for infusion at a moderately aggressive rate of 1.5 mL/kg/h. If signs of hypotension or hypoperfusion are present, a bolus dose of 10 mL/kg/h should be administered and reassessed every 1–2 h. Treatment goals include: urine output $\geq$0.5 mL/kg/h, MAP $\geq$65 mmHg, hematocrit decrease of 35%–44%, and BUN decrease $\leq$20 mg/dL. The infusion rate should be adjusted based on these dynamic parameters. Point-of-care (POC) ultrasonography can be used to assess inferior vena cava (IVC) compression rate and confirm pulmonary B-lines for early detection of fluid overload. The cumulative positive fluid balance should be limited to 3–4 L during this period unless persistent hypoperfusion is present. The goal is to improve SIRS and organ hypoperfusion, while preventing pancreatic necrosis and functional decline.

–24–48 h:

If oral intake remains impossible, transition to maintenance fluids is necessary. The fluid administration volume should be adjusted while monitoring weight and oxygenation to maintain organ perfusion while preventing excessive circulatory load and intraabdominal compartment syndrome.

After 48 h:

Early initiation of oral intake should be considered. Adjustment of maintenance fluid volume should be based on fluid loss due to complications and urine output. Continuous monitoring and observation should be conducted to determine the necessary fluid volume while preventing fluid overload.

8.5.3 Adjustment strategies based on dynamic response

Increase fluid rate if hypotension, oliguria, or rising lactate levels persist despite initial resuscitation. Decrease or pause fluids if signs of overload develop. Vasopressors, such as norepinephrine, can be administered if MAP is $<$ 65 mmHg despite adequate preload (Aggarwal et al., 2014; International Association of Pancreatology Revised Guidelines on Acute Pancreatitis 2025, 2025).

8.6 Chapter 6: Advanced monitoring and adjunctive technologies

8.6.1 Point-of-care ultrasound for volume status

POCUS has emerged as a valuable bedside tool for evaluating intravascular volume conditions in patients with AP. By evaluating the diameter and collapsibility of the IVC, clinicians can estimate CVP and guide fluid resuscitation more precisely (Furtado and Reis, 2019). In addition, lung ultrasound (LUS) can detect early signs of pulmonary edema, allowing timely adjustment of fluid rates to prevent overload (Mojoli et al., 2019). Incorporating POCUS into early resuscitation protocols enables dynamic, real-time assessment without invasive monitoring, which is particularly beneficial in resource-limited settings.

8.6.2 IVC assessment for fluid responsiveness in AP

IVC diameter variation, measured by transthoracic echocardiography, is a dynamic, noninvasive, and easily applied method for assessing fluid responsiveness in patients with acute circulatory failure admitted to the ICU and those with AP (International Association of Pancreatology Revised Guidelines on Acute Pancreatitis 2025, 2025). In assessing fluid responsiveness in AP, measurement of the IVC diameter and its respiratory fluctuations provides a useful, noninvasive, and rapid method for assessing hemodynamics. However, factors such as elevated intraperitoneal pressure and the effects of respiratory support can reduce the accuracy of IVC diameter assessment in patients with AP. Therefore, IVC assessment should be used in conjunction with other clinical findings and dynamic assessment methods rather than as a standalone indicator.

1. Measurement method
   The IVC diameter is measured using echocardiography (transthoracic or transabdominal). In this method, the IVC is visualized in two-dimensional mode through the subcostal long axis view, approximately 1–3 cm distal to the hepatic vein, from its entry to the right atrium (Furtado and Reis, 2019). Respiratory variation is assessed by evaluating the change in IVC diameter during inspiration and expiration (Furtado and Reis, 2019).
2. Indicators
   Generally, a small IVC diameter with large respiratory variation may suggest circulatory volume depletion and high fluid responsiveness (Furtado and Reis, 2019). Conversely, a large IVC diameter with small respiratory variation indicates low responsiveness to fluid loading (Furtado and Reis, 2019). However, this method has considerable limitations and can only reliably yield predictive values in a specific subgroup of patients receiving controlled invasive mechanical ventilation with a tidal volume of $\geq$8 mL/kg (Furtado and Reis, 2019). Its accuracy is further reduced by factors such as high positive end-expiratory pressure (PEEP), increased intraabdominal pressure, and right ventricular dysfunction (Furtado and Reis, 2019). In spontaneously breathing patients, negative pressure occurs during inspiration, increasing venous return and decreasing IVC diameter (Furtado and Reis, 2019). Conversely, in mechanically ventilated patients, intrathoracic pressure increases during inspiration, causing IVC diameter expansion (Charron et al., 2006).
3. Advantages and limitations of IVC assessment in AP
   1. Advantages
      - Noninvasive and can be performed at the bedside.
      - Provides rapid results

- Easily combined with other hemodynamic assessments (including blood pressure, urine output, and lactate levels)

2. Limitations
   - Increased intraabdominal pressure (ascites, intestinal distension, and peripancreatic fluid accumulation) externally compresses the IVC diameter, making it difficult to reflect true volume status (Rora Bertovi et al., 2025).
   - Respiratory fluctuations are challenging to interpret in patients on mechanical ventilation or with high PEEP (Furtado and Reis, 2019).
   - Obesity and intestinal gas can make visualization difficult (Di Nicol et al., 2023).
   - Measurements are susceptible to atrial pressure changes and right ventricular dysfunction (Furtado and Reis, 2019).

4. Clinical applications
   1. Initial assessment
      When a patient with AP presents with hypotension or decreased urine output, measuring the IVC diameter and respiratory fluctuations can quickly identify volume insufficiency. This is particularly useful for determining initial fluid management in the ICU (Furtado and Reis, 2019).
   2. Combining with fluid stress testing
      Since IVC assessment alone has limitations, its accuracy can be improved by combining it with PLR to simultaneously evaluate changes in cardiac output (Marik et al., 2011; Miller and Mandeville, 2016).
   3. Continuous monitoring
      Since hemodynamics can change during AP, regular measurement of the IVC diameter is helpful for the early detection of fluid overload and volume insufficiency. Combining this with intraabdominal pressure measurements can help prevent abdominal compartment syndrome (Li et al., 2023; Smereczyń et al., 2020).
5. Precautions specific to AP
   1. Effects of intraabdominal pressure
      In severe AP, intraabdominal pressure may increase to $\geq$20 mmHg, causing the IVC diameter to appear smaller because of external pressure (Smereczyń et al., 2020). In such cases, administering excessive fluid therapy under the mistaken assumption of volume insufficiency can further increase intraabdominal pressure and potentially worsen organ damage (Li et al., 2023).
   2. Effects of respiratory support
      Interpretation of respiratory fluctuations is reversed under mechanical ventilation; thus, the assessment must consider mechanical ventilation conditions. High PEEP, in particular, reduces venous return, and even if the IVC diameter increases, it may not indicate volume insufficiency (Furtado and Reis, 2019).
   3. Integration with cardiac function assessment

In patients with AP, cardiac function may be impaired due to myocardial depressant effects of inflammatory mediators, and the IVC diameter may reflect both volume status and cardiac function (Luo et al., 2021). Simultaneous assessment of the right heart is therefore desirable.

6. Summary
   IVC assessment in AP is a useful tool for estimating fluid responsiveness and aids in determining initial fluid strategies and monitoring decisions. However, pathological factors, such as increased intraabdominal pressure and respiratory support, can affect assessment accuracy, making it important to comprehensively evaluate these factors in combination with other dynamic assessment methods and clinical indicators. Although IVC assessment is useful, appropriate fluid management directly prevents organ dysfunction and improves prognosis. Therefore, it is essential not to rely excessively on IVC assessment alone and to perform multifaceted hemodynamic evaluations.

8.6.3 LUS in patients with AP

1. LUS is useful for the early detection of respiratory complications, optimization of fluid management, and assessment of severity in patients with AP (International Association of Pancreatology Revised Guidelines on Acute Pancreatitis 2025, 2025). In particular, AP frequently causes respiratory complications such as pulmonary edema, ARDS, and pleural effusion due to SIRS and increased capillary permeability. Therefore, LUS, which allows for noninvasive and repeated bedside assessments, has great clinical value (International Association of Pancreatology Revised Guidelines on Acute Pancreatitis 2025, 2025). Currently, several reports suggest that LUS findings may influence fluid resuscitation volume; however, no report has provided sufficient evidence for its effectiveness.
2. LUS assessment and findings (Mojoli et al., 2019; Lichtenstein et al., 2008)
   1. B-lines (indicator of pulmonary interstitial water)
      - Multiple B-lines suggest interstitial pulmonary edema and are useful early signs of fluid overload and ARDS.
      - Monitoring the appearance of B-lines from the early stages of fluid infusion in AP helps avoid excessive fluid infusion.
   2. Detection of pleural effusion
      - LUS is highly sensitive for detecting pleural effusion and can visualize even small amounts (a few milliliters).
      - It can also be used to evaluate the characteristics of pleural effusion (echoless vs. internally echo-positive) and the presence or absence of septum formation.
   3. Pulmonary consolidation
      - Useful for diagnosing atelectasis and pneumonia, and particularly valuable for evaluating secondary infections during mechanical ventilation.
   4. Profile classification (such as BLUE Protocol)

- Applicable for differentiating acute respiratory failure, including ARDS, cardiogenic pulmonary edema, pneumonia, and pneumothorax

3. Clinical significance of LUS in managing AP
   1. Optimization of fluid management
      - Sufficient fluid administration is recommended in the initial treatment of AP; however, excessive fluid administration can lead to pulmonary edema (de-Madaria et al., 2022).
      - Early detection of B-line enlargement and pleural effusion using LUS, followed by adjustment of fluid infusion rate and volume, can prevent respiratory complications (International Association of Pancreatology Revised Guidelines on Acute Pancreatitis 2025, 2025).
   2. Aid in severity assessment
      LUS findings reflect the progression of respiratory failure, and when combined with systemic severity assessments, such as the APACHE II and SOFA scores, they improve prognostic accuracy (Basu et al., 2023).
   3. Repeated bedside assessment
      - Patients with AP experience rapid changes in condition. LUS is noninvasive and can be performed repeatedly, making it suitable for monitoring in the ICU or emergency department (Lichtenstein et al., 2008).
4. Evidence and limitations
   1. Current evidence
      No large-scale randomized controlled trials have evaluated the utility of LUS specifically in patients with AP; most studies are observational or case reports. However, the value of LUS in patients admitted to the ICU and those with ARDS has been demonstrated in multiple prospective studies, suggesting its potential applicability in AP (Mojoli et al., 2019; Basu et al., 2023).
   2. Limitations
      LUS is highly operator-dependent, with substantial differences in diagnostic accuracy depending on operator experience. LUS alone may not fully capture the underlying pathology; therefore, combination with chest radiography or CT is recommended.
5. Summary
   Pulmonary echography in patients with AP is useful for the early detection of respiratory complications and individualized optimization of fluid management, making it particularly suitable for monitoring severe cases and patients under intensive care. Although large-scale evidence specific to AP is currently lacking, its use is recommended based on its established track record in the ICU and consistency with pathophysiology. A prospective study involving mechanically ventilated patients in the ICU is currently underway, and its results are anticipated (Blok et al., 2023).

8.6.4 Biomarkers for guiding resuscitation

Biomarkers such as serum lactate, BUN, and hematocrit have been studied for their usefulness in guiding fluid therapy (Tenner et al., 2024). Elevated lactate levels suggest inadequate tissue perfusion and may necessitate more aggressive resuscitation. Increased BUN and hematocrit levels may indicate hemoconcentration and insufficient fluid replacement (Tenner et al., 2024). Conversely, rapid normalization of these markers suggests either adequate resuscitation measures have been implemented, or fluid overload is imminent.

- Each biomarker

1. Lactate levels
   1. Early detection of hypoperfusion
      - Lactate levels >2 mmol/L suggest possible tissue hypoperfusion (Evans et al., 2021).
      - AP is prone to increased vascular permeability and hypovolemia due to peripancreatic inflammation, and elevated lactate levels may be observed early in the disease course (Aggarwal et al., 2014).
   2. Severity assessment
      - Persistent hyperlactemia is a predictor of multiple-organ failure, even in AP (Zeng et al., 2022).
   3. Assessment of fluid responsiveness
      - A rapid decrease in lactate levels after appropriate fluid administration suggests restoration of circulatory blood volume (Zeng et al., 2022).
      - If lactate does not decrease, fluid insufficiency, persistent hypoperfusion, or impaired oxygen utilization should be considered (Tenner et al., 2024).
   4. Target values
      - Maintain lactate levels <2 mmol/L (Zeng et al., 2022)
2. Hematocrit
   1. Assessment of hemoconcentration
      - Hematocrit >44% indicates hemoconcentration and is highly susceptible to hypovolemia (Aggarwal et al., 2014).
   2. Monitoring fluid efficacy
      - If hematocrit levels do not decrease after appropriate fluid infusion, hypovolemia may persist (Tenner et al., 2024).
3. BUN
   1. Indicator of dehydration and renal insufficiency
      - BUN ≥20 mg/dL at admission suggests dehydration or renal insufficiency (Walkowska et al., 2022).
      - In the Ministry of Health, Labor, and Welfare AP prognostic factor score in Japan, BUN ≥40 mg/dL is a criterion for severe disease (Takada et al., 2022).
   2. Assessment of fluid efficacy
      - A decrease in BUN 24 h after admission is an indicator of adequate circulatory volume recovery (Aggarwal et al., 2014).

- If BUN does not decrease, fluid insufficiency is suspected, and additional fluid administration should be considered (Aggarwal et al., 2014).

3. Prognosis prediction
   - Persistently high BUN levels are associated with an increased risk of renal damage, multiple-organ failure, and intractable disease (Tenner et al., 2024; Aggarwal et al., 2014).

8.6.4.1 Summary

Fluid intake management in AP should be based on a comprehensive assessment combining multiple indicators rather than relying on a single biomarker (International Association of Pancreatology Revised Guidelines on Acute Pancreatitis 2025, 2025). For example, monitoring temporal changes in lactate levels, BUN, and hematocrit, along with clinical symptoms and urine output, enables appropriate fluid intake management (Tenner et al., 2024; International Association of Pancreatology Revised Guidelines on Acute Pancreatitis 2025, 2025). In severe cases, although early and adequate fluid administration contributes to preventing organ failure, excessive fluid administration may lead to respiratory complications and increased intraabdominal pressure, highlighting the importance of precise management guided by biomarkers.

8.7 Chapter 7: Future directions—AI-driven fluid resuscitation

8.7.1 Overview of emerging AI applications in critical care

Fluid resuscitation remains a cornerstone of intensive care, particularly for patients with sepsis or AKI. In AP, restoring adequate tissue perfusion while avoiding complications associated with fluid overload, such as pulmonary edema, and preventing hypovolemic shock, which could worsen organ hypoperfusion, are critical in intensive care management. The diversity of critically ill patients makes the application of standardized protocols challenging for achieving optimal fluid management.

Recent developments in AI-based predictive analytics have shown promise in forecasting fluid responsiveness and identifying patients at risk of complications, such as AKI or respiratory failure (Bataille et al., 2021; Zhang et al., 2019; Tomaš et al., 2019; Liao et al., 2021; Rehm et al., 2021). Machine learning algorithms, trained on large datasets, can integrate multiple variables, such as vital signs and laboratory results, to generate individualized fluid resuscitation recommendations. These systems can operate in real time and continuously update predictions as new data become available, thereby potentially improving the safety and efficacy of early fluid therapy.

8.7.2 Ethical and regulatory considerations

The implementation of AI in fluid resuscitation raises important ethical and regulatory challenges, including safety, transparency, and clinical confidence (Mohsin Khan et al., 2025). Regulatory frameworks must ensure that AI tools are validated in diverse patient populations and clinicians retain final decision-making authority (Habli et al., 2020). In addition, continuous monitoring after implementation is essential to enable timely corrective actions.

8.8 Chapter 8: Summary and clinical recommendations

8.8.1 Key take-home messages

In AP, early, goal-directed fluid resuscitation tailored to the condition of the patient through appropriate fluid administration is essential to prevent progression to organ failure.

- Advanced monitoring tools, including established biomarkers and recently emerging technologies, such as POCUS and noninvasive cardiac output measurement, may enhance the precision of fluid management.

AI-based predictive analytics could enable personalized, real-time fluid resuscitation strategies.

Conflicts of interest

Kodai Suzuki is the CEO of Vascular Works, Inc., and has received speaker fees from Japan Blood Products Organization in 2023 and Asahi Kasei Pharma Corporation from 2022 to 2025. All other authors have nothing to declare.

References

Aggarwal, A., Manrai, M., Kochhar, R., 2014. Fluid resuscitation in acute pancreatitis. World Journal of Gastroenterology 20, 18092–18103.

Alphonsus, C.S., Rodseth, R.N., 2014. The endothelial glycocalyx: a review of the vascular barrier. Anaesthesia 69, 777–784.

Basu, S., et al., 2023. A prospective observational study to correlate lung ultrasound with clinical severity and prognosis score in patients with primary pulmonary pathology on invasive ventilatory support. International Journal of Critical Illness and Injury Science 13, 151–158.

Bataille, B., et al., 2021. Machine learning methods to improve bedside fluid responsiveness prediction in severe sepsis or septic shock: an observational study. British Journal of Anaesthesia 126, 826–834.

Biais, M., et al., 2009. Changes in stroke volume induced by passive leg raising in spontaneously breathing patients: comparison between echocardiography and Vigileo/FloTrac device. Critical Care 13, R195.

Blok, S.G., et al., 2023. Effect of lung ultrasound-guided fluid deresuscitation on duration of ventilation in intensive care unit patients (CONFIDENCE): protocol for a multicenter randomized controlled trial. Trials 24, 226.

Boxhoorn, L., et al., 2020. Acute pancreatitis. Lancet 396, 726–734.

Brinkmann, V., et al., 2004. Neutrophil extracellular traps kill bacteria. Science 303, 1532–1535.

Brown, A., Orav, J., Banks, P.A., 2000. Hemoconcentration is an early marker for organ failure and necrotizing pancreatitis. Pancreas 20, 367–372.

Cannesson, M., et al., 2008. Pleth variability index to monitor the respiratory variations in the pulse oximeter plethysmographic waveform amplitude and predict fluid responsiveness in the operating theatre. British Journal of Anaesthesia 101, 200–206.

Charron, C., Caille, V., Jardin, F., Vieillard-Baron, A., 2006. Echocardiographic measurement of fluid responsiveness. Current Opinion in Critical Care 12, 249–254.

Cuthbertson, C.M., Christophi, C., 2006. Disturbances of the microcirculation in acute pancreatitis. British Journal of Surgery 93, 518–530.

de-Madaria, E., et al., 2018. Fluid resuscitation with lactated Ringer's solution vs normal saline in acute pancreatitis: a triple-blind, randomized, controlled trial. United European Gastroenterology Journal 6, 63–72.

de-Madaria, E., et al., 2022. Aggressive or moderate fluid resuscitation in acute pancreatitis. New England Journal of Medicine 387, 989–1000.

Deshpande, S.G., Litake, M.M., 2025. Comparative study between various scoring systems in predicting the severity of acute pancreatitis. Cureus 17, e84004.

Di Nicolo, P., Tavazzi, G., Nannoni, L., Corradi, F., 2023. Inferior vena cava ultrasonography for volume status evaluation: an intriguing promise never fulfilled. Journal of Clinical Medicine 12, 2217.

Du, X.J., et al., 2011. Hydroxyethyl starch resuscitation reduces the risk of intra-abdominal hypertension in severe acute pancreatitis. Pancreas 40, 1220–1225.

Evans, L., et al., 2021. Surviving sepsis campaign: international guidelines for management of sepsis and septic shock 2021. Critical Care Medicine 49, e1063–e1143.

Foitzik, T., et al., 1995. Effect of microcirculatory perfusion on distribution of trypsinogen activation peptides in acute experimental pancreatitis. Digestive Diseases and Sciences 40, 2184–2188.

Furtado, S., Reis, L., 2019. Inferior vena cava evaluation in fluid therapy decision making in intensive care: practical implications. Revista Brasileira de Terapia Intensiva 31, 240–247.

Garg, P.K., Singh, V.P., 2019. Organ failure due to systemic injury in acute pancreatitis. Gastroenterology 156, 2008–2023.

Guo, J., et al., 2023. Comparison of early aggressive versus nonaggressive fluid resuscitation in acute pancreatitis: a meta-analysis. Therapeutic Advances in Gastroenterology 16, 17562848231192144.

Habli, I., Lawton, T., Porter, Z., 2020. Artificial intelligence in health care: accountability and safety. Bulletin of the World Health Organization 98, 251–256.

Hu, J., et al., 2020. Targeting neutrophil extracellular traps in severe acute pancreatitis treatment. Therapeutic Advances in Gastroenterology 13, 1756284820974913.

International Association of Pancreatology Revised Guidelines on Acute Pancreatitis 2025: supported and endorsed by the American Pancreatic Association, European Pancreatic

Club, Indian Pancreas Club, and Japan Pancreas Society. Pancreatology 25, 2025, 770–814.
Juvonen, P.O., et al., 1999. Splanchnic tissue perfusion in acute experimental pancreatitis. Scandinavian Journal of Gastroenterology 34, 308–314.
Keck, T., et al., 2005. Pancreatic proteases in serum induce leukocyte-endothelial adhesion and pancreatic microcirculatory failure. Pancreatology 5, 241–250.
Kellum, J.A., Song, M., Almasri, E., 2006. Hyperchloremic acidosis increases circulating inflammatory molecules in experimental sepsis. Chest 130, 962–967.
Klar, E., et al., 1994. Impact of microcirculatory flow pattern changes on the development of acute edematous and necrotizing pancreatitis in rabbit pancreas. Digestive Diseases and Sciences 39, 2639–2644.
Komara, N.L., et al., 2020. Severe acute pancreatitis: capillary permeability model linking systemic inflammation to multiorgan failure. American Journal of Physiology - Gastrointestinal and Liver Physiology 319, G573–g583.
Lee, A., et al., 2021. Lactated Ringers vs normal saline resuscitation for mild acute pancreatitis: a randomized trial. Gastroenterology 160, 955–957.e954.
Lewis, S.R., et al., 2018. Colloids versus crystalloids for fluid resuscitation in critically ill people. Cochrane Database of Systematic Reviews 8, Cd000567.
Li, X.W., et al., 2023. Comparison of clinical outcomes between aggressive and non-aggressive intravenous hydration for acute pancreatitis: a systematic review and meta-analysis. Critical Care 27, 122.
Li, H., et al., 2025. The glycocalyx: a key target for treatment of severe acute pancreatitis-associated multiple organ dysfunction syndrome. Human Cell 38, 107.
Liao, K.M., Liu, C.F., Chen, C.J., Shen, Y.T., 2021. Machine learning approaches for predicting acute respiratory failure, ventilator dependence, and mortality in chronic obstructive pulmonary disease. Diagnostics 11, 2396.
Lichtenstein, D.A., Mezière, G.A., 2008. Relevance of lung ultrasound in the diagnosis of acute respiratory failure: the BLUE protocol. Chest 134, 117–125.
Liu, T., Zhang, L., Joo, D., Sun, S.C., 2017. NF-κB signaling in inflammation. Signal Transduction and Targeted Therapy 2, 17023.
Luo, Y., et al., 2021. Comprehensive mechanism, novel markers and multidisciplinary treatment of severe acute pancreatitis-associated cardiac injury - a narrative review. Journal of Inflammation Research 14, 3145–3169.
Luo, M., et al., 2025. Signaling pathways involved in acute pancreatitis. Journal of Inflammation Research 18, 2287–2303.
Mao, E.Q., et al., 2009. Fluid therapy for severe acute pancreatitis in acute response stage. Chinese Medical Journal 122, 169–173.
Marik, P.E., Monnet, X., Teboul, J.L., 2011. Hemodynamic parameters to guide fluid therapy. Annals of Intensive Care 1, 1.
Maueröder, C., et al., 2016. Ménage-à-Trois: the ratio of bicarbonate to CO(2) and the pH regulate the capacity of neutrophils to form NETs. Frontiers in Immunology 7, 583.
Mayerle, J., et al., 2019. Genetics, cell biology, and pathophysiology of pancreatitis. Gastroenterology 156, 1951–1968.
Mentula, P., Leppäniemi, A., 2014. Position paper: timely interventions in severe acute pancreatitis are crucial for survival. World Journal of Emergency Surgery 9, 15.
Miller, A., Mandeville, J., 2016. Predicting and measuring fluid responsiveness with echocardiography. Echo Research & Practice 3, G1–g12.

Mohsin Khan, M., et al., 2025. Towards secure and trusted AI in healthcare: a systematic review of emerging innovations and ethical challenges. International Journal of Medical Informatics 195, 105780.

Mojoli, F., Bouhemad, B., Mongodi, S., Lichtenstein, D., 2019. Lung ultrasound for critically ill patients. American Journal of Respiratory and Critical Care Medicine 199, 701–714.

Norman, J., 1998. The role of cytokines in the pathogenesis of acute pancreatitis. The American Journal of Surgery 175, 76–83.

Oiwa, H., et al., 2023. Investigation of the relationship between intradialytic hypotension during hemodialysis and serum syndecan-1 concentration. Scientific Reports 13, 16753.

Pacella, D., et al., 2025. A systematic review of the predictive factors for the recurrence of acute pancreatitis. World Journal of Emergency Surgery 20, 32.

Pieńkowska, J., et al., 2016. Perfusion-CT–can we predict acute pancreatitis outcome within the first 24 hours from the onset of symptoms? PLoS One 11, e0146965.

Rada, B., 2019. Neutrophil extracellular traps. Methods in Molecular Biology 1982, 517–528.

Rehm, G.B., et al., 2021. Use of machine learning to screen for acute respiratory distress syndrome using raw ventilator waveform data. Critical Care Explorations 3, e0313.

Rora Bertovic, M., et al., 2025. Influence of increased intra-abdominal pressure on the validity of ultrasound-derived inferior vena cava measurements for estimating central venous pressure. Journal of Clinical Medicine 14, 3684.

Semler, M.W., Rice, T.W., 2016. Saline is not the first choice for crystalloid resuscitation fluids. Critical Care Medicine 44, 1541–1544.

Sinha, A., et al., 2016. Early Predictors of fluid sequestration in acute pancreatitis: a validation study. Pancreas 45, 306–310.

Smereczyński, A., Kołaczyk, K., Bernatowicz, E., 2020. Ultrasonography in the diagnosis and monitoring of intra-abdominal hypertension and abdominal compartment syndrome: ultrasonografia a nadciśnienie wewnątrzbrzuszne i zespół przedziału brzusznego. Journal of Ultrasonography 20, e201–e204.

Steen, E.H., et al., 2020. The role of the anti-inflammatory cytokine interleukin-10 in tissue fibrosis. Advances in Wound Care 9, 184–198.

Suzuki, K., Miura, T., Okada, H., 2023. The endothelial glycocalyx-All the same? No, it is not. Acute Medical-Surgical Nursing 10, e896.

Takada, T., et al., 2022. JPN clinical practice guidelines 2021 with easy-to-understand explanations for the management of acute pancreatitis. Journal of Hepato-Biliary-Pancreatic Sciences 29, 1057–1083.

Tanaka, Y., et al., 2025. Nationwide epidemiological survey of acute pancreatitis in Japan, 2021: the impact of the COVID-19 pandemic and revised clinical guidelines. Journal of Gastroenterology 60, 1437–1448.

Tenner, S., et al., 2024. American College of Gastroenterology guidelines: management of acute pancreatitis. American Journal of Gastroenterology 119, 419–437.

Tomašev, N., et al., 2019. A clinically applicable approach to continuous prediction of future acute kidney injury. Nature 572, 116–119.

Turunen, A., et al., 2024. Syndecan-1 Levels and early positive fluid balance are associated with disease severity in acute pancreatitis. Pancreas 53, e739–e747.

Walkowska, J., et al., 2022. Diagnosis and treatment of acute pancreatitis. Diagnostics 12, 1974.

Wilms, H., et al., 2014. A systematic review of goal directed fluid therapy: rating of evidence for goals and monitoring methods. Journal of Critical Care 29, 204–209.

Wu, B.U., et al., 2011. Lactated Ringer's solution reduces systemic inflammation compared with saline in patients with acute pancreatitis. Clinical Gastroenterology and Hepatology 9, 710–717.e711.

Wu, S.K., Wang, L., Wang, F., Zhang, J., 2024. Resveratrol improved mitochondrial biogenesis by activating SIRT1/PGC-1α signal pathway in SAP. Scientific Reports 14, 26216.

Xia, T., et al., 2023. Advances in the role of STAT3 in macrophage polarization. Frontiers in Immunology 14, 1160719.

Xiao, A.Y., et al., 2016. Global incidence and mortality of pancreatic diseases: a systematic review, meta-analysis, and meta-regression of population-based cohort studies. The Lancet Gastroenterology & Hepatology 1, 45–55.

Yang, K., et al., 2020. Lactate suppresses macrophage pro-inflammatory response to LPS stimulation by inhibition of YAP and NF-κB activation via GPR81-mediated signaling. Frontiers in Immunology 11, 587913.

Yunos, N.M., et al., 2012. Association between a chloride-liberal vs chloride-restrictive intravenous fluid administration strategy and kidney injury in critically ill adults. JAMA 308, 1566–1572.

Zeng, J., et al., 2022. Prognostic value of arterial lactate metabolic clearance rate in moderate and severe acute pancreatitis. Disease Markers 2022, 9233199.

Zhang, Z., Ho, K.M., Hong, Y., 2019. Machine learning for the prediction of volume responsiveness in patients with oliguric acute kidney injury in critical care. Critical Care 23, 112.

Zuckerbraun, B.S., et al., 2005. Carbon monoxide protects against the development of experimental necrotizing enterocolitis. American Journal of Physiology - Gastrointestinal and Liver Physiology 289, G607–G613.

CHAPTER

Treatment of pancreatitis 2: Nutritional strategies

9

Ryo Kamidani[1, 2], Kunihiro Shirai[3]

[1]*Department of Emergency and Disaster Medicine, Gifu University Graduate School of Medicine, Gifu, Japan;* [2]*Saitama Children's Medical Center, Department of Pediatric Critical Care, Saitama, Japan;* [3]*Kagawa Prefectural Central Hospital, Department of Emergency Medicine, Kagawa, Japan*

9.1 Introduction

Nutritional strategies in acute pancreatitis (AP) have evolved from the traditional concept of "pancreatic rest" to an evidence-based approach emphasizing early nutritional support as a cornerstone of therapy. For decades, patients were kept nil per os (NPO) to avoid stimulation of pancreatic exocrine secretion (Banks & Freeman, 2006; Forsmark & Baillie, 2007). However, this practice was founded on theoretical assumptions rather than on robust evidence. Prolonged fasting promotes intestinal mucosal atrophy, increases bacterial translocation, and enhances systemic inflammation, all of which may aggravate pancreatic and extrapancreatic injuries (Spanier et al., 2011; Arvanitakis et al., 2024; De Lucia et al., 2023). Consequently, clinical practice has shifted from prolonged fasting to strategies that prioritize maintaining gut integrity through early nutritional support.

The gastrointestinal tract plays a central role in the pathophysiology of severe AP. Disruption of the gut barrier leads to the translocation of enteric bacteria and endotoxins, contributing to infected pancreatic necrosis, which is a major cause of morbidity and mortality (Cicalese et al., 2001; Dervenis et al., 2003). By sustaining splanchnic perfusion and supporting the mucosal structure, enteral nutrition (EN) influences immune pathways and reduces the likelihood of microbial passage into systemic circulation. Consequently, recent guidelines, including those from the ESPEN guideline on clinical nutrition in acute and chronic pancreatitis (ESPEN 2020), the AGA clinical practice update (AGA 2020), and the JPN clinical practice guidelines 2021 with easy-to-understand explanations for the management of acute pancreatitis (JPN guidelines for acute pancreatitis 2021), uniformly recommend the early initiation of EN for patients predicted to have moderate to severe AP (Arvanitakis et al., 2020, 2024; Baron et al., 2020; Takada et al., 2022).

The metabolic response to inflammation in AP is characterized by marked hypermetabolism, protein catabolism, and insulin resistance. Energy expenditure may increase by 20%–40%, whereas nitrogen loss can exceed 20 g/day in severe cases (Bouffard et al., 1989; Lakananurak and Gramlich, 2020). This hypercatabolic state

Advanced Insights into Acute Pancreatitis. https://doi.org/10.1016/B978-0-443-44736-5.00012-7

leads to the rapid depletion of lean body mass and immune dysfunction, predisposing patients to sepsis and delayed recovery. Thus, adequate and timely nutritional support is essential not only for maintaining nutritional status but also for modulating systemic inflammation and improving outcomes.

Recent randomized controlled trials and meta-analyses have confirmed that early EN, preferably within 24–48 h after admission, significantly reduces infectious complications, organ failure, and mortality compared with delayed feeding or total parenteral nutrition (TPN) (Qi et al., 2018; Wu et al., 2018; Feng et al., 2017). Both gastric and postpyloric tube feeding are generally feasible options with comparable tolerance and safety in most patients. When EN is contraindicated or not tolerated, parenteral nutrition (PN) remains a necessary adjunct that requires careful metabolic monitoring to prevent hyperglycemia, hypertriglyceridemia, and catheter-related infections.

Along with the evolution of feeding routes, advances in formulation science have enabled the development of specialized EN and PN products tailored to the pathophysiological demands of AP. Semielemental or peptide-based formulas, medium-chain triglyceride (MCT)-rich emulsions, and immune-modulating nutrients, such as glutamine, arginine, and omega-3 fatty acids, are increasingly used to optimize metabolic control and inflammatory modulation (Hegazi and DeWitt, 2014; Pearce et al., 2006; Lei et al., 2015). However, controversies remain regarding the routine use of immunonutrition and probiotics, as some trials have shown neutral or even adverse effects in critically ill populations (Endo et al., 2018).

In summary, nutritional support in AP is no longer viewed as a passive caloric replacement but as an active therapeutic intervention that influences the disease trajectory. Understanding the indications, timing, route, and composition of nutritional support is fundamental to improving outcomes. This chapter reviews the current evidence and practical strategies for nutritional management in AP, focusing on adult patients in critical care settings and highlighting the evolving role of targeted formulations and emerging precision nutrition approaches.

9.2 Indications for nutritional support

Patients with moderate to severe AP typically experience persistent systemic inflammatory response syndrome (SIRS), progressive protein catabolism, prolonged intolerance of oral intake, and an increased risk of infectious complications. Nutritional support is a central component of the therapy for these patients. The indications for initiating nutritional support are determined by the predicted disease severity, anticipated duration of oral intolerance, risk of infectious complications, and the need to maintain gastrointestinal integrity (Fig. 9.1). International guidelines, including ESPEN 2020 , AGA 2020, and the JPN guidelines for acute pancreatitis 2021, concur that nutritional support should be initiated within 24–48 h in patients who are unable to tolerate oral intake or who are predicted to develop moderate to severe disease (Arvanitakis et al., 2020, 2024; Baron et al., 2020; Takada et al., 2022).

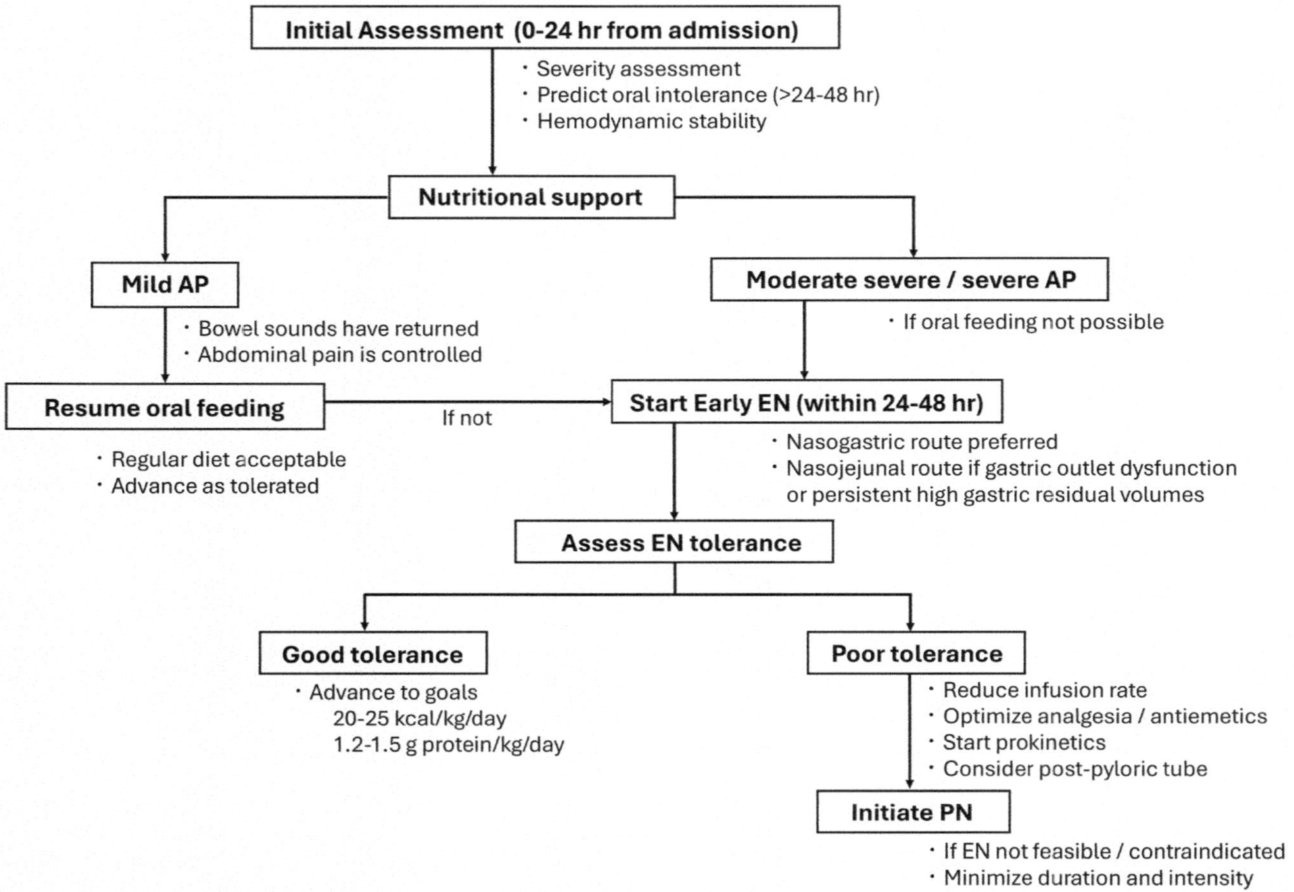

FIGURE 9.1 Algorithm for initiating and advancing nutritional support in acute pancreatitis.

This figure summarizes a stepwise algorithm for nutritional support in acute pancreatitis. After assessing severity, predicted oral intolerance, and hemodynamic stability, patients who can tolerate oral feeding advance their diet as symptoms improve. If oral intake is not possible, enteral nutrition is initiated through the nasogastric or, when needed, nasojejunal route and advanced toward caloric and protein goals. Management of feeding intolerance and indications for parenteral nutrition are also outlined.

- Author-created figure adapted from ESPEN 2020, AGA 2020, and the JPN guidelines for acute pancreatitis 2021.

Table 9.1 The revised Atlanta 2012 classification for grading severity of acute pancreatitis.

Grade of severity	Criteria
Mild	No organ failure No local or systemic complications
No variables present	Organ failure that resolves within 48 h (transient organ failure) and/or Local or systemic complications
Total	Persistent single or multiple organ failure (>48 h)

Adapted from the Revised Atlanta Classification 2012 (Sarr, M.G., 2013. 2012 revision of the Atlanta Classification of acute pancreatitis. Polish Archives of Internal Medicine 123 (3), 118–124, https://doi.org/10.20452/pamw.1627).

The predicted severity of AP is closely linked to the need for timely nutritional support. Several scoring systems, including the Revised Atlanta Classification (Table 9.1), APACHE II, BISAP (Table 9.2), and the Japanese prognostic factor scoring system, help identify patients at high risk of organ failure, necrotizing pancreatitis, and prolonged fasting (Takada et al., 2022; Sarr, 2013; Harshit Kumar and Singh Griwan, 2018; Zhang et al., 2014; Wu et al., 2008). Although most patients with mild AP can resume oral feeding within a few days, clinical features such as persistent abdominal pain or ileus, ongoing vomiting, high nasogastric residuals, sustained opioid requirements, or radiographic evidence of evolving necrotizing pancreatitis indicate the likelihood of prolonged oral intolerance. In such cases, EN should be initiated if intake is not feasible within the first 24–48 h.

Table 9.2 The beside index for severity in acute pancreatitis (BISAP) scoring system.

Individual components
B—BUN >25 mg/dL
I—Impaired mental status (Glasgow Coma score <15)
S—Systemic inflammatory response syndrome (SIRS) SIRS is defined as two or more of the following: Temperature of <36 or >38°C Respiratory rate >20 breaths/min or Pa CO_2 < 32 mmHg Pulse >90 beats/min WBC <4000 or >12,000 cells/mm^3 or >10% immature bands
A—Age >60 years
P—Pleural effusion detected on imaging

Adapted from the BISAP scoring system (Wu, B.U., Johannes, R.S., Sun, X., Tabak, Y., Conwell, D.L., Banks, P.A., 2008. The early prediction of mortality in acute pancreatitis: a large population-based study. Gut 57 (12), 1698–1703, https://doi.org/10.1136/gut.2008.152702).

In severe AP, disruption of the intestinal mucosal barrier, alterations in the gut microbiota, and increased intestinal permeability promote bacterial translocation and subsequent infected pancreatic necrosis. Early nutritional support mitigates these effects by maintaining luminal nutrient flow, preserving tight junction integrity, and supporting splanchnic perfusion. Multiple randomized controlled trials and meta-analyses have demonstrated that early nutritional support reduces the incidence of infected pancreatic necrosis, nutrition-related infectious complications, sepsis, and multiple organ dysfunction (Arvanitakis et al., 2020; Qi et al., 2018; Marik and Zaloga, 2004). Moreover, patients with moderate-to-severe AP exhibit a pronounced hypermetabolic state, and failure to meet the increased energy and protein requirements leads to an accelerated loss of lean body mass, impaired immune function, and a higher risk of organ dysfunction.

Even when factors external to the pancreas, such as major comorbid illness (e.g., trauma, sepsis, chronic malnutrition), postoperative states, preexisting gastrointestinal dysfunction, or alcohol use disorder with baseline malnutrition, compromise feeding ability, early EN should be initiated regardless of pancreatitis severity to prevent further deterioration of nutritional reserves. Conversely, conditions such as uncontrolled shock requiring escalating vasopressor support, suspected acute mesenteric ischemia or bowel perforation, severe intestinal obstruction, or abdominal compartment syndrome with impaired perfusion are absolute or relative contraindications to enteral feeding. In such situations, PN may be considered while reassessing tolerance, as the patient's hemodynamic status stabilizes.

9.3 Enteral nutrition

Enteral nutrition (EN) has become the preferred method of nutritional support in AP based on a solid physiological rationale and high-quality clinical evidence. Far beyond merely providing calories and protein, EN preserves the mucosal architecture and decreases bacterial translocation, thereby suppressing secondary complications. These physiological advantages have underpinned the paradigm shift from "pancreatic rest" toward early gut-directed therapy.

A substantial body of evidence, including randomized controlled trials and meta-analyses, supports the initiation of EN within 24–48 h in patients who cannot tolerate oral intake or are at risk of moderate to severe disease (Arvanitakis et al., 2020, 2024; Baron et al., 2020; Takada et al., 2022; Li et al., 2013). In mild pancreatitis, oral intake can be safely resumed immediately intestinal motility recovers, shortening the length of hospital stay without adverse events (Horibe et al., 2016, 2020; Ramírez-Maldonado et al., 2021). Compared with delayed initiation or TPN, early EN has consistently been shown to reduce the incidence of infection, organ failure, and mortality (Wu et al., 2018; Petrov et al., 2008). The concern that EN might exacerbate pancreatic stimulation has largely been disproven, as both the nasogastric and nasojejunal routes are generally safe, well-tolerated, and do not worsen abdominal symptoms or biochemical markers of disease activity. These findings provide a rationale for early EN as a first-line therapy in most patients with AP.

Historically, nasojejunal feeding has been preferred based on the theoretical reduction of pancreatic stimulation; however, accumulating data indicate that nasogastric feeding and nasojejunal feeding are broadly equivalent in terms of tolerance, safety, and clinical outcomes (Dutta et al., 2020; Chang et al., 2013). Accordingly, many guidelines now recommend nasogastric feeding as the practical first-choice route, limiting nasojejunal feeding to patients with severe gastric outlet dysfunction, high residual volume, or persistent feeding intolerance (Arvanitakis et al., 2020, 2024).

The choice of enteral formula is another important aspect of clinical management. For most patients, a standard polymeric formula is an appropriate starting point. In cases of intolerance to polymeric formulas or when malabsorption or gastrointestinal dysfunction is suspected, semielemental or peptide-based formulas containing hydrolyzed proteins and MCTs may be considered, as they require less pancreatic enzymatic activity. Several studies have suggested that peptide-based or MCT-rich formulas improve tolerance and reduce abdominal symptoms, although their superiority over polymeric formulas has not been consistently demonstrated (Endo et al., 2018; Rinninella et al., 2017; Tiengou et al., 2006). Given the relatively limited clinical benefit and higher cost, routine use of elemental formulas is not recommended (Arvanitakis et al., 2020).

Immune-modulating nutrients such as glutamine, arginine, and omega-3 fatty acids have been extensively studied in critical illness and AP. Although some evidence suggests that omega-3-enriched formulas may reduce systemic inflammation and infectious complications, and glutamine supplementation may support enterocyte integrity and nitrogen balance, the evidence remains inconsistent, and current guidelines do not endorse routine use of immunonutrition in AP (Arvanitakis et al., 2020; Lei et al., 2015; Lasztity et al., 2005; Asrani et al., 2013; Arutla et al., 2019). It may be considered selectively in research settings or in highly catabolic patients, but it does not substitute for early standard EN.

Practical considerations for EN delivery include initiating at a low infusion rate and gradually advancing toward target goals, while monitoring clinical and gastrointestinal parameters (e.g., abdominal distension, nausea, vomiting, stool patterns, and gastric residuals) for tolerance. The typical initial caloric target is 20–25 kcal/kg/day, and the protein requirement to counter hypercatabolism is 1.2–1.5 g/kg/day (McClave et al., 2016). Hemodynamic stability is essential for the safe initiation of EN; modest vasopressor support is not an absolute contraindication, and feeding may commence cautiously in patients whose perfusion parameters are improving. Feeding interruptions should be minimized, and strategies such as prokinetic agents, postpyloric tube placement, or formula adjustment should be employed to enhance tolerance.

In summary, EN has been established as a foundational therapy for the management of AP. By preserving gut integrity, reducing systemic inflammation, and preventing severe infectious complications, it has a direct impact on patient outcomes. The paradigm shift from pancreatic rest to early gut-protective EN is one of the most significant advances in modern AP treatment.

9.4 Parenteral nutrition

In the nutritional management of AP, PN, widely used historically because of concerns over pancreatic stimulation, has now been repositioned as a secondary strategy reserved for patients in whom EN cannot be implemented, is not tolerated despite appropriate optimization, or is contraindicated due to gastrointestinal pathology (Arvanitakis et al., 2020, 2024; Baron et al., 2020; Takada et al., 2022). Specifically, these scenarios include persistent intolerance to EN despite postpyloric tube placement and prokinetic therapy, high-output fistulas, severe ileus or bowel obstruction, suspected mesenteric ischemia, and abdominal compartment syndrome with impaired perfusion. In hemodynamically unstable patients requiring escalating vasopressor support, PN may serve as a temporary bridge until EN is safe.

Randomized trials and meta-analyses have consistently demonstrated that PN, compared with early EN, is associated with increased rates of infectious complications, including catheter-related bloodstream infections, intensified systemic inflammation, and progression to organ dysfunction (Yi et al., 2012; Liu and Gao, 2021). These adverse outcomes are thought to be derived from the loss of enteral stimulation, impairment of gut barrier function, and systemic effects of intravenous nutrition (particularly hyperglycemia). Therefore, the guidelines recommend PN only when adequate EN cannot be achieved within the first few days of illness (Arvanitakis et al., 2020, 2024; Baron et al., 2020; Takada et al., 2022).

Careful metabolic management is required when PN is administered. Hyperglycemia independently worsens AP outcomes; therefore, glycemic control is essential (Lu et al., 2021b; Yang et al., 2022; Guan et al., 2025). Lipid emulsions, including mixed-oil or fish-oil-containing formulations, can be used safely; however, in the presence of severe hypertriglyceridemia, dose adjustment or temporary cessation is necessary. Electrolytes, micronutrients, and trace elements must be rigorously monitored, and adjustments must be made to prevent complications such as refeeding syndrome.

EN should be initiated as early as gastrointestinal function is allowed. Even minimal enteral intake provides trophic stimulation, improves mucosal integrity, and reduces infectious complications; thus, once EN achieves approximately 60%–70% of the caloric requirements, PN can be progressively tapered (McClave et al., 2016; Heidegger et al., 2013; Wilkinson and Dickerson, 2016). In modern clinical practice, PN plays an important but deliberately limited role, functioning mainly as a bridge therapy when EN cannot be promptly achieved.

9.5 Nutritional formulations and emerging therapies

Several EN formulations are available for patients with AP, and the choice of formula should be guided by digestive capacity, tolerance, metabolic demands, and the severity of gastrointestinal dysfunction. Standard polymeric formulations containing intact proteins and long-chain triglycerides remain the first-line option for most patients because of their wide availability, cost-effectiveness, and generally good tolerance. Their effectiveness is supported by systematic reviews and meta-analyses demonstrating that in routine cases of AP, specialized enteral formulations do not provide consistent advantages over standard polymeric formulas in terms of tolerance, infectious complications, or mortality (Arvanitakis et al., 2020; Petrov et al., 2009; Poropat et al., 2015).

Semielemental or peptide-based formulas contain hydrolyzed proteins and MCTs, thereby reducing the pancreatic digestive demand. These formulas may be advantageous in cases of malabsorption, significant gastrointestinal dysfunction, high gastric residual volume, and intolerance to polymeric feeds. Although several trials have suggested improved tolerance and reduced abdominal symptoms with peptide-based or MCT-rich formulas, the evidence is heterogeneous, and large randomized trials have not demonstrated clear superiority over standard polymeric formulas (Arvanitakis et al., 2020; Tiengou et al., 2006; Poropat et al., 2015). Consequently, their use is typically targeted at selected patients rather than being routinely applied. Elemental formulas composed of free amino acids are considered only in specific cases owing to their high osmolarity, cost, and limited supporting evidence (Puri et al., 2021).

Immunomodulatory nutrients, including glutamine, arginine, nucleotides, and omega-3 polyunsaturated fatty acids, have attracted considerable interest because of their potential roles in modulating inflammation, preserving mucosal function, and improving the immune response. Omega-3 –enriched formulations and fish oil-containing parenteral lipid emulsions may attenuate systemic inflammation, reduce infectious complications, and favorably influence lipid mediators (Lasztity et al., 2005; Wang et al., 2008; Pan et al., 2017; Al-Leswas et al., 2020). Glutamine supplementation has demonstrated potential benefits for enterocyte integrity, nitrogen balance, and oxidative stress (Asrani et al., 2013; Pan et al., 2017; Singh et al., 2014). However, high-quality evidence regarding AP remains limited and heterogeneous, and international guidelines do not endorse routine immunonutrition (Arvanitakis et al., 2020). Consequently, omega-3–enriched formulas and lipid emulsions, as well as glutamine-containing regimens, are best regarded as adjunctive options reserved for research settings or patients with high catabolism and demonstrated deficiencies, rather than as central components of standard therapy.

Emerging therapies include microbiome-modulating strategies such as targeted probiotics, synbiotics, and prebiotics aimed at restoring gut homeostasis and reducing bacterial translocation. Although early studies suggested potential benefits, particularly in reducing infectious complications, subsequent trials have produced conflicting results. One notable trial reported increased mortality with certain probiotic combinations (Oláh et al., 2002, 2007; Besselink et al., 2008; Malik et al., 2023). Consequently, the routine use of probiotics is not recommended in AP, and ongoing research is exploring safer and more targeted microbiome interventions (Lu et al., 2021a).

Technologies to assess energy expenditure, such as indirect calorimetry, may allow for more individualized nutritional prescriptions and optimize macronutrient delivery while minimizing the risks of overfeeding and hyperglycemia (Lu et al., 2021a; Valainathan et al., 2019). Similarly, pharmacological agents targeting intestinal permeability, immune modulation, and oxidative stress are promising avenues, although they remain under investigation (Moggia et al., 2017; Szatmary et al., 2022).

Overall, although specialized formulations and novel therapies offer intriguing possibilities, standard polymeric EN remains the foundation for nutritional support in AP. Emerging approaches should be integrated cautiously and guided by evolving evidence and individualized patient needs.

9.6 Monitoring and complications

Effective nutritional therapy for AP requires structured and continuous monitoring to ensure adequate nutrient delivery, early detection of intolerance, and timely recognition of nutrition-related complications. This is particularly important in patients with moderate or severe disease in whom hypermetabolism, organ dysfunction, and sustained catabolic stress increase the risk of both underfeeding and overfeeding. Monitoring should encompass the clinical, gastrointestinal, hemodynamic, and biochemical domains, with the intensity of surveillance adjusted according to disease severity and care settings (general ward vs. intensive care unit).

Clinical monitoring begins with the regular assessment of pain, abdominal distension, nausea, vomiting, bowel sounds, and stool characteristics. These parameters provide early clues to the gastrointestinal tolerance of EN and the development of complications, such as bowel obstruction, pyloric dysfunction, or abdominal compartment syndrome. Vital signs must be checked frequently to detect hemodynamic instability or sepsis. Tachycardia, hypotension, escalating vasopressor use, or new-onset fever may indicate worsening systemic inflammation, infection, or intolerance to ongoing nutritional support. Similarly, respiratory patterns and oxygenation are crucial, as progressive abdominal distension or aspiration may aggravate respiratory failure in patients with preexisting pulmonary compromise.

In patients receiving EN, tolerance should be assessed at least once daily, with particular attention to abdominal symptoms, frequency of vomiting, and gastric residual volume (GRV). Mild increases in GRV alone do not mandate cessation of EN; however, persistent large residuals, vomiting, or marked abdominal distention should prompt interventions such as slowing the infusion rate, optimizing analgesia and antiemetic therapy, initiating prokinetic agents, or considering postpyloric tube placement. Diarrhea may reflect hyperosmolar feeds, antibiotic-associated dysbiosis, exocrine pancreatic insufficiency, or enteric infections (e.g., *Clostridioides difficile*). This evaluation should distinguish benign osmotic diarrhea from clinically significant fluid and electrolyte losses that require intervention. Conversely, constipation or ileus may necessitate adjustment of opioid dosing, laxative use, or temporary modification of enteral feeding rates.

Biochemical monitoring is essential for guiding both EN and PN. During the acute phase, serum electrolytes, urea, creatinine, glucose, and liver enzymes should be measured daily or nearly daily to enable timely adjustment of nutritional prescriptions and the detection of complications (da Silva et al., 2020; Friedli et al., 2017). Particular attention should be paid to phosphate, potassium, and magnesium levels, especially in patients with severe malnutrition or chronic undernutrition who are at risk of refeeding syndrome. If early hypophosphatemia or combined electrolyte derangements are observed, caloric delivery should be reduced, and aggressive replacement should be instituted without delay to prevent progression to overt refeeding syndrome (da Silva et al., 2020). As AP frequently causes stress hyperglycemia and transient endocrine dysfunction, close surveillance of blood glucose levels is mandatory. Uncontrolled hyperglycemia is associated with increased infection rates and worse clinical outcomes. Maintaining safe, moderate glycemic control (e.g., 140–180 mg/dL) is recommended in most critically ill patients, including those with AP (Yang et al., 2022; Dungan et al., 2009). In patients receiving lipid-containing formulations, whether EN or PN, serum triglyceride concentrations should be monitored regularly, and doses should be reduced or temporarily withheld if marked hypertriglyceridemia develops, particularly in those with hypertriglyceridemia-associated pancreatitis.

Longitudinal assessment of nutritional status should include monitoring of body weight, fluid-adjusted cumulative balance, and, where feasible over longer time frames, indices of muscle mass or function (e.g., mid-arm circumference and handgrip strength). In critically ill patients where a more precise adjustment of energy and protein delivery is desired, nitrogen balance studies or indirect calorimetry may be considered, although these tools are not universally available. The trajectories of inflammatory markers such as C-reactive protein, procalcitonin, and interleukin-6 should be interpreted cautiously. Although they may reflect the interaction between systemic inflammation and nutritional interventions, they are nonspecific and should not be used in isolation to judge feeding tolerance or the success of nutrition therapy (Bharadwaj et al., 2016; Liu et al., 2023).

EN in AP is associated with a distinct spectrum of complications, including aspiration pneumonia, worsening abdominal pain, bowel obstruction, and, rarely, intestinal ischemia. The aspiration risk can be mitigated by head-of-bed elevation, avoidance of excessive gastric overdistension, and use of postpyloric feeding in patients at a high risk of aspiration or with impaired airway protection. New-onset or worsening abdominal pain, rising lactate levels, or signs of peritonism during EN should prompt urgent evaluation of bowel ischemia, perforation, or necrosis, particularly in patients with severe disease or in those receiving vasopressor support. Abdominal compartment syndrome, although uncommon, is a serious complication.

Progressive abdominal distension, increased intraabdominal pressure, oliguria, and respiratory compromise warrant rapid multidisciplinary assessment, with temporary reduction or interruption of EN considered as part of the management strategy.

PN has a profile of complications, notably catheter-related bloodstream infections, thrombosis, metabolic disturbances, and hepatobiliary dysfunction. Strict aseptic techniques during catheter insertion and handling, daily inspection of the insertion site, and prompt evaluation for unexplained fever or sepsis are essential. Metabolic complications include hyperglycemia, hypertriglyceridemia, and electrolyte abnormalities, and prolonged therapy, cholestasis, and PN-associated liver disease may develop. These risks underscore the need to limit the duration and intensity of PN, preferentially use the enteral route whenever feasible, and taper PN as soon as enteral or oral intake provides the majority of the caloric requirements.

Finally, safe and effective nutritional management depends on meticulous documentation and regular multidisciplinary reviews. Integrating inputs from gastroenterologists, intensivists, surgeons, dietitians, pharmacists, and nursing staff facilitates early problem detection and timely adjustment of nutritional strategies. Structured protocols and checklists that specify nutritional targets, monitoring parameters, and predefined responses to intolerance or complications can promote adherence to best practices and reduce unwarranted variation in care. Thus, vigilant monitoring is a proactive tool for maximizing the benefits and minimizing the risks of nutritional support in patients with AP.

9.7 Pediatric and special populations

Pediatric and other special populations pose unique challenges to nutritional management in AP because of differences in etiology, metabolic reserves, comorbid conditions, and long-term consequences. In children, the evidence base remains relatively limited compared with that in adults, and most recommendations are derived from small studies and expert consensus. Nevertheless, recent pediatric guidelines and consensus statements broadly align with adult data in emphasizing early, enteral, and individualized nutritional support while simultaneously underscoring the need to preserve growth and neurodevelopmental trajectories (Abu-El-Haija et al., 2018). In children, the etiologic profiles of AP differ from those in adults, with biliary disease, drug-induced injury, trauma, systemic diseases, genetic factors, and structural abnormalities being prominent causes. Children have lower absolute nutrient reserves and are, therefore, more vulnerable to growth failure and micronutrient deficiencies, particularly in the setting of recurrent or chronic disease. The joint ESPGHAN–NASPGHAN position paper highlights that even at first presentation, and especially in severe or necrotizing diseases, children may develop malnutrition and fat-soluble vitamin deficiencies (A, D, E, and K) and recommends regular follow-up of anthropometry, dietary intake, and selected laboratory parameters (Abu-El-Haija et al., 2018).

A review of early EN in pediatric AP suggested that in mild to moderately severe disease, oral feeding or tube feeding with normal-fat diets is safe and may reduce the length of stay and complications (Bukowski et al., 2022). Recent ESPGHAN–NASPGHAN position papers and expert consensus documents recommend initiating oral or gastric EN within 48 h of admission in the absence of clear contraindications and do not mandate strict low-fat diets for most children with AP (Abu-El-Haija et al., 2018). As in adults, feeding advancement should be guided by clinical tolerance rather than serum amylase or lipase activity (Abu-El-Haija et al., 2018). In children predicted to have moderately severe or severe disease, or in those unable to maintain adequate oral intake within 24–48 h of onset, EN should be preferred over PN whenever feasible. Postpyloric feeding, as in adults, should be reserved for children with marked gastric outlet dysfunction, persistently high residual gastric volumes, or a high risk of aspiration. PN is considered a last resort in pediatric AP and is indicated only when EN is contraindicated or remains intolerable despite optimization. When PN is required, prescriptions should be weight- and age-adjusted with careful monitoring for refeeding syndrome and catheter-related complications (Abu-El-Haija et al., 2018).

The general principles of pediatric nutrition apply to the formula selection for pediatric AP. Age-appropriate polymeric formulas are recommended for most children, and human breast milk, wherever available, is the preferred option for infants. Semielemental or peptide-based formulas should be reserved for children with significant malabsorption, short bowel syndrome, or persistent intolerance to polymeric feed (Abu-El-Haija et al., 2018; Martínez-Vázquez et al., 2025; Matthai et al., 2020). Evidence that specialized formulas confer superior clinical outcomes over standard pediatric formulas is limited, and most guidelines prioritize practicality, tolerance, and cost-effectiveness over theoretical reductions in pancreatic stimulation (Arvanitakis et al., 2020, 2024; Petrov et al., 2009; Abu-El-Haija et al., 2018).

In addition to pediatric patients, several other subgroups require individualized nutritional strategies. Overweight and obesity are highly prevalent among adults hospitalized for AP. Obesity is a well-established risk factor for severe disease, high rates of pancreatic necrosis, and poor clinical outcomes (Martínez et al., 2006; Dobszai et al., 2019; Khetpal et al., 2025). Observational studies and a randomized trial involving obese patients have indicated that early EN should not be withheld solely because of obesity and may attenuate some of its adverse effects, possibly by preserving gut barrier function, reducing bacterial translocation, and dampening systemic inflammation (Jin et al., 2020; Ammori et al., 1999). In obese patients, energy targets are usually calculated based on the adjusted or ideal body weight to avoid overfeeding, whereas protein targets are kept relatively high to preserve lean body mass. As metabolic syndrome and hypertriglyceridemia frequently coexist in this population, meticulous glycemic control and regular monitoring of serum triglyceride levels are particularly important.

Pregnancy represents another special context in which both maternal and fetal considerations shape nutritional management. Gallstone disease remains the leading cause of AP during pregnancy, whereas hypertriglyceridemia-induced pancreatitis is of particular concern during late pregnancy. Case series and narrative reviews suggest that strict dietary fat restriction, incorporation of MCTs, and in selected cases, temporary PN can be used to control triglyceride levels while maintaining adequate maternal–fetal nutrition (Mizushima et al., 1998; Wong et al., 2015; Loh et al., 2005; Gupta et al., 2022). As in nonpregnant adults, EN is always preferred when feasible. However, if severe nausea and vomiting or ileus prevent adequate enteral intake and threaten fetal growth, the threshold for initiating PN may be lowered. In such situations, collaboration among the obstetrics, endocrinology, and nutrition teams is essential.

Patients with preexisting organ dysfunction, such as chronic kidney disease, progressive liver disease, or heart failure; frail elderly individuals; chronic alcoholics; and patients with chronic malnutrition are at a particularly high risk for complications and refeeding syndrome when initiating nutritional support. In these groups, careful screening for malnutrition, conservative initiation of caloric intake, and strict monitoring of phosphate, potassium, and magnesium levels are required. The existing consensus recommendations on refeeding syndrome advocate a cautious, stepwise escalation of energy intake together with proactive electrolyte replacement in high-risk patients; this approach can be applied directly to the management of AP (Arvanitakis et al., 2020; da Silva et al., 2020). In the context of renal or hepatic impairment, it may be necessary to adjust protein and fluid prescriptions, ideally in collaboration with nephrology and hepatology specialists.

Across these diverse populations, core principles remain consistent: prioritize EN at the earliest possible time and at an appropriate dose; reserve PN only for clear indications; and individualize the prescription of energy, protein, and micronutrients according to age, comorbidities, and metabolic risk. Applying these fundamental principles from a perspective that considers the needs of specific populations, such as children, pregnant women, obese patients, and those with severe complications, enables nutritional support to contribute optimally to recovery while minimizing iatrogenic complications.

9.8 Future directions

Future research on nutritional management in AP is likely to move beyond the dichotomy between EN and PN toward a more refined phenotype-based metabolic approach. Although early EN is established as the standard treatment for moderate-to-severe AP, significant challenges remain regarding the optimal timing and volume of administration, patient selection criteria, and methods for integrating evolving technologies and adjunctive therapies into routine clinical practice.

A key priority is to refine individualized energy and protein prescriptions. Current clinical practice relies on weight-based estimates; however, the energy requirements in AP can fluctuate significantly over time. Where indirect calorimetry is available, its wider adoption enables precise adjustments, helping to avoid harmful metabolic effects from both undernutrition and overnutrition. Future studies should evaluate whether nutrition management based on indirect calorimetry improves clinically meaningful outcomes in AP, particularly in patients with persistent organ dysfunction and prolonged ICU stay.

The second focus is on improved phenotypic analysis of patients to guide nutritional strategies. AP is a heterogeneous syndrome rather than a single phenotype; therefore, patients should be stratified based on etiology (biliary, alcoholic, hypertriglyceridemia, etc.), severity, metabolic risk factors (sarcopenia, obesity, diabetes, etc.), and concomitant organ dysfunction. It is crucial to explore whether individualized nutritional approaches, such as early initiation, different macronutrient patterns, and stepwise titration, can modify organ failure, infection, and recovery outcomes. The association between AP and obesity/metabolic syndromes warrants particular attention. Establishing methods to meet protein requirements optimally while avoiding excessive nonprotein calorie intake and worsening glucose metabolism abnormalities or hypertriglyceridemia is an urgent clinical challenge.

The composition of nutritional formulations remains an active area of research. Although standard polymer-based EN is considered safe and suitable for most patients with AP, robust trials with sufficient power are needed to determine whether specialized formulations meaningfully improve tolerability, reduce pain, or impact key endpoints in specific subgroups. Similarly, immunomodulatory nutrients such as glutamine, arginine, and omega-3 polyunsaturated fatty acids possess biologically compelling mechanisms of action; however, their clinical value in AP remains uncertain. Future studies should move beyond small, single-center trials to incorporate standardized formulations, well-defined patient populations, and robust outcomes rather than merely surrogate markers.

The gut-pancreas-microbiome axis is arguably the most rapidly evolving field. Experimental and clinical data suggest that gut microbiota dysbiosis, barrier dysfunction, and bacterial translocation are key contributors to the transition from sterile to infectious pancreatic necrosis and to the development of systemic complications (Patel et al., 2021; Tan et al., 2015; Fritz et al., 2010). The initial enthusiasm for broad-spectrum probiotic and synbiotic supplementation is waning owing to conflicting trial results and safety concerns (including reports of increased mortality and intestinal ischemia with specific formulations). Attention is now shifting toward more targeted and safer microbiome-directed interventions: approaches combining defined microbial consortia, postbiotics, selective bacteriophages, and small-molecule compounds that modulate the microbiome with barrier function enhancement strategies. Careful design of trials using modern microbiome profiling to stratify patients based on baseline microbial signatures and rigorous monitoring of safety are essential to determine whether these approaches can reduce infectious necrosis and sepsis without reproducing past adverse events.

Finally, implementation science and healthcare systems research play a crucial role in translating existing evidence into practice. Despite strong evidence supporting early EN, many patients with AP in real-world clinical settings still experience prolonged fasting or delayed nutritional intake. It is essential to understand barriers to guideline-aligned nutritional management, such as clinicians' beliefs about "pancreatic rest," concerns about aspiration, and logistical constraints regarding enteral tube placement, and to validate interventions that overcome these barriers. Digital tools such as clinical decision support systems, standardized order sets, and real-time dashboards displaying nutritional adequacy and tolerance can promote adherence to best practices.

9.9 Conclusion

Early enteral and individualized nutritional therapy has evolved from a secondary consideration to a pillar of AP management. EN is now firmly established as the preferred treatment modality for most patients. In mild AP, oral intake can be resumed promptly once intestinal motility recovers, whereas in moderately severe to severe AP, early EN within 24–48 h of admission is recommended. In most cases, standard polymeric formulations and straightforward tube-feeding routes are sufficient. PN should be reserved for clearly defined indications and minimized in terms of duration and intensity. Although advanced formulations and emerging biologically targeted therapies are promising, their use must be guided by evolving evidence and thoughtful patient selection. Ultimately, a structured multidisciplinary approach to nutritional support provides a practical, evidence-based strategy to improve the outcomes of this complex and heterogeneous disease.

References

Abu-El-Haija, M., Uc, A., Werlin, S.L., Freeman, A.J., Georgieva, M., Jojkić-Pavkov, D., Kalnins, D., Kochavi, B., Koot, B.G.P., Van Biervliet, S., Walkowiak, J., Wilschanski, M., Morinville, V.D., 2018. Nutritional considerations in pediatric pancreatitis: a position paper from the NASPHAN pancreas committee and ESPHAN cystic fibrosis/pancreas working group. Journal of Pediatric Gastroenterology and Nutrition 67 (1), 131–143. https://doi.org/10.1097/MPG.0000000000002023.

Al-Leswas, D., Eltweri, A.M., Chung, W.Y., Arshad, A., Stephenson, J.A., Al-Taan, O., Pollard, C., Fisk, H.L., Calder, P.C., Garcea, G., Metcalfe, M.S., Dennison, A.R., 2020. Intravenous omega-3 fatty acids are associated with better clinical outcome and less inflammation in patients with predicted severe acute pancreatitis: a randomised double blind controlled trial. Clinical Nutrition 39 (9), 2711–2719. https://doi.org/10.1016/j.clnu.2018.04.003.

Ammori, B.J., Leeder, P.C., King, R.F.G.J., Barclay, G.R., Martin, I.G., Larvin, M., McMahon, M.J., 1999. Early increase in intestinal permeability in patients with severe acute pancreatitis: correlation with endotoxemia, organ failure, and mortality. Journal of Gastrointestinal Surgery 3 (3), 252–262. https://doi.org/10.1016/S1091-255X(99)80067-5.

Arutla, M., Raghunath, M., Deepika, G., Jakkampudi, A., Murthy, H.V.V., Rao, G.V., Reddy, D.N., Talukdar, R., 2019. Efficacy of enteral glutamine supplementation in patients with severe and predicted severe acute pancreatitis— a randomized controlled trial.

Indian Journal of Gastroenterology 38 (4), 338–347. https://doi.org/10.1007/s12664-019-00962-7.

Arvanitakis, M., Ockenga, J., Bezmarevic, M., Gianotti, L., Krznarić, Ž., Lobo, D.N., Löser, C., Madl, C., Meier, R., Phillips, M., Rasmussen, H.H., Van Hooft, J.E., Bischoff, S.C., 2020. ESPEN guideline on clinical nutrition in acute and chronic pancreatitis. Clinical Nutrition 39 (3), 612–631. https://doi.org/10.1016/j.clnu.2020.01.004.

Arvanitakis, M., Ockenga, J., Bezmarevic, M., Gianotti, L., Krznarić, Ž., Lobo, D.N., Löser, C., Madl, C., Meier, R., Phillips, M., Rasmussen, H.H., Van Hooft, J.E., Bischoff, S.C., 2024. ESPEN practical guideline on clinical nutrition in acute and chronic pancreatitis. Clinical Nutrition 43 (2), 395–412. https://doi.org/10.1016/j.clnu.2023.12.019.

Asrani, V., Chang, W.K., Dong, Z., Hardy, G., Windsor, J.A., Petrov, M.S., 2013. Glutamine supplementation in acute pancreatitis: a meta-analysis of randomized controlled trials. Pancreatology 13 (5), 468–474. https://doi.org/10.1016/j.pan.2013.07.282.

Banks, P.A., Freeman, M.L., October 2006. Practice parameters committee of the American College of gastroenterology. Practice guidelines in acute pancreatitis. American Journal of Gastroenterology 101 (10), 2379–2400. https://doi.org/10.1111/j.1572-0241.2006.00856.x.

Baron, T.H., DiMaio, C.J., Wang, A.Y., Morgan, K.A., 2020. American gastroenterological association clinical practice update: management of pancreatic necrosis. Gastroenterology 158 (1), 67. https://doi.org/10.1053/j.gastro.2019.07.064.

Besselink, M.G., van Santvoort, H.C., Buskens, E., Boermeester, M.A., van Goor, H., Timmerman, H.M., Nieuwenhuijs, V.B., Bollen, T.L., van Ramshorst, B., Witteman, B.J., Rosman, C., Ploeg, R.J., Brink, M.A., Schaapherder, A.F., Dejong, C.H., Wahab, P.J., van Laarhoven, C.J., van der Harst, E., van Eijck, C.H., Cuesta, M.A., Akkermans, L.M., Gooszen, H.G., 2008. Probiotic prophylaxis in predicted severe acute pancreatitis: a randomised, double-blind, placebo-controlled trial. The Lancet 371 (9613), 651–659. https://doi.org/10.1016/S0140-6736(08)60207-X.

Bharadwaj, S., Ginoya, S., Tandon, P., Gohel, T.D., Guirguis, J., Vallabh, H., Jevenn, A., Hanouneh, I., 2016. Malnutrition: laboratory markers vs nutritional assessment. Gastroenterology Report 4 (4), 272–280. https://doi.org/10.1093/gastro/gow013.

Bouffard, Y.H., Delafosse, B.X., Annat, G.J., Viale, J.P., Bertrand, O.M., Motin, J.P., 1989. Energy expenditure during severe acute pancreatitis. Journal of Parenteral and Enteral Nutrition 13 (1), 26–29. https://doi.org/10.1177/014860718901300126.

Bukowski, J.S., Dembiński, Ł., Dziekiewicz, M., Banaszkiewicz, A., 2022. Early enteral nutrition in paediatric acute pancreatitis—a review of published studies. Nutrients 14 (16). https://doi.org/10.3390/nu14163441. Available from: http://www.mdpi.com/journal/nutrients/.

Chang, Y.s., Fu, H.q., Xiao, Y.m., Liu, J.c., 2013. Nasogastric or nasojejunal feeding in predicted severe acute pancreatitis: a meta-analysis. Critical Care 17 (3). https://doi.org/10.1186/cc12790China. Available from: http://ccforum.com/content/17/3/R118.

Cicalese, L., Sahai, A., Sileri, P., Rastellini, C., Subbotin, V., Ford, H., Lee, K., 2001. Acute pancreatitis and bacterial translocation. Digestive Diseases and Sciences 46 (5), 1127–1132. https://doi.org/10.1023/A:1010786701289.

da Silva, J.S.V., Seres, D.S., Sabino, K., Adams, S.C., Berdahl, G.J., Citty, S.W., Cober, M.P., Evans, D.C., Greaves, J.R., Gura, K.M., Michalski, A., Plogsted, S., Sacks, G.S., Tucker, A.M., Worthington, P., Walker, R.N., Ayers, P., 2020. ASPEN consensus recommendations for refeeding syndrome. Nutrition in Clinical Practice 35 (2), 178–195. https://doi.org/10.1002/ncp.10474.

De Lucia, S.S., Candelli, M., Polito, G., Maresca, R., Mezza, T., Schepis, T., Pellegrino, A., Zileri Dal Verme, L., Nicoletti, A., Franceschi, F., Gasbarrini, A., Nista, E.C., 2023. Nutrition in acute pancreatitis: from the old paradigm to the new evidence. Nutrients 15 (8). https://doi.org/10.3390/nu15081939. Available from: http://www.mdpi.com/journal/nutrients/.

Dervenis, C., Hatzitheoklitos, E., Smailis, D., 2003. Bacterial translocation and its prevention in acute pancreatitis. Journal of Hepato-Biliary-Pancreatic Surgery 10 (6), 415–418. https://doi.org/10.1007/s00534-002-0727-5.

Dobszai, D., Mátrai, P., Gyöngyi, Z., Csupor, D., Bajor, J., Erőss, B., Mikó, A., Szakó, L., Meczker, Á., Hágendorn, R., Márta, K., Szentesi, A., Hegyi, P., 2019. Body-mass index correlates with severity and mortality in acute pancreatitis: a meta-analysis. World Journal of Gastroenterology 25 (6), 729–743. https://doi.org/10.3748/wjg.v25.i6.729.

Dungan, K.M., Braithwaite, S.S., Preiser, J.C., 2009. Stress hyperglycaemia. The Lancet 373 (9677), 1798–1807. https://doi.org/10.1016/S0140-6736(09)60553-5.

Dutta, A.K., Goel, A., Kirubakaran, R., Chacko, A., Tharyan, P., 2020. Nasogastric versus nasojejunal tube feeding for severe acute pancreatitis. Cochrane Database of Systematic Reviews 2020 (3). https://doi.org/10.1002/14651858.CD010582.pub2. Available from: https://www.cochranelibrary.com/cdsr/table-of-contents.

Endo, A., Shiraishi, A., Fushimi, K., Murata, K., Otomo, Y., 2018. Comparative effectiveness of elemental formula in the early enteral nutrition management of acute pancreatitis: a retrospective cohort study. Annals of Intensive Care 8 (1). https://doi.org/10.1186/s13613-018-0414-6.

Feng, P., He, C., Liao, G., Chen, Y., 2017. Early enteral nutrition versus delayed enteral nutrition in acute pancreatitis: a PRISMA-compliant systematic review and meta-analysis. Medicine (United States) 96 (46). https://doi.org/10.1097/MD.0000000000008648. Available from: http://journals.lww.com/md-journal.

Forsmark, C.E., Baillie, J., AGA Institute Clinical Practice and Economics Committee; AGA Institute Governing Board, May 2007. AGA Institute technical review on acute pancreatitis. Gastroenterology 132 (5), 2022–2044. https://doi.org/10.1053/j.gastro.2007.03.065.

Friedli, N., Stanga, Z., Sobotka, L., Culkin, A., Kondrup, J., Laviano, A., Mueller, B., Schuetz, P., 2017. Revisiting the refeeding syndrome: results of a systematic review. Nutrition 35, 151–160. https://doi.org/10.1016/j.nut.2016.05.016.

Fritz, S., Hackert, T., Hartwig, W., Rossmanith, F., Strobel, O., Schneider, L., Will-Schweiger, K., Kommerell, M., Büchler, M.W., Werner, J., 2010. Bacterial translocation and infected pancreatic necrosis in acute necrotizing pancreatitis derives from small bowel rather than from colon. The American Journal of Surgery 200 (1), 111–117. https://doi.org/10.1016/j.amjsurg.2009.08.019.

Guan, Y., Liu, G., Tang, F., Wu, X., Shi, J., Huang, Q., 2025. Stress hyperglycemia in acute pancreatitis: from mechanisms to prognostic implications. Life Sciences 365, 123469. https://doi.org/10.1016/j.lfs.2025.123469.

Gupta, M., Liti, B., Barrett, C., Thompson, P.D., Fernandez, A.B., 2022. Prevention and management of hypertriglyceridemia-induced acute pancreatitis during pregnancy: a systematic review. The American Journal of Medicine 135 (6), 709–714. https://doi.org/10.1016/j.amjmed.2021.12.006.

Harshit Kumar, A., Singh Griwan, M., 2018. A comparison of Apache II, BISAP, Ranson's score and modified CTSI in predicting the severity of acute pancreatitis based on the

2012 revised Atlanta Classification. Gastroenterology Report 6 (2), 127–131. https://doi.org/10.1093/gastro/gox029.

Hegazi, R.A., DeWitt, T., 2014. Enteral nutrition and immune modulation of acute pancreatitis. World Journal of Gastroenterology 20 (43), 16101–16105. https://doi.org/10.3748/wjg.v20.i43.16101.

Heidegger, C.P., Berger, M.M., Graf, S., Zingg, W., Darmon, P., Costanza, M.C., Thibault, R., Pichard, C., 2013. Optimisation of energy provision with supplemental parenteral nutrition in critically ill patients: a randomised controlled clinical trial. The Lancet 381 (9864), 385–393. https://doi.org/10.1016/S0140-6736(12)61351-8.

Horibe, M., Nishizawa, T., Suzuki, H., Minami, K., Yahagi, N., Iwasaki, E., Kanai, T., 2016. Timing of oral refeeding in acute pancreatitis: a systematic review and meta-analysis. United European Gastroenterology Journal 4 (6), 725–732. https://doi.org/10.1177/2050640615612368.

Horibe, M., Iwasaki, E., Nakagawa, A., Matsuzaki, J., Minami, K., Machida, Y., Tamagawa, H., Takimoto, Y., Ueda, M., Katayama, T., Kawasaki, S., Matsushita, M., Seino, T., Fukuhara, S., Kanai, T., 2020. Efficacy and safety of immediate oral intake in patients with mild acute pancreatitis: a randomized controlled trial. Nutrition 74, 110724. https://doi.org/10.1016/j.nut.2020.110724.

Jin, Z., Wang, Z., Wang, J., 2020. Early enteral nutrition prevent acute pancreatitis from deteriorating in obese patients. Journal of Clinical Gastroenterology 54 (2), 184–191. https://doi.org/10.1097/MCG.0000000000001117.

Khetpal, N., Sharbatji, M., Asfari, M.M., Ahmad, S., 2025. Outcomes of acute pancreatitis hospitalizations with obesity. Digestive Diseases and Sciences 70 (4), 1350–1359. https://doi.org/10.1007/s10620-025-08880-9.

Lakananurak, N., Gramlich, L., 2020. Nutrition management in acute pancreatitis: clinical practice consideration. World Journal of Clinical Cases 8 (9), 1561–1573. https://doi.org/10.12998/WJCC.V8.I9.1561.

Lasztity, N., Hamvas, J., Biró, L., Németh, E., Marosvölgyi, T., Decsi, T., Pap, A., Antal, M., 2005. Effect of enterally administered n-3 polyunsaturated fatty acids in acute pancreatitis - a prospective randomized clinical trial. Clinical Nutrition 24 (2), 198–205. https://doi.org/10.1016/j.clnu.2004.12.008.

Lei, Q.C., Wang, X.Y., Xia, X.F., Zheng, H.Z., Bi, J.C., Tian, F., Li, N., 2015. The role of omega-3 fatty acids in acute pancreatitis: a meta-analysis of randomized controlled trials. Nutrients 7 (4), 2261–2273. https://doi.org/10.3390/nu7042261.

Li, J.Y., Yu, T., Chen, G.C., Yuan, Y.H., Zhong, W., Zhao, L.N., Chen, Q.K., 2013. Enteral nutrition within 48 hours of admission improves clinical outcomes of acute pancreatitis by reducing complications: a meta-analysis. PLoS One 8 (6). https://doi.org/10.1371/journal.pone.0064926China. Available from: http://www.plosone.org/article/fetchObject.action?uri=info%3Adoi%2F10.1371%2Fjournal.pone.0064926&representation=PDF.

Liu, M., Gao, C., 2021. A systematic review and meta-analysis of the effect of total parenteral nutrition and enteral nutrition on the prognosis of patients with acute pancreatitis. Annals of Palliative Medicine 10 (10), 10779–10788. https://doi.org/10.21037/apm-21-2469.

Liu, Q., Fan, X., Cui, W., Wang, X., Zhang, Z., Wang, N., Qiao, L., 2023. Serum cholinesterase, C-reactive protein, interleukin 6, and procalcitonin levels as predictors of mortality in patients in the intensive care unit. Turkish Journal of Anaesthesiology and Reanimation 51 (5), 408–413. https://doi.org/10.4274/TJAR.2023.231349.

Loh, J.A., Rickels, M.R., Williams, J., Iqbal, N., 2005. Total parenteral nutrition in management of hyperlipidemic pancreatitis during pregnancy . Endocrine Practice. Endocrine Practice 11 (5), 325–330. https://doi.org/10.4158/EP.11.5.325.

Lu, W.W., Chen, X., Ni, J.L., Zhu, S.L., Fei, A.H., Wang, X.S., 2021a. The role of gut microbiota in the pathogenesis and treatment of acute pancreatitis: a narrative review. Annals of Palliative Medicine 10 (3), 3445–3451. https://doi.org/10.21037/apm-21-429.

Lu, Y., Zhang, Q., Lou, J., 2021b. Blood glucose-related indicators are associated with in-hospital mortality in critically ill patients with acute pancreatitis. Scientific Reports 11 (1). https://doi.org/10.1038/s41598-021-94697-1.

Malik, A., Amjad, W., Farooq, U., Saleem, S.S., Malik, M.I., 2023. Evaluation of the effect of probiotics on septic complications in patients with severe acute pancreatitis. A systematic review and meta-analysis. Przegląd Gastroenterologiczny 18 (3), 281–291. https://doi.org/10.5114/PG.2022.118164.

Marik, P.E., Zaloga, G.P., 2004. Meta-analysis of parenteral-nutrition versus enteral nutrition in patients with acute pancreatitis. British Medical Journal 328 (7453), 1407–1410. https://doi.org/10.1136/bmj.38118.593900.55.

Martínez, J., Johnson, C.D., Sánchez-Payá, J., de Madaria, E., Robles-Díaz, G., Pérez-Mateo, M., 2006. Obesity is a definitive risk factor of severity and mortality in acute pancreatitis: an updated meta-analysis. Pancreatology 6 (3), 206–209. https://doi.org/10.1159/000092104.

Martínez-Vázquez, J., Martínez-Vargas, E., Mora-Jiménez, J., Arguedas-Chacón, S., García-Montero, J., Chaves-Romero, N., Murillo-Cubero, J., Zavaleta-Monestel, E., 2025. Enteral infant formulas: a narrative review of historical evolution, nutritional composition, and clinical use in pediatrics. Cureus 17 (4). https://doi.org/10.7759/cureus.82692.

Matthai, J., Mohan, N., Viswanathan, M.S., Shanmugam, N., Bharadia, L., Bhatnagar, S., Srikanth, K.P., 2020. Therapeutic enteral formulas in children. Indian Pediatrics 57 (4), 343–348. https://doi.org/10.1007/s13312-020-1787-9.

McClave, S.A., Taylor, B.E., Martindale, R.G., Warren, M.M., Johnson, D.R., Braunschweig, C., McCarthy, M.S., Davanos, E., Rice, T.W., Cresci, G.A., Gervasio, J. M., Sacks, G.S., Roberts, P.R., Compher, C., 2016. Guidelines for the provision and assessment of nutrition support therapy in the adult critically ill patient: society of critical care medicine (SCCM) and American society for parenteral and enteral nutrition (A.S.P.E.N.). Journal of Parenteral and Enteral Nutrition 40 (2), 159–211. https://doi.org/10.1177/0148607115621863.

Mizushima, T., Ochi, K., Matsumura, N., Ichimura, M., Ishibashi, T., Tsuboi, K., Harada, H., 1998. Prevention of hyperlipidemic acute pancreatitis during pregnancy with medium-chain triglyceride nutritional support. International Journal of Pancreatology 23 (3), 187–192. https://doi.org/10.1007/bf02788396.

Moggia, E., Koti, R., Belgaumkar, A.P., Fazio, F., Pereira, S.P., Davidson, B.R., Gurusamy, K. S., 2017. Pharmacological interventions for acute pancreatitis. Cochrane Database of Systematic Reviews 2017 (4). https://doi.org/10.1002/14651858.CD011384.pub2. Available from: http://as.wiley.com/WileyCDA/Brand/id-6.html.

Oláh, A., Belágyi, T., Issekutz, A., Gamal, M.E., Bengmark, S., 2002. Randomized clinical trial of specific lactobacillus and fibre supplement to early enteral nutrition in patients with acute pancreatitis. British Journal of Surgery 89 (9), 1103–1107. https://doi.org/10.1046/j.1365-2168.2002.02189.x.

Oláh, A., Belágyi, T., Pótó, L., Romics, L., Bengmark, S., 2007. Synbiotic control of inflammation and infection in severe acute pancreatitis: a prospective, randomized, double blind study. Hepato-Gastroenterology 54 (74), 590–594.

Pan, L.L., Li, J., Shamoon, M., Bhatia, M., Sun, J., June 30, 2017. Recent advances on nutrition in treatment of acute pancreatitis. Frontiers in Immunology 8, 762. https://doi.org/10.3389/fimmu.2017.00762. Erratum in: Front Immunol. 2018 Apr 23;9:849. doi: 10.3389/fimmu.2018.00849.

Patel, B.K., Patel, K.H., Bhatia, M., Iyer, S.G., Madhavan, K., Moochhala, S.M., 2021. Gut microbiome in acute pancreatitis: a review based on current literature. World Journal of Gastroenterology 27 (30), 5019–5036. https://doi.org/10.3748/wjg.v27.i30.5019.

Pearce, C.B., Sadek, S.A., Walters, A.M., Goggin, P.M., Somers, S.S., Toh, S.K., Johns, T., Duncan, H.D., 2006. A double-blind, randomised, controlled trial to study the effects of an enteral feed supplemented with glutamine, arginine, and omega-3 fatty acid in predicted acute severe pancreatitis. Journal of the Pancreas 7 (4), 361–371.

Petrov, M.S., Van Santvoort, H.C., Besselink, M.G.H., Van Der Heijden, G.J.M.G., Windsor, J.A., Gooszen, H.G., 2008. Enteral nutrition and the risk of mortality and infectious complications in patients with severe acute pancreatitis: a meta-analysis of randomized trials. Archives of Surgery 143 (11), 1111–1117. https://doi.org/10.1001/archsurg.143.11.1111Netherlands.

Petrov, M.S., Loveday, B.P.T., Pylypchuk, R.D., McIlroy, K., Phillips, A.R.J., Windsor, J.A., 2009. Systematic review and meta-analysis of enteral nutrition formulations in acute pancreatitis. British Journal of Surgery 96 (11), 1243–1252. https://doi.org/10.1002/bjs.6862NewZealand.

Poropat, G., Giljaca, V., Hauser, G., Štimac, D., 2015. Enteral nutrition formulations for acute pancreatitis. Cochrane Database of Systematic Reviews 2015 (3). https://doi.org/10.1002/14651858.CD010605.pub2. Available from: http://as.wiley.com/WileyCDA/Brand/id-6.html.

Puri, P., Dhiman, R.K., Taneja, S., Tandon, P., Merli, M., Anand, A.C., Arora, A., Acharya, S. K., Benjamin, J., Chawla, Y.K., Dadhich, S., Duseja, A., Eapan, C.E., Goel, A., Kalra, N., Kapoor, D., Kumar, A., Madan, K., Nagral, A., Pandey, G., Rao, P.N., Saigal, S., Saraf, N., Saraswat, V.A., Saraya, A., Sarin, S.K., Sharma, P., Shalimar, Shukla, A., Sidhu, S.S., Singh, N., Singh, S.P., Srivastava, A., Wadhawan, M., 2021. Nutrition in chronic liver disease: consensus statement of the Indian national association for study of the liver. Journal of Clinical and Experimental Hepatology 11 (1), 97–143. https://doi.org/10.1016/j.jceh.2020.09.003.

Qi, D., Yu, B., Huang, J., Peng, M., 2018. Meta-analysis of early enteral nutrition provided within 24 hours of admission on clinical outcomes in acute pancreatitis. Journal of Parenteral and Enteral Nutrition 42 (7), 1139–1147. https://doi.org/10.1002/jpen.1139. Available from: http://onlinelibrary.wiley.com/journal/10.1002/(ISSN)1941-2444.

Ramírez-Maldonado, E., López Gordo, S., Pueyo, E.M., Sánchez-García, A., Mayol, S., González, S., Elvira, J., Memba, R., Fondevila, C., Jorba, R., 2021. Immediate oral refeeding in patients with mild and moderate acute pancreatitis: a multicenter, randomized controlled trial (PADI trial). Annals of Surgery 274 (2), 255–263. https://doi.org/10.1097/SLA.0000000000004596.

Rinninella, E., Annetta, M.G., Serricchio, M.L., Dal Lago, A.A., Miggiano, G.A.D., Mele, M. C., 2017. Nutritional support in acute pancreatitis: from physiopathology to practice. An evidence-based approach. European Review for Medical and Pharmacological Sciences 21 (2), 421–432.

Sarr, M.G., 2013. 2012 revision of the Atlanta Classification of acute pancreatitis. Polish Archives of Internal Medicine 123 (3), 118–124. https://doi.org/10.20452/pamw.1627.

Singh, N., Mishra, S.K., Sachdev, V., Sharma, H., Upadhyay, A.D., Arora, I., Saraya, A., 2014. Effect of oral glutamine supplementation on gut permeability and endotoxemia in patients with severe acute pancreatitis: a randomized controlled trial. Pancreas 43 (6), 867–873. https://doi.org/10.1097/MPA.0000000000000124.

Spanier, B.W.M., Bruno, M.J., Mathus-Vliegen, E.M.H., 2011. Enteral nutrition and acute pancreatitis: a review. Gastroenterology Research and Practice 2011, 1–9. https://doi.org/10.1155/2011/857949.

Szatmary, P., Grammatikopoulos, T., Cai, W., Huang, W., Mukherjee, R., Halloran, C., Beyer, G., Sutton, R., 2022. Acute pancreatitis: diagnosis and treatment. Drugs 82 (12), 1251–1276. https://doi.org/10.1007/s40265-022-01766-4.

Takada, T., Isaji, S., Mayumi, T., Yoshida, M., Takeyama, Y., Itoi, T., Sano, K., Iizawa, Y., Masamune, A., Hirota, M., Okamoto, K., Inoue, D., Kitamura, N., Mori, Y., Mukai, S., Kiriyama, S., Shirai, K., Tsuchiya, A., Higuchi, R., Hirashita, T., 2022. JPN clinical practice guidelines 2021 with easy-to-understand explanations for the management of acute pancreatitis. Journal of Hepato-Biliary-Pancreatic Sciences 29 (10), 1057–1083. https://doi.org/10.1002/jhbp.1146. Available from: http://onlinelibrary.wiley.com/journal/10.1002/(ISSN)1868-6982.

Tan, C., Ling, Z., Huang, Y., Cao, Y., Liu, Q., Cai, T., Yuan, H., Liu, C., Li, Y., Xu, K., 2015. Dysbiosis of intestinal microbiota associated with inflammation involved in the progression of acute pancreatitis. Pancreas 44 (6), 868–875. https://doi.org/10.1097/MPA.0000000000000355.

Tiengou, L.E., Gloro, R., Pouzoulet, J., Bouhier, K., Read, M.H., Arnaud-Battandier, F., Plaze, J.M., Blaizot, X., Dao, T., Piquet, M.A., 2006. Semi-elemental formula or polymeric formula: is there a better choice for enteral nutrition in acute pancreatitis? Randomized comparative study. Journal of Parenteral and Enteral Nutrition 30 (1), 1–5. https://doi.org/10.1177/014860710603000101.

Valainathan, S., Boukris, A., Arapis, K., Schoch, N., Goujon, G., Konstantinou, D., Bécheur, H., Pelletier, A.L., 2019. Energy expenditure in acute pancreatitis evaluated by the Harris–Benedict equation compared with indirect calorimetry. Clinical Nutrition ESPEN 33, 57–59. https://doi.org/10.1016/j.clnesp.2019.07.007.

Wang, X., Li, W., Li, N., Li, J., 2008. ω-3 fatty acids - supplemented parenteral nutrition decreases hyperinflammatory response and attenuates systemic disease sequelae in severe acute pancreatitis: a randomized and controlled study. Journal of Parenteral and Enteral Nutrition 32 (3), 236–241. https://doi.org/10.1177/0148607108316189.

Wilkinson, R.E., Dickerson, R.N., 2016. "New" indications for parenteral nutrition. Hospital Pharmacy 51 (10), 795–797. https://doi.org/10.1310/hpj5110-795. Available from: http://archive.hospital-pharmacy.com/doi/pdf/10.1310/hpj5110-795.

Wong, B., Ooi, T.C., Keely, E., 2015. Severe gestational hypertriglyceridemia: a practical approach for clinicians. Obstetric Medicine 8 (4), 158–167. https://doi.org/10.1177/1753495X15594082.

Wu, B.U., Johannes, R.S., Sun, X., Tabak, Y., Conwell, D.L., Banks, P.A., 2008. The early prediction of mortality in acute pancreatitis: a large population-based study. Gut 57 (12), 1698–1703. https://doi.org/10.1136/gut.2008.152702.

Wu, P., Li, L., Sun, W., 2018. Efficacy comparisons of enteral nutrition and parenteral nutrition in patients with severe acute pancreatitis: a meta-analysis from randomized controlled trials. Bioscience Reports 38 (6). https://doi.org/10.1042/BSR20181515. Available from: http://www.bioscirep.org/content/38/6/BSR20181515.full-text.pdf.

Yang, X., Zhang, R., Jin, T., Zhu, P., Yao, L., Li, L., Cai, W., Mukherjee, R., Du, D., Fu, X., Xue, J., Martina, R., Liu, T., Pendharkar, S., Phillips, A.R., Singh, V.K., Sutton, R., Windsor, J.A., Deng, L., Xia, Q., Huang, W., 2022. Stress hyperglycemia is independently associated with persistent organ failure in acute pancreatitis. Digestive Diseases and Sciences 67 (5), 1879–1889. https://doi.org/10.1007/s10620-021-06982-8.

Yi, F., Ge, L., Zhao, J., Yuan, L., Zhou, F., Chen, Z., Zhu, Y., Xia, B., 2012. Meta-analysis: total parenteral nutrition versus total enteral nutrition in predicted severe acute pancreatitis. Internal Medicine 51 (6), 523–530. https://doi.org/10.2169/internalmedicine.51.6685.

Zhang, J., Shahbaz, M., Fang, R., Liang, B., Gao, C., Gao, H., Ijaz, M., Peng, C., Wang, B., Niu, Z., Niu, J., 2014. Comparison of the BISAP scores for predicting the severity of acute pancreatitis in Chinese patients according to the latest Atlanta classification. Journal of Hepato-Biliary-Pancreatic Sciences 21 (9), 689–694. https://doi.org/10.1002/jhbp.118.

CHAPTER

10 Treatment of pancreatitis 3: Follow-up procedures and long-term care

Mitsuru Okuno[1], Tsuyoshi Mukai[1], Keisuke Iwata[2], Fumiya Kataoka[1], Eiichi Tomita[1], Hisataka Moriwaki[1], Masahito Shimizu[3]

[1]*Department of Gastroenterology, Matsunami General Hospital, Gifu, Japan;* [2]*Department of Gastroenterology, Gifu Municipal Hospital, Gifu, Japan;* [3]*First Department of Internal Medicine, Gifu University Hospital, Gifu, Japan*

ABBREVIATIONS

AP	acute pancreatitis
CBD	common bile duct
CBDS	common bile duct stone
CE-CT	contrast-enhanced computed tomography
CT	computed tomography
DPDS	disconnected pancreatic duct syndrome
EHL	electrohydraulic lithotripsy
ENPD	endoscopic nasopancreatic drainage
ERCP	endoscopic retrograde cholangiopancreatography
ERP	endoscopic retrograde pancreatography
ESWL	extracorporeal shock wave lithotripsy
EUS	endoscopic ultrasonography
EUS-FNA	EUS-fine needle aspiration
EUS-PD	EUS-pancreatic drainage
IPMN	intraductal papillary mucinous neoplasm
LAMS	lumen apposing metal stent
MRCP	magnetic resonance cholangiopancreatography
MRI	magnetic resonance imaging
PFC	pancreatic fluid collection
POPS	peroral pancreatoscopy
RCT	randomized controlled trial
WON	walled-off necrosis

10.1 Foreword

Acute pancreatitis (AP) is caused by alcohol consumption, gallstones, pancreatic stones, tumors, and postoperative strictures. When AP occurs, treatment is

Advanced Insights into Acute Pancreatitis. https://doi.org/10.1016/B978-0-443-44736-5.00005-X

administered based on the specific cause. As AP recurrence can be fatal, the goal of management (follow-up) after AP is to prevent the recurrence of pancreatitis, and in the case of tumors, to detect them at an early stage.

Computed tomography (CT) and magnetic resonance imaging (MRI) are common imaging modalities used during follow-up. However, these modalities may be difficult to detect in cases of small stones or tumors because of the detection limits of CT and MRI. Endoscopic follow-ups, such as endoscopic ultrasonography (EUS) and/or endoscopic retrograde pancreatography (ERP), are important for evaluating the causes of AP (stone and/or tumor) and obtaining pathological samples (pancreatic juice cytology).

Stent management is necessary in cases where endoscopic intervention is performed for AP treatment. It needs to consider the timing of stent exchange and removal in accordance with the condition of the patient. The pancreas also has exocrine and endocrine functions, including secretion of digestive enzymes and insulin. AP can cause a decline in exoendocrine function owing to pancreatic tissue damage. This chapter describes the follow-up methods for AP, including the identification of the causes of AP and follow-up methods for each disease.

10.2 Alcoholic AP

Alcoholic pancreatitis is a major cause of AP, and continued alcohol use is a risk factor for the development of AP and chronic pancreatitis. Alcohol abstinence can decrease the risk of pancreatitis; however, the goal of alcohol abstinence is generally difficult to achieve because of alcohol dependence.

The "harm reduction" method, in which the harm caused by drinking is reduced as much as possible by reducing the amount of alcohol consumed, is employed when alcohol abstinence is difficult. When a trial for harm reduction is difficult, referral to a specialized institution should be considered.

10.3 Hypertriglyceridemia

The incidence of AP caused by hypertriglyceridemia is approximately 2.3% of all cases of AP (Masamune et al., 2020). Hypertriglyceridemia is associated with the development of pancreatitis, and the incidence of pancreatitis is 15%–20% when triglyceride levels exceed 1000 mg/dL (Pascual et al., 2019). Hypertriglyceridemia is caused by genetic abnormalities (primary hypertriglyceridemia) or other factors including obesity, alcohol consumption, pregnancy, and drug administration (secondary hypertriglyceridemia).

Although the control of hypertriglyceridemia is key to preventing AP recurrence, guidance on the target triglyceride level is lacking. One report has indicated a triglyceride level of 500 mg/dL as an independent factor for persistent organ failure (Lu et al., 2020). Maintaining low triglyceride levels is important, although triglyceride levels should be maintained below 500 mg/dL to prevent AP recurrence and reduce organ damage in the event of AP recurrence.

Bezafibrate is a pharmacological drug used to treat hypertriglyceridemia. Outpatient follow-up is important to confirm compliance with oral medications. In addition, because it has some adverse events, including rhabdomyolysis, creatine kinase levels should be assessed using blood tests during follow-up.

10.4 Gallstone pancreatitis

Gallstone pancreatitis is caused by bile duct stones or biliary sludge that affects the main papilla, resulting in pancreatic duct obstruction or papillary edema. Therefore, an endoscopic papillary approach is necessary to remove the stones or biliary sludge. When CT or MRI detects common bile duct (CBD) stones (CBDS) at the onset of AP, CBDS can be removed. However, whether the CBDS has passed from the main papilla or remains in the CBD remains unclear (cf. CT imaging studies do not detect CBDS (Wang et al., 1988; Moon et al., 2005). If the CBDS remains in the CBD, it can cause AP recurrence. Endoscopic retrograde cholangiopancreatography (ERCP) can be performed while evaluating the presence of CBDS; however, considering ERCP-related adverse events, performing ERCP in all cases to evaluate CBDS presence is impractical. EUS allows for a detailed evaluation of the CBD with minimal invasion. Some randomized controlled trials (RCTs) have been conducted in which ERCP was preceded or followed by EUS evaluation for the presence of CBDS (Liu et al., 2005; Polkowski et al., 2007; Lee et al., 2008). EUS can prevent ERCP in the absence of CBDS, contributing to a reduction in ERCP-related adverse events. Therefore, even after improvement in AP, evaluating the CBD using EUS is important to prevent the recurrence of gallstone pancreatitis with minimal invasion.

Figure 10.1 shows a case in which the CBDS was difficult to visualize using CT. EUS can be used to visualize a small CBDS in the distal CBD. Particularly for small CBDS, EUS is superior because of its high spatial resolution and is useful for deciding the indications for ERCP.

MRI is also used to evaluate the presence of CBDS; however, small CBDS can be difficult to detect. A prospective study evaluating the abilities of MRI and EUS for small CBDS showed that EUS was superior to MRI (Suzuki et al., 2022). In addition, MRI may be difficult to visualize because of peripancreatic inflammation, although EUS evaluation should be considered after the improvement of AP.

The presence or absence of gallstones may result in CBDS. Without cholecystectomy after AP in gallstone cases, the rate of rehospitalization is high owing to gallstone-related complications, including recurrent gallstone pancreatitis (Yadav, O'Connell and Papachristou, 2012).

As for the timing of cholecystectomy (gallbladder stone removal), RCTs are comparing the early surgery group (approximately 5 days after pancreatitis) or the delayed surgery group in patients with mild AP. Early surgery was associated with fewer gallstone-related adverse events, and postoperative adverse events were similar between early and delayed surgery (da Costa et al., 2019; Noel et al., 2018; Jee et al., 2018). By contrast, as for the operative time, the early surgery

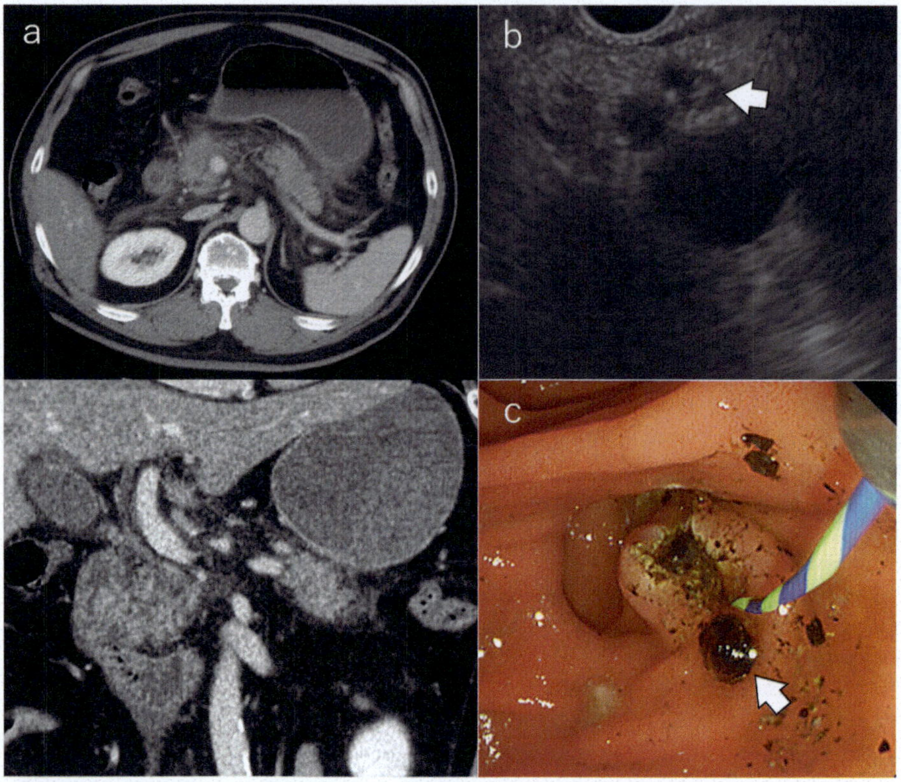

FIGURE 10.1

(A) Computed tomography reveals acute pancreatitis; however, it does not reveal a common bile stone. (B) Endoscopic ultrasonography can reveal a small common bile stone (arrow). (C) After endoscopic sphincterotomy, a small common bile stone is removed (arrow).

group was significantly longer than the delayed surgery group (120 min in the early group/88 min in the delayed surgery group [$P < .01$]) (Noel et al., 2018). Although the early surgery group had longer operative times, no significant differences in hospital stay or adverse events were observed between the two groups (Lyu et al., 2018). Therefore, early cholecystectomy may be useful in avoiding AP recurrence, if early surgery is feasible.

However, cholecystectomy after severe AP, which can lead to pancreatic fluid collection (PFC), can be complicated by surgical infection. Early surgery after severe AP is reportedly associated with a higher risk of postoperative infection and adverse events than delayed surgery (Nealon et al., 2004). In addition, in cases of severe AP, endoscopic treatment for the PFC, which must be performed 4 weeks after the onset of AP, is occasionally required. Therefore, considering the timing of cholecystectomy, it should be performed after the improvement of abdominal inflammation, and delayed surgery should be considered 4–6 weeks after severe AP.

10.5 Pancreatic cancer/intraductal papillary mucinous neoplasm

AP occurs in early pancreatic cancer and may be caused by the obstruction of the MPD by pancreatic cancer. A possible relationship between AP and (early) pancreatic cancer is suggested. A study comparing AP and control groups that matched the backgrounds of approximately 250,000 patients reported that the AP group had a significantly higher incidence of pancreatic cancer than the control group (hazard ratio = 2.02) (Kirkegård et al., 2018). A meta-analysis of the relationship between pancreatic cancer and AP has also reported a correlation with the occurrence of pancreatic cancer (odds ratio = 2.12) (Bansal and Sonnenberg, 1995; Duell et al., 2012).

A similar mechanism can occur in IPMN. The main cause of pancreatitis in IPMN cases is obstruction of the MPD by mucins or tumors. Several reports have indicated that AP is common in cases of IPMN with high-grade dysplasia or invasive cancer (Tsutsumi et al., 2011; Morales-Oyarvide et al., 2015). Therefore, the follow-up of patients after AP treatment is important considering the onset of malignant diseases.

Follow-up imaging for early pancreatic cancer or small IPMN nodules using contrast-enhanced CT (CE-CT) and magnetic resonance cholangiopancreatography (MRCP) is important. CE-CT is useful for identifying tumors and evaluating nodules in cystic lesions with contrast enhancement. MRCP is useful for evaluating MPD strictures (Fig. 10.2A) and cystic lesions, including IPMN (Correa-Gallego et al., 2023). Therefore, because these examinations have different functions, they can be used alternately during follow-up. In addition, EUS enables the visualization of microlesions that are difficult to evaluate using CE-CT and MRCP, and the detailed evaluation of cysts in IPMN. If EUS is feasible, it should be included in imaging examinations. EUS can detect small lesions in IPMNs. However, debris can accumulate inside a cyst, making it difficult to determine whether it is a nodule. Contrast-enhanced EUS using ultrasound contrast agents (cf. Sonazoid) can confirm the contrast effect if it is a nodule, which is useful for determining the malignancy (Fig. 10.3). Figure 10.3A shows blood flow through the lesion in the pancreatic cyst by contrast-enhanced EUS. No flow in the lesion was observed, suggesting the accumulation of debris. By contrast, flow was detected in the lesion, suggesting the presence of a tumor (Fig. 10.3B).

EUS-fine-needle aspiration (EUS-FNA) is useful for diagnosing pancreatic cancer (Van Riet et al., 2021); however, when the tumor cannot be detected by EUS, EUS-FNA cannot be performed, and a pathological diagnosis cannot be made. When a tumor cannot be identified by imaging, especially in early pancreatic cancer, the effectiveness of multiple pancreatic juice cytologies has also been reported. The transpapillary endoscopic nasopancreatic drainage (ENPD) tube allows the collection of pancreatic juices with pancreatic cancer cells floating in pancreatic juice (Fig. 10.2B). However, the sensitivity of single pancreatic juice cytology using

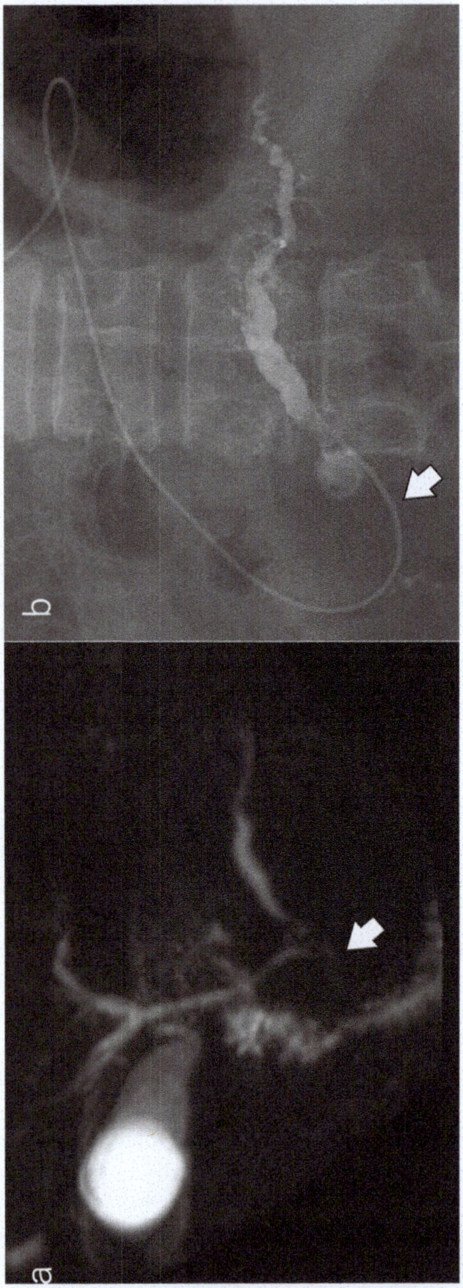

FIGURE 10.2

(A) Magnetic resonance cholangiopancreatography shows the main pancreatic head duct stricture (arrow) of the patient after acute pancreatitis. (B) Endoscopic nasopancreatic drainage is placed into the main pancreatic duct for the collection of multiple pancreatic juices. The main pancreatic head duct stricture (arrow) is also evaluated.

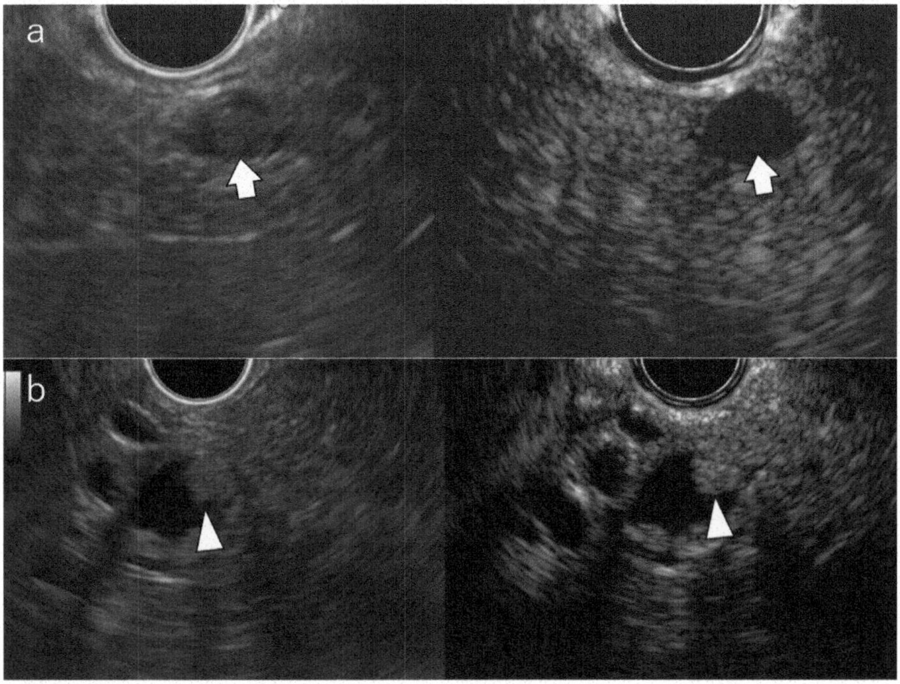

FIGURE 10.3

The evaluation of the lesion in the pancreatic cyst. (A) Endoscopic ultrasonography (EUS) detects lesions in pancreatic cysts. Contrast-enhanced EUS (right panel) was used to evaluate blood flow in the lesion. No flow lesions were detected in the pancreatic cyst, suggesting accumulation of debris (arrow). (B) Contrast-enhanced EUS (right panel) was used to evaluate blood flow in the lesion. Flow is observed in the pancreatic cyst lesion, suggesting a tumor (arrowhead).

this method is insufficient. Multiple pancreatic juice cytology (75%) has a higher diagnostic accuracy than single pancreatic juice cytology (38%) (Nakamura et al., 2023; Ikemoto et al., 2021). Therefore, performing multiple cytological examinations after ENPD tube placement enables an accurate pathological diagnosis and confirmation of the causes of pancreatic duct stricture and AP.

10.6 Pancreatic stone (chronic pancreatitis)

Pancreatitis can recur because of pancreatic duct strictures and pancreatic stones, which are associated with chronic pancreatitis. Stent placement for MPD strictures can improve and prevent the recurrence of pancreatitis and symptoms associated with chronic pancreatitis, including abdominal pain. In clinical practice, the plastic stent can be gradually thickened to 10-Fr plastic stent to dilate the MPD stricture.

Stent placement should be continued for >1 year with regular exchanges. If MPD dilation is inadequate or the plastic stent is difficult to remove, multiple plastic or metallic stents should be considered. Multiple PS stent placements are effective in improving MPD strictures and abdominal symptoms because of sufficient pancreatic duct dilation (Sofi et al., 2021; Costamagna et al., 2006). The efficacy of metallic stents for pancreatic ducts has also been reported, with a high efficacy of 82.5% in improving abdominal symptoms (Shen et al., 2014). When abdominal symptoms do not improve after stent placement, surgical treatment, including Frey's procedure, pancreatectomy, and pancreatic duct jejunal anastomosis, is performed. Surgical treatment can achieve complete resolution in 62.5% of patients with abdominal symptoms (Binmoeller et al., 1995). Other reports have also indicated that endoscopic pancreatic stent placement had limited long-term results; however, surgical treatment was effective in these patients (Smits et al., 1995).

An analysis of the effect of pancreatic stent placement before pancreatic surgery reported no differences in postoperative adverse events or abdominal symptoms (Boerma et al., 2002). Therefore, considering the clinical strategy, endoscopic stent placement is the first choice for pancreatic strictures and stones; however, surgical treatment is recommended for cases in which endoscopic treatment is difficult to control for abdominal symptoms.

Extracorporeal shock wave lithotripsy (ESWL) is an established treatment for pancreatic duct stones. The effectiveness of ESWL was reported in 52.9% of patients who had complete pain control, 33.4% who had partial pain control, and the stone disappearance rate was 70.7%. The stone recurrence rate was 18.8% and the incidence of pancreatitis was 4.2% (Moole et al., 2016).

ESWL can crush hard pancreatic stones. Although endoscopic pancreatic stone removal carries the risk of stone basket impaction, the stones can be removed through transpapillary treatment after crushing. To avoid basket impaction, it is important for performing ESWL to crush the stone sufficiently sized (≤5-mm) for stone removal.

When ESWL is not effective, the electrohydraulic lithotripsy (EHL) for pancreatic stones using a peroral pancreatoscopy (POPS) is also an effective method (Fig. 10.4). Figure 10.4A shows complete obstruction by a pancreatic stone in the MPD (arrow). After insertion of a 9-Fr POPS into the MPD for visualization of the pancreatic stone, the stone was crushed and removed using an EHL. After setting the EHL probe on the pancreatic stone, the stone was crushed (Fig. 10.4B). After crushing the stone, it can be removed using a basket or balloon catheter through the main papilla.

A comparison between POPS and ESWL reported no difference in the effectiveness of pancreatic stone treatment (Siranart et al., 2025). ESWL occasionally requires multiple sessions, which may result in a prolonged hospital stay. Considering the length of hospitalization and ability to directly remove stones using the transpapillary approach, POPS may be a useful option for the treatment of pancreatic stones.

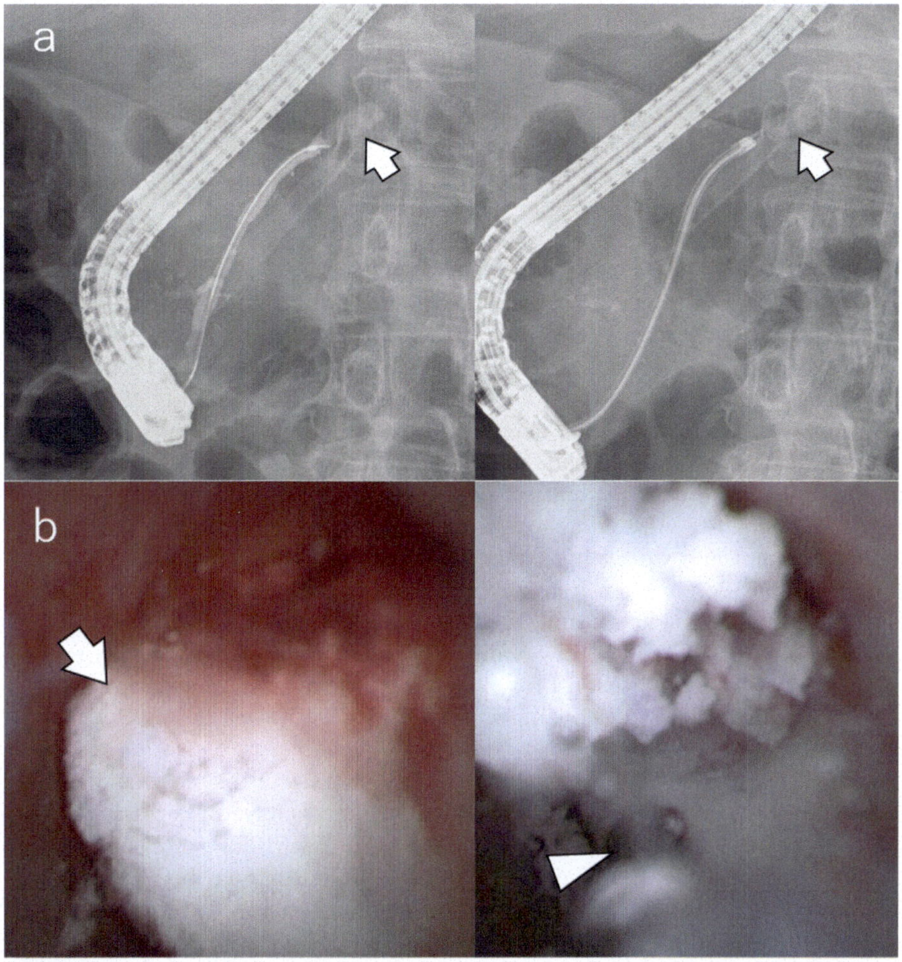

FIGURE 10.4

Treatment of a pancreatic stone. (A) Endoscopic retrograde pancreatography reveals that the pancreatic stone completely obstructs the main pancreatic duct (MPD) (arrow). Peroral pancreatoscopy (POPS) is performed to crush pancreatic stones using electrohydraulic lithotripsy (EHL). (B) POPS detected a pancreatic stone in the MPD (arrow). After setting the EHL probe (arrowhead), the pancreatic stone was crushed using the EHL.

10.7 Follow-up after pancreatic fluid collection treatment

PFC can occur after severe AP. To drain and treat the PFC, a plastic stent and/or lumen-apposing metal stent (LAMS) is inserted into the PFC. After stent placement, pancreatic fluid leakage from the ruptured MPD into the PFC cavity occurred. Therefore, a stent must be placed to ensure drainage and access to the PFC cavity in cases of reinfection with fluid retention. The timing of plastic stent removal was better after several months, considering reintervention and access to the cavity. Figure 10.5 shows the PFC recurrence. After treatment for walled-off necrosis (WON) using LAMS placement and endoscopic necrosectomy, a plastic stent was placed in the WON cavity. The PS was removed after 2 months. However, relapse of the PFC was observed (Fig. 10.5D). This may be attributable to the early timing of the removal, the cavity not disappearing, and the presence of a ruptured MPD. Therefore, the timing of the stent removal should be carefully considered. In cases of extensive pancreatic necrosis or damage to the MPD, pancreatic juice may not drain through the transpapillary route, and the pancreatic fluid may leak into the PFC cavity.

Disconnected pancreatic duct syndrome (DPDS) can occur in 16%–23% of WON cases (Lawrence et al., 2008; Nealon et al., 2009). DPDS is defined as a condition in which damage and necrosis of the MPD result in a loss of connection between the head and tail MPD. Leakage of pancreatic fluid from the ruptured position into the abdominal cavity (Fig. 10.6A) (Kozarek et al., 1991). DPDS is difficult to diagnose and can be overlooked, leading to inappropriate treatment and worsening prognosis. Therefore, the possibility of DPDS should be considered when treating patients with severe AP accompanied by pancreatic necrosis (Larsen and Kozarek, 2016). Evaluating the presence of DPDS is difficult, although DPDS can be suspected based on CT or MRCP findings (Fig. 10.6A). ERP is useful for evaluating the presence or absence of MPD rupture and DPDS and subsequent treatment selection (Rana et al., 2015; Bang et al., 2016).

Most cases of DPDS do not improve with conservative treatment and require endoscopic or surgical treatment. The endoscopic approach involves placing a stent to bridge the head and tail ends of the ruptured MPD (Fig. 10.6B) or placing a stent from the ruptured MPD to the fluid collection cavity (Nadkarni et al., 2015; Varadarajulu et al., 2013). However, although the transpapillary approach is effective in partial MPD ruptures with a success rate of 92%, in cases of complete MPD rupture, the success rate is only 20% because of the difficulty in accessing the caudal pancreas (Jang et al., 2016). Figure 10.6 shows a case of partial MPD rupture. Although partial MPD rupture cases have a high success rate, regular pancreatic duct stent replacement and dilation of the ruptured area are necessary because of fibrosis and narrowing of the ruptured area.

Surgical treatment was selected when endoscopic treatment was ineffective. Distal pancreatectomy, which includes ruptured MPD, is commonly performed, and the spleen is removed simultaneously owing to the extensive inflammation caused by AP (Maatman et al., 2020). Alternatively, when bypass surgery, including pancreaticoduodenal anastomosis of the distal pancreas, is available, pancreatic function can be preserved, and the incidence of adverse events is lower than that of pancreatic resection (Fischer et al., 2014; Dokmak et al., 2021). The surgical method was selected based on the patient's condition.

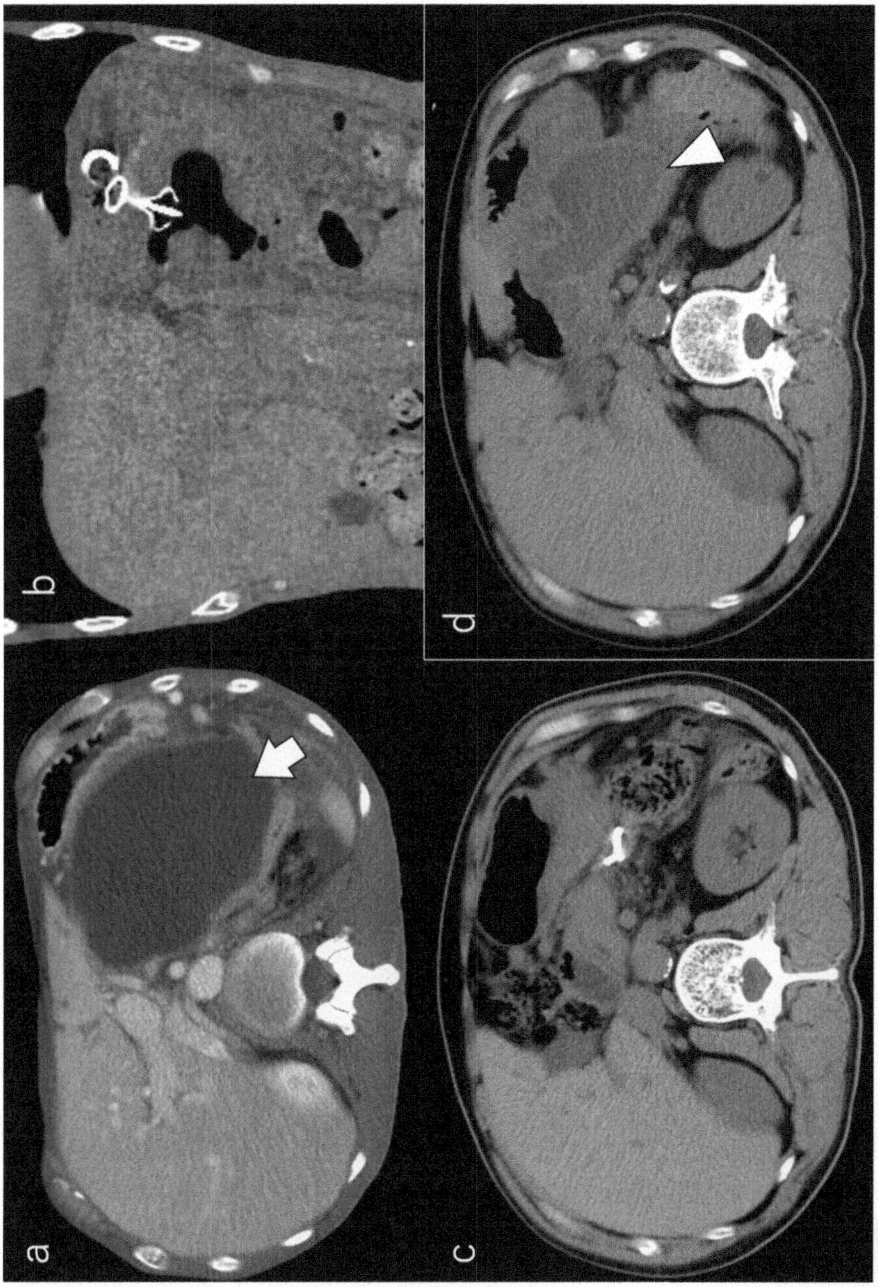

FIGURE 10.5

Follow-up of a pancreatic fluid collection. (A) Computed tomography revealed pancreatic fluid collection (PFC) in the tail caused by acute pancreatitis. (B) A lumen-apposing metal stent is placed between the gastric wall and PFC. (C) After the PFC treatment, a plastic stent (PS) was placed in the PFC cavity. The PS was removed after 2 months. (D) Relapse of the PFC was observed after PS removal. The PFC was then subjected to PS replacement.

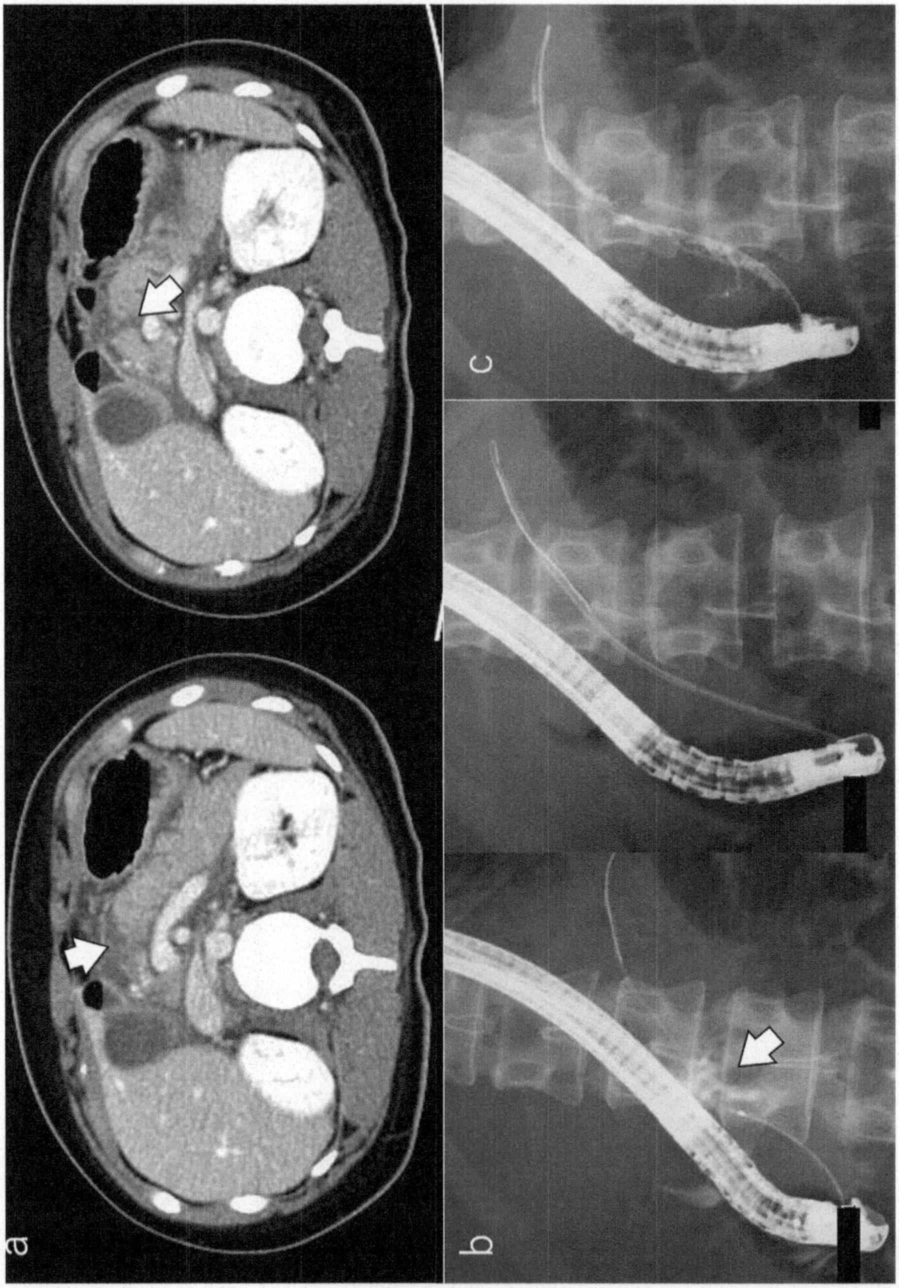

FIGURE 10.6

Disconnected pancreatic duct syndrome after acute pancreatitis. (A) Computed tomography image showing leakage of pancreatic juice from the disconnected head and body of the pancreatic duct (arrow) after acute pancreatitis. (B) Endoscopic retrograde pancreatography reveals leakage of the contrast medium from the head and body of the pancreas (arrow). A plastic stent is placed to bridge the disconnected pancreatic duct. (C) Six months later, the disconnected pancreatic ducts improved. The contrast medium did not leak.

10.8 Pancreaticointestinal anastomosis stricture after partial pancreatectomy

After partial pancreatectomy, such as pancreaticoduodenectomy, stricture of the pancreaticointestinal anastomosis can occur, causing obstructive pancreatitis. To treat the obstructive pancreatitis, endoscopic stent placement to the pancreatic duct anastomosis is required. ERP is typically performed by enteroscopy. However, endoscopic stent placement can be difficult because of complete anastomotic obstruction and/or intestinal deformation. To resolve these difficulties, the use of a transparent long cap and gel immersion endoscope has been reported (Kasai et al., 2022; Shimatani et al., 2023, 2024; Okuno et al., 2023).

However, when finding an anastomosis and stent placement is difficult even with these methods, EUS-pancreatic drainage (EUS-PD) is a good salvage technique. EUS-PD was performed using a 19-gauge or 22-gauge needle puncture of the MPD, followed by the placement of a guidewire into the MPD. After dilating the puncture route, a plastic stent was placed in the MPD (Fig. 10.7). After obstructive pancreatitis is treated using EUS-PD, routine stent exchange is required to manage the drainage route and avoid stent occlusion.

Stent removal in the EUS-PD route may cause MPD reobstruction; therefore, stent removal may be possible by reopening the pancreaticointestinal anastomosis. When the guidewire from the EUS-PD route passes through the anastomotic stricture, stent placement is possible using the rendezvous method (Fig. 10.8). Although anastomotic dilation is difficult to perform with multiple stents through the EUS-PD route, the rendezvous method can easily detect the anastomosis (Fig. 10.8A) and allows multiple stent placements (Fig. 10.8B), which will be sufficient for anastomotic dilation. If a stent is placed from the anastomosis to the distal end of the MPD, the EUS-PD route can be occluded naturally. After stent placement using the ERP procedure, the anastomosis can be managed using enteroscopy.

The timing of stent removal should be considered in each case to confirm the improvement in pancreatic stricture length and enlargement of the anastomosis. After removal, the patient should be followed up with CT and blood tests to evaluate the occurrence of anastomosis reobstruction.

10.9 Exo/endocrine pancreas function

AP can damage pancreatic tissue, causing postpancreatitis exocrine and endocrine dysfunction. In addition, long-term functional decline may occur, which can lead to chronic pancreatitis with 5–15 an occurrence rate (Takeyama, 2009; Ahmed Ali et al., 2016; Sankaran et al., 2015).

Concerning exocrine dysfunction, 62% of the hospitalized patients with AP and 35% of the follow-up patients demonstrated exocrine dysfunction. The frequency of exocrine dysfunction is higher in severe than in mild pancreatitis, and the frequency of alcohol-induced AP is higher than that of gallstones (Huang et al., 2019).

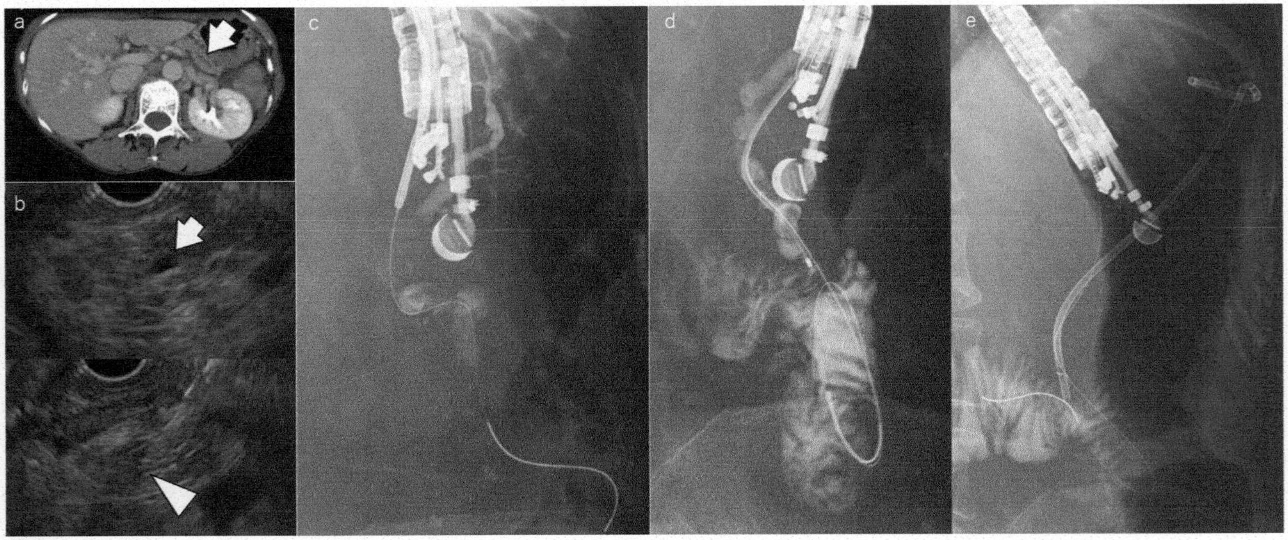

FIGURE 10.7

Treatment and follow-up of the obstructive pancreatitis after postpancreatectomy by endoscopic ultrasound-guided pancreatic duct drainage. (A) Computed tomography after pancreatectomy revealed obstructive pancreatitis and a dilated main pancreatic duct (arrow). (B) Using endoscopic ultrasonography, the slightly dilated main pancreatic duct (arrow) is punctured with a 22-gauge needle (arrowhead). (C) A 0.018-inch guidewire is inserted into the pancreatic duct. (D) The puncture route was dilated, and an additional 0.025-inch guidewire was placed in the pancreatic duct. e. A 7-Fr plastic stent is placed into the main pancreatic duct.

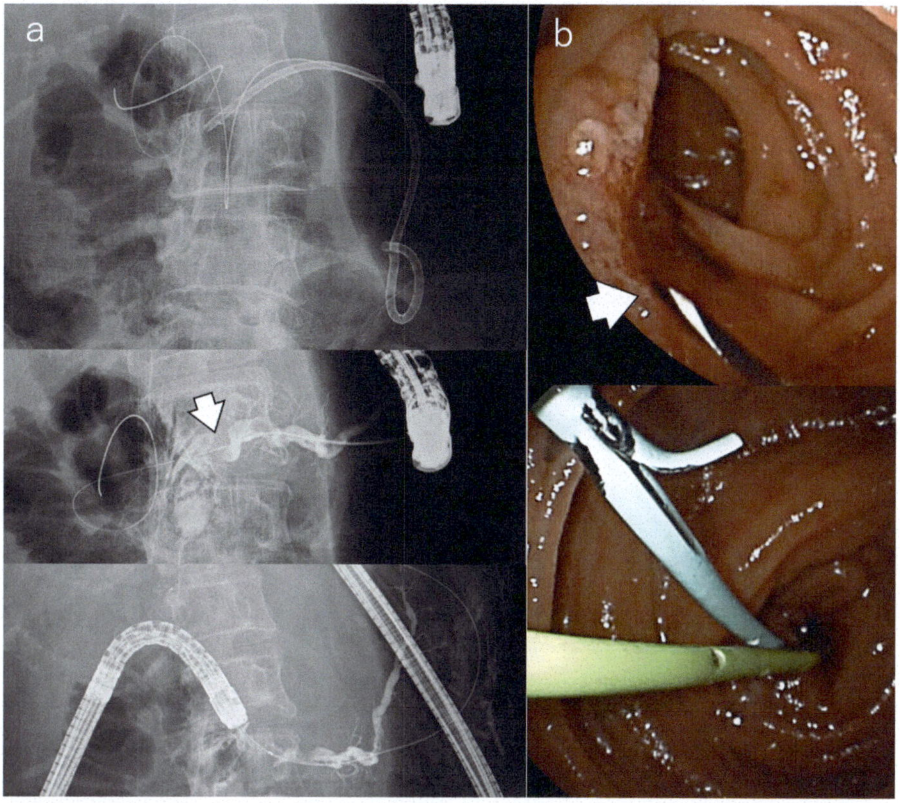

FIGURE 10.8

Pancreatic stent placement using the rendezvous method through endoscopic ultrasound-guided pancreatic duct drainage (EUS-PD) route. (A) The guidewire was placed in the intestine through the EUS-PD route for stent placement using the rendezvous method. (B) A guidewire was located in the intestine through a pancreaticointestinal anastomotic stricture. Double plastic stents were placed at the anastomotic stricture using an enteroscopy.

Exocrine dysfunction due to alcoholic pancreatitis or severe pancreatitis may cause gastrointestinal symptoms, including long-term postpancreatitis diarrhea (Hollemans et al., 2018). Inadequate food absorption leads to undernutrition, vitamin deficiency, and weight loss. Therefore, long-term follow-up and administration of digestive agents, such as Pancrelipase, should be considered. Malabsorption of fat leads to malabsorption of fat-soluble vitamins, resulting in decreased levels of fat-soluble vitamins (vitamins A, D, K, and E) (Nakamura et al., 1996). In the multicenter study on chronic pancreatitis, serum vitamin A and E levels were significantly reduced. However, nutritional markers including fat-soluble vitamins and prealbumin can improve in patients taking pancreatic digestive enzymes (Greer et al., 2019). Vitamin D deficiency is relatively common (Lankisch et al., 1986). Patients with chronic pancreatitis have decreased bone density and serum vitamin D (25(OH) D) levels, which are markers of osteoporosis (Greer et al., 2019; Sikkens et al., 2013; Duggan et al., 2015). No high-level evidence reports have investigated whether osteoporosis treatment is necessary for the treatment of pancreatic exocrine insufficiency. However, vitamin D and bisphosphonate preparations are expected to be effective. Therefore, when fatty stools are observed, fat-soluble vitamins should be measured.

Regarding endocrine dysfunction, the incidence of developing diabetes after AP is 15% within 1 year, and the risk increases further after 5 years (Das et al., 2014; Shen et al., 2015). Alcoholic AP is more common than other etiologies of pancreatitis in terms of the incidence of diabetes (Takeyama, 2009).

Thus, exo/endocrine dysfunction is highly possible, especially after alcoholic or severe pancreatitis, and follow-up observations including blood tests and consideration of medications are necessary.

References

Ahmed Ali, U., Issa, Y., Hagenaars, J.C., Bakker, O.J., van Goor, H., Nieuwenhuijs, V.B., Bollen, T.L., van Ramshorst, B., Witteman, B.J., Brink, M.A., Schaapherder, A.F., Dejong, C.H., Spanier, B.W.M., Heisterkamp, J., van der Harst, E., van Eijck, C.H., Besselink, M.G., Gooszen, H.G., van Santvoort, H.C., Boermeester, M.A., 2016. Risk of recurrent pancreatitis and progression to chronic pancreatitis after a first episode of acute pancreatitis. Clinical Gastroenterology and Hepatology 14 (5), 738–746. https://doi.org/10.1016/j.cgh.2015.12.040.

Bang, J., Navaneethan, U., Hasan, M., Hawes, R., Varadarajulu, S., 2016. EUS correlates of disconnected pancreatic duct syndrome in walled-off necrosis. Endoscopy International Open 04 (08), E883. https://doi.org/10.1055/s-0042-112586.

Bansal, P., Sonnenberg, A., 1995. Pancreatitis is a risk factor for pancreatic cancer. Gastroenterology 109 (1), 247–251. https://doi.org/10.1016/0016-5085(95)90291-0.

Binmoeller, K.F., Jue, P., Seifert, H., Nam, W.C., Izbicki, J., Soehendra, N., 1995. Endoscopic pancreatic stent drainage in chronic pancreatitis and a dominant stricture: long-term results. Endoscopy 27 (9), 638–644. https://doi.org/10.1055/s-2007-1005780.

Boerma, D., Van Gulik, T.M., Rauws, E.A.J., Obertop, H., Gouma, D.J., 2002. Outcome of pancreaticojejunostomy after previous endoscopic stenting in patients with chronic pancreatitis. European Journal of Surgery 168 (4), 223–228. https://doi.org/10.1080/11024150260102834.

Correa-Gallego, C., Miyasaka, Y., Hozaka, Y., Nishino, H., Kawamoto, M., Vieira, D.L., Ohtsuka, T., Wolfgang, C., 2023. Surveillance after resection of non-invasive intraductal papillary mucinous neoplasms (IPMN). A systematic review. Pancreatology 23 (3), 258–265. https://doi.org/10.1016/j.pan.2023.02.008.

Costamagna, G., Bulajic, M., Tringali, A., Pandolfi, M., Gabbrielli, A., Spada, C., Petruzziello, L., Familiari, P., Mutignani, M., 2006. Multiple stenting of refractory pancreatic duct strictures in severe chronic pancreatitis: long-term results. Endoscopy 38 (03), 254–259. https://doi.org/10.1055/s-2005-921069.

da Costa, D.W., Schepers, N.J., Bouwense, S.A., Hollemans, R.A., van Santvoort, H.C., Bollen, T.L., Consten, E.C., van Goor, H., Hofker, S., Gooszen, H.G., Boerma, D.,

Besselink, M.G., 2019. Predicting a 'difficult cholecystectomy' after mild gallstone pancreatitis. HPB 21 (7), 827–833. https://doi.org/10.1016/j.hpb.2018.10.015.

Das, S.L.M., Singh, P.P., Phillips, A.R.J., Murphy, R., Windsor, J.A., Petrov, M.S., 2014. Newly diagnosed diabetes mellitus after acute pancreatitis: a systematic review and meta-analysis. Gut 63 (5), 818–831. https://doi.org/10.1136/gutjnl-2013-305062.

Dokmak, S., Tetart, A., Aussilhou, B., Choquet, A., Rebours, V., Vullierme, M.P., Soubrane, O., Ruszniewski, P., Lévy, P., Sauvanet, A., 2021. French reconnection: a conservative pancreato-enteric reconnection for disconnected pancreatic duct syndrome. Pancreatology 21 (1), 282–290. https://doi.org/10.1016/j.pan.2020.10.045.

Duell, E.J., Lucenteforte, E., Olson, S.H., Bracci, P.M., Li, D., Risch, H.A., Silverman, D.T., Ji, B.T., Gallinger, S., Holly, E.A., Fontham, E.H., Maisonneuve, P., Bueno-de-Mesquita, H.B., Ghadirian, P., Kurtz, R.C., Ludwig, E., Yu, H., Lowenfels, A.B., Seminara, D., Petersen, G.M., La Vecchia, C., Boffetta, P., 2012. Pancreatitis and pancreatic cancer risk: a pooled analysis in the international pancreatic cancer case-control consortium (PanC4). Annals of Oncology 23 (11), 2964–2970. https://doi.org/10.1093/annonc/mds140.

Duggan, S.N., Purcell, C., Kilbane, M., O'Keane, M., McKenna, M., Gaffney, P., Ridgway, P.F., Boran, G., Conlon, K.C., 2015. An association between abnormal bone turnover, systemic inflammation, and osteoporosis in patients with chronic pancreatitis: a case-matched study. American Journal of Gastroenterology 110 (2), 336–345. https://doi.org/10.1038/ajg.2014.430.

Fischer, T.D., Gutman, D.S., Hughes, S.J., Trevino, J.G., Behrns, K.E., 2014. Disconnected pancreatic duct syndrome: disease classification and management strategies. Journal of the American College of Surgeons 219 (4), 704–712. https://doi.org/10.1016/j.jamcollsurg.2014.03.055.

Greer, J.B., Greer, P., Sandhu, B.S., Alkaade, S., Wilcox, C.M., Anderson, M.A., Sherman, S., Gardner, T.B., Lewis, M.D., Guda, N.M., Muniraj, T., Conwell, D., Cote, G.A., Forsmark, C.E., Banks, P.A., Tang, G., Stello, K., Gelrud, A., Brand, R.E., Slivka, A., Whitcomb, D.C., Yadav, D., 2019. Nutrition and inflammatory biomarkers in chronic pancreatitis patients. Nutrition in Clinical Practice 34 (3), 387–399. https://doi.org/10.1002/ncp.10186.

Hollemans, R.A., Hallensleben, N.D.L., Mager, D.J., Kelder, J.C., Besselink, M.G., Bruno, M.J., Verdonk, R.C., van Santvoort, H.C., 2018. Pancreatic exocrine insufficiency following acute pancreatitis: systematic review and study level meta-analysis. Pancreatology 18 (3), 253–262. https://doi.org/10.1016/j.pan.2018.02.009.

Huang, W., de la Iglesia-García, D., Baston-Rey, I., Calviño-Suarez, C., Lariño-Noia, J., Iglesias-Garcia, J., Shi, N., Zhang, X., Cai, W., Deng, L., Moore, D., Singh, V.K., Xia, Q., Windsor, J.A., Domínguez-Muñoz, J.E., Sutton, R., 2019. Exocrine pancreatic insufficiency following acute pancreatitis: systematic review and meta-analysis. Digestive Diseases and Sciences 64 (7), 1985–2005. https://doi.org/10.1007/s10620-019-05568-9.

Ikemoto, J., Serikawa, M., Hanada, K., Eguchi, N., Sasaki, T., Fujimoto, Y., Sugiyama, S., Yamaguchi, A., Noma, B., Kamigaki, M., Minami, T., Okazaki, A., Yukutake, M., Ishii, Y., Mouri, T., Shimizu, A., Tsuboi, T., Arihiro, K., Chayama, K., 2021. Clinical analysis of early-stage pancreatic cancer and proposal for a new diagnostic algorithm: a multicenter observational study. Diagnostics 11 (2), 287. https://doi.org/10.3390/diagnostics11020287.

Jang, J.W., Kim, M.H., Oh, D., Cho, D.H., Song, T.J., Park, D.H., Lee, S.S., Seo, D.W., Lee, S.K., Moon, S.H., 2016. Factors and outcomes associated with pancreatic duct disruption in patients with acute necrotizing pancreatitis. Pancreatology 16 (6), 958–965. https://doi.org/10.1016/j.pan.2016.09.009.

Jee, S.L., Jarmin, R., Lim, K.F., Raman, K., 2018. Outcomes of early versus delayed cholecystectomy in patients with mild to moderate acute biliary pancreatitis: a randomized prospective study. Asian Journal of Surgery 41 (1), 47–54. https://doi.org/10.1016/j.asjsur.2016.07.010.

Kasai, T., Shimatani, M., Mitsuyama, T., Yamashina, T., Kano, M., Yuba, T., Naganuma, M., 2022. Endoscopic approach for a pinhole-like benign stenosis in a pancreaticojejunal anastomosis using a double-balloon endoscope with a clear long cap attachment. Endoscopy 54 (S 02), E977. https://doi.org/10.1055/a-1883-9514.

Kirkegård, J., Cronin-Fenton, D., Heide-Jørgensen, U., Mortensen, F.V., 2018. Acute pancreatitis and pancreatic cancer risk: a nationwide matched-cohort study in Denmark. Gastroenterology 154 (6), 1729–1736. https://doi.org/10.1053/j.gastro.2018.02.011.

Kozarek, R.A., Ball, T.J., Patterson, D.J., Freeny, P.C., Ryan, J.A., William Traverso, L., 1991. Endoscopic transpapillary therapy for disrupted pancreatic duct and peripancreatic fluid collections. Gastroenterology 100 (5), 1362–1370. https://doi.org/10.1016/0016-5085(91)90790-r.

Lankisch, P.G., Lembcke, B., Wemken, G., Creutzfeldt, W., 1986. Functional reserve capacity of the exocrine pancreas. Digestion 35 (3), 175–181. https://doi.org/10.1159/000199364.

Larsen, M., Kozarek, R.A., 2016. Management of disconnected pancreatic duct syndrome. Current Treatment Options in Gastroenterology 14 (3), 348–359. https://doi.org/10.1007/s11938-016-0098-7.

Lawrence, C., Howell, D.A., Stefan, A.M., Conklin, D.E., Lukens, F.J., Martin, R.F., Landes, A., Benz, B., 2008. Disconnected pancreatic tail syndrome: potential for endoscopic therapy and results of long-term follow-up. Gastrointestinal Endoscopy 67 (4), 673–679. https://doi.org/10.1016/j.gie.2007.07.017.

Lee, Y.T., Chan, F.K.L., Leung, W.K., Chan, H.L.Y., Wu, J.C.Y., Yung, M.Y., Ng, E.K.W., Lau, J.Y.W., Sung, J.J.Y., 2008. Comparison of EUS and ERCP in the investigation with suspected biliary obstruction caused by choledocholithiasis: a randomized study. Gastrointestinal Endoscopy 67 (4), 660–668. https://doi.org/10.1016/j.gie.2007.07.025.

Liu, C.L., Fan, S.T., Lo, C.M., Tso, W.K., Wong, Y., Poon, R.T.P., Lam, C.M., Wong, B.C., Wong, J., 2005. Comparison of early endoscopic ultrasonography and endoscopic retrograde cholangiopancreatography in the management of acute biliary pancreatitis: a prospective randomized study. Clinical Gastroenterology and Hepatology 3 (12), 1238–1244. https://doi.org/10.1016/S1542-3565(05)00619-1.

Lu, Z., Li, M., Guo, F., Zhang, G., Song, S., Liu, N., Wang, D., 2020. Timely reduction of triglyceride levels is associated with decreased persistent organ failure in hypertriglyceridemic pancreatitis. Pancreas 49 (1), 105–110. https://doi.org/10.1097/mpa.0000000000001463.

Lyu, Y.X., Cheng, Y.X., Jin, H.F., Jin, X., Cheng, B., Lu, D., 2018. Same-admission versus delayed cholecystectomy for mild acute biliary pancreatitis: a systematic review and meta-analysis. BMC Surgery 18 (1). https://doi.org/10.1186/s12893-018-0445-9. Available from: www.biomedcentral.com/bmcsurg.

Maatman, T.K., Roch, A.M., Lewellen, K.A., Heimberger, M.A., Ceppa, E.P., House, M.G., Nakeeb, A., Schmidt, C.M., Zyromski, N.J., 2020. Disconnected pancreatic duct syndrome: spectrum of operative management. Journal of Surgical Research 247, 297–303. https://doi.org/10.1016/j.jss.2019.09.068.

Masamune, A., Kikuta, K., Hamada, S., Tsuji, I., Takeyama, Y., Shimosegawa, T., Okazaki, K., Matsumoto, R., Kume, K., Miura, S., Takikawa, T., Hongo, S., Tanaka, Y., Sakai, T., Ueno, M., Unno, J., Toki, M., Nishimura, M., Kitamura, K., Kobashigawa, K., Ashida, K., Takahashi, K., Watanabe, T., Sato, Y., Satoh, A., Yoshida, N., Tanaka, K., Kondo, H., Suzuki, N., Sato, H., Mitoro, A., Minami, K., Iwasaki, E., Okuwaki, K., Mukai, T., Kudo, Y., Azuma, S., Takeda, Y., Setsuda, M., Tsuchida, K., Irisawa, A., Motoya, D., Shirahata, N., Iwata, K., Oza, N., Kawaguchi, Y., Hoshino, A., Hirano, A., Sakakihara, I., Watanabe, N., Funayama, H., Nakamura, Y., Yamamoto, T., Uchida, H.H., Iizawa, Y., Imamura, M., Sekine, M., Inui, K., Yamamoto, S., Sato, S., Ishii, S., Yakushijin, T., Yamamoto, K., Masaka, T., Yoshida, M., Arizumi, T., Goto, M., Takenaka, M., Miyakawa, H., Otsuka, Y., Nishi, M., Suwa, K., Sakagami, J., Satoh, M., Koga, H., Takagi, T., Nagahama, M., Hatayama, K., Kaneko, N., Mori, M., Nakai, Y., Iwashita, Y., Iwatsuka, K., Hayashi, K., Terai, S., Kaneto, H., Nishikawa, T., Uchi, M., Honjo, M., Takagi, T., Mizukami, K., Tamura, T., Gushima, T., Sato, A., Shimizu, K., Aruga, Y., Nonogaki, K., Kaino, M., Iwashita, T., Tsuruoka, M., Miura, M., Hasegawa, H., Takai, T., Yokota, H., Fujisawa, T., Uetsuki, K., Kobayashi, T., yamaji, N., Itoi, T., Muraki, T., Watanabe, T., Inatomi, O., Ushijima, T., Takehara, Y., Kawamura, H., Fukasawa, M., Hagiwara, H., Tamura, H., Moriya, T., Kuwatani, M., Kawaji, Y., Kamisawa, T., Abue, M., Fukunishi, M., Onoyama, H., Mochizuki, S., Sumimoto, K., Koyabu, M., Nishio, A., Sakao, M., Imamura, Y., Ohta, H., Nishimura, N., Shiihara, M., Anan, H., Motoya, M., Chiba, M., Deguchi, T., Kofunato, Y., Sawada, N., Katada, K., Matsumoto, S., Higashide, S., Okano, H., Ishida, M., Yamazaki, M., Tamura, N., Muraoka, T., Uza, N., Kodama, Y., Muro, S., Fujimori, N., Ikeda, T., Shimokawahara, N., Otsu, N., Yano, Y., Sugawara, S., Takaura, K., Ogura, Y., Furuuchi, T., Shibasaki, M., Kishimoto, M., Ishida, M., 2020. Clinical practice of acute pancreatitis in Japan: an analysis of nationwide epidemiological survey in 2016. Pancreatology 20 (4), 629–636. https://doi.org/10.1016/j.pan.2020.04.013.

Moole, H., Jaeger, A., Bechtold, M.L., Forcione, D., Taneja, D., Puli, S.R., 2016. Success of extracorporeal shock wave lithotripsy in chronic calcific pancreatitis management: a meta-analysis and systematic review. Pancreas 45 (5), 651–658. https://doi.org/10.1097/MPA.0000000000000512.

Moon, J.H., Cho, Y.D., Cha, S.W., Cheon, Y.K., Ahn, H.C., Kim, Y.S., Kim, Y.S., Lee, J.S., Lee, M.S., Lee, H.K., Shim, C.S., Kim, B.S., 2005. The detection of bile duct stones in suspected biliary pancreatitis: comparison of MRCP, ERCP, and intraductal US. American Journal of Gastroenterology 100 (5), 1051–1057. https://doi.org/10.1111/j.1572-0241.2005.41057.x.

Morales-Oyarvide, V., Mino-Kenudson, M., Ferrone, C.R., Gonzalez-Gonzalez, L.A., Warshaw, A.L., Lillemoe, K.D., Fernández-Del Castillo, C., 2015. Acute pancreatitis in intraductal papillary mucinous neoplasms: a common predictor of malignant intestinal subtype. Surgery 158 (5), 1219–1225. https://doi.org/10.1016/j.surg.2015.04.029.

Nadkarni, N.A., Kotwal, V., Sarr, M.G., Vege, S.S., 2015. Disconnected pancreatic duct syndrome endoscopic stent or surgeon's knife? Pancreas 44 (1), 16–22. https://doi.org/10.1097/MPA.0000000000000216.

Nakamura, T., Takebe, K., Imamura, K.I., Tando, Y., Yamada, N., Arai, Y., Terada, A., Ishii, M., Kikuchi, H., Suda, T., 1996. Fat-soluble vitamins in patients with chronic pancreatitis (pancreatic insufficiency). Acta Gastro-Enterologica Belgica 59 (1), 10–14.

Nakamura, S., Ishii, Y., Serikawa, M., Hanada, K., Eguchi, N., Sasaki, T., Fujimoto, Y., Yamaguchi, A., Sugiyama, S., Noma, B., Kamigaki, M., Minami, T., Okazaki, A., Yukutake, M., Mouri, T., Tatsukawa, Y., Ikemoto, J., Arihiro, K., Oka, S., 2023. Diagnostic ability and safety of repeated pancreatic juice cytology using an endoscopic naso-pancreatic drainage catheter for pancreatic ductal adenocarcinoma: a multicenter prospective study. Diagnostics 13 (16), 2696. https://doi.org/10.3390/diagnostics13162696.

Nealon, W.H., Bawduniak, J., Walser, E.M., Pitt, H.A., Behrns, K.E., Stain, S.C., 2004. Appropriate timing of cholecystectomy in patients who present with moderate to severe gallstone-associated acute pancreatitis with peripancreatic fluid collections. Annals of Surgery 239 (6), 741–751. https://doi.org/10.1097/01.sla.0000128688.97556.94.

Nealon, W.H., Bhutani, M., Riall, T.S., Raju, G., Ozkan, O., Neilan, R., 2009. A unifying concept: pancreatic ductal anatomy both predicts and determines the major complications resulting from pancreatitis. Journal of the American College of Surgeons 208 (5), 790–799. https://doi.org/10.1016/j.jamcollsurg.2008.12.027.

Noel, R., Arnelo, U., Lundell, L., Hammarqvist, F., Jumaa, H., Enochsson, L., Sandblom, G., 2018. Index versus delayed cholecystectomy in mild gallstone pancreatitis: results of a randomized controlled trial. HPB 20 (10), 932–938. https://doi.org/10.1016/j.hpb.2018.03.016.

Okuno, N., Hara, K., Haba, S., Kuwahara, T., Kuraishi, Y., Yanaidani, T., 2023. Gel immersion radial incision and cutting for pancreaticojejunostomy anastomotic stricture. Endoscopy 55 (S 01), E696. https://doi.org/10.1055/a-2058-8461.

Pascual, I., Sanahuja, A., García, N., Vázquez, P., Moreno, O., Tosca, J., Peña, A., Garayoa, A., Lluch, P., Mora, F., 2019. Association of elevated serum triglyceride levels with a more severe course of acute pancreatitis: cohort analysis of 1457 patients. Pancreatology 19 (5), 623–629. https://doi.org/10.1016/j.pan.2019.06.006.

Polkowski, M., Regula, J., Tilszer, A., Butruk, E., 2007. Endoscopic ultrasound versus endoscopic retrograde cholangiography for patients with intermediate probability of bile duct stones: a randomized trial comparing two management strategies. Endoscopy 39 (04), 296–303. https://doi.org/10.1055/s-2007-966264.

Rana, S.S., Bhasin, D.K., Sharma, R., Gupta, R., 2015. Factors determining recurrence of fluid collections following migration of intended long term transmural stents in patients with walled off pancreatic necrosis and disconnected pancreatic duct syndrome. Endoscopic Ultrasound 4 (3), 208–212. https://doi.org/10.4103/2303-9027.162999.

Sankaran, S.J., Xiao, A.Y., Wu, L.M., Windsor, J.A., Forsmark, C.E., Petrov, M.S., 2015. Frequency of progression from acute to chronic pancreatitis and risk factors: a meta-analysis. Gastroenterology 149 (6), 1490. https://doi.org/10.1053/j.gastro.2015.07.066.

Shen, Y., Liu, M., Chen, M., Li, Y., Lu, Y., Zou, X., 2014. Covered metal stent or multiple plastic stents for refractory pancreatic ductal strictures in chronic pancreatitis: a systematic review. Pancreatology 14 (2), 87–90. https://doi.org/10.1016/j.pan.2013.12.005.

Shen, H.N., Yang, C.C., Chang, Y.H., Lu, C.L., Li, C.Y., 2015. Risk of diabetes mellitus after first-attack acute pancreatitis: a national population-based study. American Journal of Gastroenterology 110 (12), 1698–1706. https://doi.org/10.1038/ajg.2015.356.

Shimatani, M., Mitsuyama, T., Yamashina, T., 2023. New treatment for completely obstructed pancreaticojejunal anastomosis using gel-immersion double-balloon ERP with long cap. Clinical Gastroenterology and Hepatology 21 (11), A25. https://doi.org/10.1016/j.cgh.2023.04.009.

Shimatani, M., Mitsuyama, T., Yamashina, T., Takeo, M., Horitani, S., Saito, N., Matsumoto, H., Orino, M., Kano, M., Yuba, T., Takayama, T., Nakagawa, T., Takayama, S., 2024. Advanced technical tips and recent insights in ERCP using balloon-assisted endoscopy. DEN Open 4 (1). https://doi.org/10.1002/deo2.301.

Sikkens, E.C.M., Cahen, D.L., Koch, A.D., Braat, H., Poley, J.W., Kuipers, E.J., Bruno, M.J., 2013. The prevalence of fat-soluble vitamin deficiencies and a decreased bone mass in patients with chronic pancreatitis. Pancreatology 13 (3), 238–242. https://doi.org/10.1016/j.pan.2013.02.008.

Siranart, N., Kozai, L., Simadibrata, D.M., Pornananrat, N., Roongphornchai, P., Pajareya, P., Worapongpaiboon, R., Phutinart, S., Dendumrongsup, W., Chumpangern, Y., Jaroenlapnopparat, A., Vantanasiri, K., Tantitanawat, K., 2025. Per-oral pancreatoscopy-guided lithotripsy versus extracorporeal shock wave lithotripsy in pancreatic stone: a meta-analysis. Digestive Diseases and Sciences 70 (7), 2506–2520. https://doi.org/10.1007/s10620-025-08952-w.

Smits, M.E., Badiga, S.M., Rauws, E.A.J., Tytgat, G.N.J., Huibregtse, K., 1995. Long-term results of pancreatic stents in chronic pancreatitis. Gastrointestinal Endoscopy 42 (5), 461–467. https://doi.org/10.1016/S0016-5107(95)70051-X.

Sofi, A.A., Khan, M.A., Ahmad, S., Khan, Z., Peerzada, M.M., Sunguk, J., Vargo, J., 2021. Comparison of clinical outcomes of multiple plastic stents and covered metal stent in refractory pancreatic ductal strictures in chronic pancreatitis- a systematic review and meta-analysis. Pancreatology 21 (5), 854–861. https://doi.org/10.1016/j.pan.2021.03.017.

Suzuki, M., Sekino, Y., Hosono, K., Yamamoto, K., Kawana, K., Nagase, H., Kubota, K., Nakajima, A., 2022. Endoscopic ultrasound versus magnetic resonance cholangiopancreatography for the diagnosis of computed tomography-negative common bile duct stone: prospective randomized controlled trial. Digestive Endoscopy 34 (5), 1052–1059. https://doi.org/10.1111/den.14193. Available from: http://onlinelibrary.wiley.com/journal/10.1111/(ISSN)1443-1661.

Takeyama, Y., 2009. Long-term prognosis of acute pancreatitis in Japan. Clinical Gastroenterology and Hepatology 7 (11), S15. https://doi.org/10.1016/j.cgh.2009.08.022.

Tsutsumi, K., Ohtsuka, T., Oda, Y., Sadakari, Y., Mori, Y., Aishima, S., Takahata, S., Nakamura, M., Mizumoto, K., Tanaka, M., 2011. A history of acute pancreatitis in intraductal papillary mucinous neoplasms of the pancreas is a potential predictive factor for malignant papillary subtype. Pancreatology 10 (6), 707–712. https://doi.org/10.1159/000320696.

Van Riet, P.A., Erler, N.S., Bruno, M.J., Cahen, D.L., 2021. Comparison of fine-needle aspiration and fine-needle biopsy devices for endoscopic ultrasound-guided sampling of solid lesions: a systemic review and meta-analysis. Endoscopy 53 (4), 411–423. https://doi.org/10.1055/a-1206-5552.

Varadarajulu, S., Rana, S.S., Bhasin, D.K., 2013. Endoscopic therapy for pancreatic duct leaks and disruptions. Gastrointestinal Endoscopy Clinics of North America 23 (4), 863–892. https://doi.org/10.1016/j.giec.2013.06.008.

Wang, S.S., Lin, X.Z., Tsai, Y.T., Lee, S.D., Pan, H.B., Chou, Y.H., Su, C.H., Lee, C.H., Shiesh, S.C., Lin, C.Y., Lin, H.C., 1988. Clinical significance of ultrasonography, computed tomography, and biochemical tests in the rapid diagnosis of gallstone-related pancreatitis: a prospective study. Pancreas 3 (2), 153–158. https://doi.org/10.1097/00006676-198804000-00007.

Yadav, D., O'Connell, M., Papachristou, G.I., 2012. Natural history following the first attack of acute pancreatitis. American Journal of Gastroenterology 107 (7), 1096–1103. https://doi.org/10.1038/ajg.2012.126.

CHAPTER

Pharmacotherapy for pancreatitis 11

Keiko Suzuki[1,2] and Akio Suzuki[3,4]

[1]*Gifu University Graduate School of Medicine, Department of Emergency and Disaster Medicine, Gifu, Japan;* [2]*Gifu Pharmaceutical University, Practical Pharmaceutical Research Center, Gifu, Japan;* [3]*Gifu University Hospital, Department of Pharmacy, Gifu, Japan;* [4]*Gifu Pharmaceutical University, Laboratory of Advanced Medical Pharmacy, Gifu, Japan*

11.1 Fluid resuscitation

11.1.1 Overview

Acute pancreatitis (AP) is characterized by pancreatic, peripancreatic, and systemic edema, predominantly caused by increased vascular permeability secondary to systemic inflammatory responses and endothelial injury (Singh and Garg, 2016; Dumnicka et al., 2017). These pathophysiological changes result in third-space fluid sequestration and subsequent intravascular volume depletion. Intravascular hypovolemia leads to pancreatic ischemia, exacerbation of systemic inflammation, and an increased risk of complications, including pancreatic necrosis, organ dysfunction, infectious events, and mortality (Takeda et al., 2005; Yegneswaran et al., 2011; Besselink et al., 2009; Mofidi et al., 2006). Several animal studies have demonstrated that regional hypoperfusion of the pancreas is associated with necrosis and may be corrected by fluid resuscitation (Kusterer et al., 1991; Juvonen et al., 1999; Kinnala et al., 2002). These findings are supported by clinical evidence demonstrating that early and adequate fluid resuscitation improves pancreatic perfusion and reduces the risk of progression to severe disease (Trikudanathan et al., 2012; de-Madaria et al., 2011). Maintenance of adequate intravascular volume is therefore a cornerstone of early management to preserve pancreatic microcirculation and reduce the risk of progression to severe disease. Current clinical practice guidelines recommend initiating fluid resuscitation as early as possible after the diagnosis of AP (Tenner et al., 2024; Crockett et al., 2018; Párniczky et al., 2025; Takada et al., 2022).

11.1.2 The type of fluid

The type of fluid recommended for initial resuscitation is an isotonic crystalloid solution, typically lactated Ringer's solution (Tenner et al., 2024; Párniczky et al., 2025; Takada et al., 2022). A meta-analysis showed that these significantly

Advanced Insights into Acute Pancreatitis. https://doi.org/10.1016/B978-0-443-44736-5.00001-2

decreased the odds of moderately severe/severe AP, the rate of intensive care unit (ICU) admission, local complications, and the length of hospital stay compared to normal saline (Hong et al., 2024). In contrast, colloid solutions are generally not recommended due to a lack of clear benefit, including no improvement in mortality compared with fluid resuscitation without hydroxyethyl starch (HES) and a significant increase in potential risk of multiple organ failure (Di Martino et al., 2021; Dawson et al., 2023).

11.1.3 Infusion rate of initial fluid resuscitation

The optimal infusion rate of initial fluid resuscitation to maximize clinical benefit while avoiding iatrogenic fluid overload has long been debated. High-rate fluid therapy has the potential to cause fluid overload, precipitating or exacerbating heart and respiratory failure. A randomized clinical trial of 249 patients with AP compared goal-directed aggressive resuscitation (20 mL/kg bolus followed by 3 mL/kg/h) with moderate resuscitation (10 mL/kg bolus in hypovolemic patients or no bolus in normovolemic patients, followed by 1.5 mL/kg/h) using lactated Ringer's solution (de-Madaria et al., 2022). The trial found no significant difference between the two groups in the incidence of moderately severe or severe pancreatitis (aggressive resuscitation: 22.1% vs. moderate resuscitation: 17.3%; adjusted relative risk [RR]: 1.30; 95% confidence interval [CI]: 0.78–2.18; $P = .32$), and aggressive resuscitation was associated with a higher incidence of fluid overload (aggressive resuscitation: 20.5% vs. moderate resuscitation: 6.3%; adjusted RR: 2.85; 95% CI: 1.36–5.94, $P = .004$). Moreover, a systematic review and a meta-analysis both revealed that aggressive fluid resuscitation is associated with higher mortality compared with moderate fluid replacement in patients with AP (Ding and Chen, 2023; Dawson et al., 2023). Therefore, moderate fluid resuscitation (1.5 mL/kg/h) with an individualized goal-directed approach is recommended (Párniczky et al., 2025). The strategy for fluid resuscitation should be tailored to the individual risk profile (e.g., age, comorbidities) and, in cases requiring ICU admission, to individual resuscitation goals using a combination of clinical signs, physical examination, vital signs (e.g., mean arterial pressure (MAP) 65–85 mmHg, heart rate $<$100/min), urine output (e.g., $>$0.5 mL/kg/h), biochemical markers (e.g., blood urea nitrogen (BUN) $<$ 20 mg/dL, hematocrit $<$44%), imaging results, and advanced (non-)invasive targets (Párniczky et al., 2025).

11.2 Antimicrobial therapy

11.2.1 Overview

In AP, infectious complications of hospital-acquired infections, such as bacteremia, pneumonia, cholangitis, urinary tract infection, or later infection of infected pancreatic necrosis (IPN), are a major determinant of mortality (Boxhoorn et al., 2020). Among these, IPN is particularly critical, with a high mortality rate of about 30% (Banks et al., 2006). AP is initiated by autodigestion of the pancreas due to premature activation of digestive enzymes, leading to a sterile inflammatory response in the early phase, but this can become infected over time (Besselink et al., 2009; Banks et al., 2013). This is thought to result from bacterial translocation from the gut to the adjacent necrotic pancreatic parenchyma (Banks et al., 2013). It has been reported that hospital-acquired infections, such as bacteremia and pneumonia, tend to occur relatively early in hospitalization (median day 7), whereas IPN typically develops later, after 2 weeks of admission (median day 26) (Besselink et al., 2009).

Contemporary clinical guidelines unanimously recommend against the use of routine prophylactic antibiotics, even in patients with necrotizing pancreatitis, due to the absence of demonstrated benefit in preventing infection or reducing mortality (Tenner et al., 2024; Crockett et al., 2018; Párniczky et al., 2025; Takada et al., 2022). On the other hand, antibiotic therapy should be initiated as soon as possible when infection is suspected or confirmed on the basis of clinical, laboratory, or imaging evidence (Howell and Davis, 2017).

11.2.2 Indications for antimicrobial therapy

Early in the clinical course of AP, patients may also develop systemic inflammatory response syndrome (SIRS), which can be difficult to distinguish from sepsis because both present with fever, tachycardia, tachypnea, and leukocytosis. When infection is suspected, antibiotics should be initiated while the source is being investigated.

On the other hand, IPN may be suspected if any of the following are present: 1. Positive microbiologic culture from body fluids, for example, blood, sputum, bile, urine, and drain fluid, and 2. presence of gas bubbles within the pancreatic/peripancreatic necrotic collection on a computed tomography (CT) scan (Párniczky et al., 2025). Procalcitonin levels help guide whether to initiate antibiotic therapy in suspected infection. Comparisons of fine-needle aspiration (FNA) and clinical or imaging/blood-based assessments have shown similar diagnostic accuracy for IPN (86% vs. 87%, $P = .83$) (Van Baal et al., 2014), with FNA sensitivity and specificity of 88%–98% and 90%–100%, and clinical/imaging assessments of 45%–100% and 55%–93%, respectively (Chen et al., 2017; Ding et al., 2019; Ji et al., 2016; Rau et al., 1998; He et al., 2017; Moran et al., 2016). CT–or endoscopic ultrasound (EUS)-guided FNA for Gram stain and culture is not recommended (Tenner et al., 2024; Crockett et al., 2018; Párniczky et al., 2025; Takada et al., 2022) and may carry the risk of contaminating sterile collections.

11.2.3 Antimicrobial management

The causative organisms of IPN commonly include Enterobacterales, Enterococcus species, *Staphylococcus aureus*, Staphylococcus epidermidis, anaerobic bacteria, and Candida species (Gilbert et al., 2025). When IPN is suspected, initiation of broad-spectrum intravenous antimicrobial therapy with sufficient penetration into necrotic pancreatic tissue should be initiated promptly. Recommended antibiotics for empiric therapy include carbapenems (e.g., meropenem, imipenem/cilastatin), piperacillin/tazobactam, or a third-generation cephalosporin (e.g., cefepime) or fluoroquinolone (e.g., levofloxacin, moxifloxacin) combined with an anti-anaerobic drug such as metronidazole (Wolbrink et al., 2020). Antifungal therapy is not recommended empirically and should be restricted to confirmed or a high suspicion of fungal infection, which is generally observed in patients with a prolonged ICU stay, prior broad-spectrum antibiotic exposure, presence of central venous catheters, or receipt of total parenteral nutrition (Párniczky et al., 2025; Baron et al., 2020). Echinocandins are recommended as preemptive therapy in patients with a high suspicion of fungal involvement or when Candida is detected in necrotic tissue (Pappas et al., 2016).

Critically ill patients show highly variable antibiotic pharmacokinetics due to alterations in fluid status, serum albumin levels, organ function, and microvascular perfusion, and antibiotic dosing needs to be carefully optimized based on the patient's clinical condition (Roberts and Lipman, 2009). For example, in sepsis or SIRS, augmented renal clearance and an increased volume of distribution often necessitate full, nondose-reduced loading doses and standard or higher maintenance dosing. In contrast, the later development of organ dysfunction, including renal and hepatic impairment, or the need for continuous renal replacement therapy (CRRT) may require dose reduction or an extension of dosing intervals (or both). Accordingly, antibiotic dosing needs to be carefully optimized based on the patient's clinical condition. Table 11.1 shows dose adjustments for antibiotics in the presence of hepatic and renal dysfunction, while Table 11.2 shows recommended antibiotic dosing during blood purification therapies (Gilbert et al., 2025).

The duration of antimicrobial therapy should be determined by the nature of the underlying infection and the adequacy of source control rather than by any predefined course length. In IPN, antibiotics are generally continued for at least 2 weeks, and discontinuation is considered after 5–7 days if cultures become negative and clinical signs of infection have resolved (Sahu et al., 2025). For hospital-acquired infections complicating AP—such as cholangitis, pneumonia, or urinary tract infection—management should follow established, disease-specific guidelines.

Table 11.1 Dosage adjustments of common antimicrobial agents in renal and hepatic impairment.[a]

	Half-life (hour) (normal renal function)	Standard dosage (normal renal function)	Ccr (mL/min)	Dosage adjustment for hepatic impairment
Meropenem	1	1 g i.v. every 8 h	Ccr ≥50: Do not adjust dosage Ccr 26–50: 1 g every 12 h Ccr 10–25:0.5 g every 12 h Ccr <10:0.5 g every 24 h	Child-Pugh A–C: Do not adjust dosage
Imipenem/ Cilastatin	1	500 mg – 1 g i.v. every 6 h	500 mg every 6 h: Ccr ≥90: Do not adjust dosage Ccr <60–90:400 mg every 6 h Ccr <30–60:300 mg every 6 h Ccr <15–30:200 mg every 6 h 1 g every 6 h: Ccr ≥90: Do not adjust dosage Ccr <60–90:750 mg every 8 h Ccr <30–60:500 mg every 6 h Ccr <15–30:500 mg every 12 h	Child-Pugh A–C: Do not adjust dosage
Levofloxacin	7	750 mg p.o. or i.v. every 24 h	Ccr >50: Do not adjust dosage Ccr 20–49:750 mg every 48 h Ccr <20:750 mg once, then 500 mg every 48 h	Child-Pugh A–C: Do not adjust dosage
Moxifloxacin	10–14	400 mg p.o. or i.v. every 24 h	Do not adjust dosage	Child-Pugh A–C: Do not adjust dosage
Metronidazole	6–14	7.5 mg/kg p.o. or i.v. every 6 h	Ccr ≥10: Do not adjust dosage Ccr <10:7.5 mg/kg every 12 h	Child-Pugh A–B: Do not adjust dosage Child-Pugh C: 50% reduction
Tazobactam/ Piperacillin (Long-term)	PIPC: 1 TAZ: 1	Long-term i.v. 4.5 g i.v. (over 4 h or longer) every 8 h	Long-term i.v. Ccr >20: Do not adjust dosage Ccr <20:4.5 g (over 4 h or longer) every 12 h	Child-Pugh A–C: Do not adjust dosage
Cefepime	2	2 g i.v. every 8 h	Ccr >60: Do not adjust dosage Ccr 30–60:2 g every 12 h Ccr 11–29:2 g every 24 h Ccr <10:1 g every 24 h	Child-Pugh A–C: Do not adjust dosage
Micafungin	15–17	100 mg i.v. every 24 h	Do not adjust dosage	
Caspofungin	13	50 mg i.v. every 24 h	Do not adjust dosage	

[a] i.v., *intravenous;* p.o., *per os;* Ccr, *creatinine clearance. Based on information from The Sanford Guide to Antimicrobial Therapy.*

Table 11.2 Dosing recommendations for antimicrobial agents in continuous and intermittent renal replacement therapies.[a]

	HD	CAPD	CRRT	SLED
Meropenem	0.5 g every 24 h (Administer after HD on dialysis days)	0.5 g every 24 h	CVVH, CVVHD: 0.75–1 g every 8 h (Stasi et al., 2023)	0.5 g every 8 h
Imipenem/Cilastatin	Target dose 500 mg Every 6 h: 200 mg every 6 h (Administer after HD) Target dose 1 g Every 6 h: 500 mg every 12 h (administer after HD)	No data	0.5–1 g every 12 h (depends on MIC and purification volume) (Fish et al., 2005)	No data
Levofloxacin	750 mg once, then 500 mg every 48 h	750 mg once, then 500 mg every 48 h	750 mg once, then 500 mg every 48 h	No data
Moxifloxacin	Do not adjust dosage	Do not adjust dosage	Do not adjust dosage	Do not adjust dosage
Metronidazole	Do not adjust dosage	Do not adjust dosage	Do not adjust dosage	No data
Tazobactam/Piperacillin (long-term)	3.375 g every 12 h (over 4 h or longer) every 12 h	No data	4.5 g (over 4 h or longer) every 8 h	No data
Cefepime	1 g every 24 h (administer after HD)	2 g every 48 h	2 g every 12–24 h	No data
Micafungin	Do not adjust dosage	Do not adjust dosage	Consider 150–200 mg every 24 h (Honore et al., 2018)	No data
Caspofungin	Do not adjust dosage	Do not adjust dosage	Do not adjust dosage	No data

[a] HD, *hemodialysis;* CAPD, *continuous ambulatory peritoneal dialysis;* CRRT, *continuous renal replacement therapy;* SLED, *sustained low-efficiency dialysis;* CVVH, *continuous veno-venous hemofiltration;* CVVHD, *continuous veno-venous hemodialysis. Based on information from The Sanford Guide to Antimicrobial Therapy.*

11.3 Analgesic therapy

11.3.1 Overview

Abdominal pain is the most common and burdensome symptom of AP (Földi et al., 2022). Severe abdominal pain arises from a combination of pancreatic inflammation, capsular distension, peripancreatic edema, and neuropathic irritation (Wu et al., 2023). Effective analgesia may not only prevent a reduction in quality of life (QOL) but may also prevent respiratory compromise, reduce sympathetic overactivation, and facilitate early mobilization, all factors that significantly influence clinical trajectory (Herranz Prinz et al., 2022). Pain management is therefore a central component of the early treatment of AP.

Pharmacological therapies include nonopioid analgesic acetaminophen, nonsteroidal anti-inflammatory drugs (NSAIDs), opioids, and local anesthetics. Major international and national guidelines for AP do not mandate a single "first-line" analgesic (Párniczky et al., 2025; Takada et al., 2022). Table 11.3 summarizes the evidence regarding the effectiveness of various analgesic agents in AP. It is therefore recommended to tailor analgesic choice—nonopioid or opioid, route, and dosing—according to pain intensity, organ function, and evolving clinical course.

11.3.2 Opioids

Opioids exert their analgesic effect by inhibiting neurotransmitter release in the spinal cord and modulating pain pathways through mu (μ), kappa (κ), and delta (δ) receptors (Kanjhan, 1995). Clinically used opioids are categorized as full agonists (e. g., morphine, fentanyl), partial agonists (e.g., buprenorphine), and agonist–antagonists (e.g., pentazocine). Opioids remain an important option for managing moderate to severe pain in AP because of their potent analgesic effects and rapid onset of action, particularly for visceral pain (Pandanaboyana et al., 2024). Accordingly, their use is recommended to achieve effective pain control in AP patients with severe pain (Párniczky et al., 2025; Takada et al., 2022). Two meta-analyses have shown that opioids and nonsteroidal anti-inflammatory drug (NSAIDs) provide comparable efficacy and safety in mild AP, although opioid analgesics are associated with a lower need for rescue analgesics (Thavanesan et al., 2022; Nelson et al., 2022). Similarly, two randomized controlled trials that compared intravenous opioid analgesics (either pentazocine or buprenorphine) with intravenous diclofenac (an NSAID) showed that buprenorphine is more effective and equally safe for analgesia in AP patients, including those with moderately severe or severe disease (Saini et al., 2024; Mahapatra et al., 2019).

Opioids have been avoided in AP due to concerns about sphincter of Oddi spasm, but recent evidence shows that they can be used safely without increasing sphincter-related adverse events (Gülen et al., 2016). Additionally, opioids have adverse effects such as respiratory depression, constipation, and potential dependence, but their short-term use in AP is generally considered safe based on

Table 11.3 Pharmacologic analgesia for acute pancreatitis: Summary of reported evidence.

Drug/comparison	Dose/regimen	Key findings	References
NSAIDs (WHO analgesic ladder)	Low-potency NSAIDs first; escalate to high-potency NSAIDs or NSAIDs + opioids	Mild–moderate AP pain often controlled by NSAIDs alone; step-up approach recommended	Banks et al. (2006), Banks (1997), Mayerle et al. (2005)
Morphine	Morphine 10 mg SC every 4 h	Historically criticized for sphincter of Oddi dysfunction, but no significant worsening of AP course proven	Peiró et al. (2008), Helm et al. (1988)
Metamizole versus morphine	Metamizole 2 g/8 h IV versus morphine 10 mg/4 h SC	Metamizole produced slightly more frequent and faster pain relief than SC morphine	Peiró et al. (2008)
Buprenorphine versus Pethidine (Historical reference to pethidine)	Not fully stated (Long-acting buprenorphine versus pethidine)	Historically, pethidine was recommended as the first-line analgesic for AP. Blamey et al. Demonstrated that buprenorphine provided analgesia comparable to pethidine, while posing a lower risk of physical opioid dependence.	Helm et al. (1988)
Buprenorphine versus procaine	Continuous IV buprenorphine versus continuous IV procaine	Buprenorphine required fewer rescue analgesics; lower VAS pain scores in first 2 days	Jakobs et al. (2000)
Procaine versus pentazocine	Procaine 2 g/24 h continuous IV versus pentazocine IV every 6 h	Procaine group required more rescue analgesics (98% vs. 44%); pain scores higher over first 3 days in pentazocine group	Kahl et al. (2004)
Overall opioids versus nonopioids	Various regimens	No difference in AP-related complications or major adverse events between opioids and nonopioid analgesics; opioids may reduce rescue analgesic need	Basurto Ona et al. (2013), Jakobs et al. (2000), Stevens et al. (2002), Thompson (2001)

previous reports (Basurto Ona et al., 2013). However, opioids should be avoided in patients with respiratory depression, altered sensorium, and paralytic ileus (Basurto Ona et al., 2013).

11.3.3 NSAIDs and other anesthetics

NSAIDs, including diclofenac, ketorolac, dexketoprofen, and metamizole, control pain by inhibiting cyclooxygenase (COX)-dependent prostaglandin formation (Vane and Botting, 1995). They have been used for managing pain in AP, and are particularly useful in patients in whom opioid use may be contraindicated or who are at risk of opioid-related side effects.

A systematic review of the efficacy of NSAIDs in treating AP and its systemic complications reported that NSAIDs are effective in suppressing proinflammatory cytokines, relieving pain, ameliorating systemic complications, and reducing mortality, and that serious adverse events are rarely reported (Wu et al., 2020). Moreover, a previous meta-analysis of five studies involving 227 participants reported that, among three studies which included 162 participants, opioids did not significantly differ from nonopioid analgesics in terms of the requirement for rescue analgesia, pain intensity, clinical outcomes, or adverse events in the management of pain associated with AP (Basurto Ona et al., 2013). However, NSAIDs should be used with caution in patients with renal impairment, gastric ulcers, or those at risk of bleeding.

Acetaminophen was previously thought to act via COX inhibition, but is now believed to produce analgesia in the central nervous system through its metabolite AM404, which modulates TRPV1 and CB1 receptors (Ohashi and Kohno, 2020). Due to its favorable safety profile, acetaminophen is often prescribed for patients in whom NSAIDs are contraindicated, including those with gastric ulcers, bronchial asthma, pregnant or nursing women, and children (Ohashi and Kohno, 2020; Meng et al., 2013). Although there is no evidence supporting the use of parenterally administered local anesthetics for pain management in AP, local anesthetics (e.g., procaine and bupivacaine) are occasionally used (Stigliano et al., 2017).

11.3.4 Severity-based analgesia

Analgesic administration in AP should be guided by both disease severity and the patient's ability to tolerate oral intake. Patients with severe pain will require intravenous analgesia and patient-controlled analgesia should be considered (Stigliano et al., 2017). Epidural analgesia may be considered for patients with severe pain and co-morbid cardiovascular or respiratory disease (Párniczky et al., 2025). Pain severity should be assessed using standardized, objective tools, such as the Numeric Rating Scale (NRS) or Visual Analog Scale (VAS), to guide titration, evaluate treatment response, and avoid unnecessary escalation (Párniczky et al., 2025).

11.4 Protease inhibitors

11.4.1 Overview

AP is an inflammatory disorder of the abdomen which is initiated by the ectopic activation of trypsin within pancreatic acinar cells, and which subsequently triggers a cascade of digestive enzyme activation leading to autodigestion (Frossard et al., 2008; Afghani et al., 2015). The release of trypsin and damage-associated molecular patterns (DAMPs) from injured acinar cells induces pancreatic and peripancreatic necrosis and contributes to organ dysfunction, including hepatic and renal impairment, primarily through vascular endothelial injury (Afghani et al., 2015; Kang et al., 2014). Given that premature intrapancreatic activation of digestive enzymes—particularly trypsin—is therefore central to the pathophysiology of AP and drives autodigestion, inflammation, and systemic complications, protease inhibitors, including ulinastatin, gabexate mesilate, and nafamostat mesilate, have long been investigated as potential disease-modifying agents. However, robust evidence supporting their routine therapeutic use in AP remains lacking (Párniczky et al., 2025). The following section summarizes current evidence for individual agents.

11.4.2 Ulinastatin

Ulinastatin is a human urine–derived serine protease inhibitor with anti-inflammatory, immunomodulatory, and organ-protective properties. It also suppresses neutrophil infiltration, the release of inflammatory cytokines such as tumor necrosis factor-α (TNF-α) and interleukin-6 (IL-6), and enzyme-mediated proteolysis, and thereby reduces systemic inflammation and related complications (He and Zhang, 2020; Zhao, Ma and Wang, 2022; Feng et al., 2022).

Recent meta-analyses support the potential utility of ulinastatin in severe AP. A systematic review and meta-analysis of 12 studies (1244 patients) demonstrated a significant reduction in mortality (relative risk [RR] 0.36; 95% CI 0.20–0.65), along with improvements in white blood cell count, C-reactive protein level, TNF-α level, serum IL-6, and time to resolution of abdominal pain (Bandyopadhyay et al., 2025). However, no consistent benefit was observed in the duration of hospitalization or APACHE II score. Moreover, two meta-analyses evaluating combination therapy with ulinastatin plus a somatostatin analogue (somatostatin or octreotide) reported reduced complication rates [e.g., acute respiratory distress syndrome (ARDS), acute kidney injury (AKI), multiple organ dysfunction syndrome (MODS)], faster symptom resolution and pain relief, shorter hospitalization, and improved biochemical markers compared with control therapy (Horváth et al., 2022; Zhu et al., 2025). Nevertheless, persistent limitations, including heterogeneity in study design, variable dosing regimens, concomitant therapies, and potential publication bias, limit confidence in these findings, as well as the question of whether ulinastatin should be adopted, either alone or as part of combination therapy, as a standard disease-modifying treatment for severe AP remains uncertain.

11.4.3 Gabexate mesilate

Gabexate mesilate is a synthetic protease inhibitor that was developed to improve drug distribution into the pancreas relative to the previously used aprotinin. Gabexate mesilate is primarily a synthetic serine protease inhibitor that targets trypsin and other serine proteases, which improves pancreatic microcirculation in the early phase of AP (Chen et al., 1996), while modulating inflammatory cytokines such as TNF-α, IL-6, and IL-10 (Chen et al., 2006).

A meta-analysis of 17 randomized controlled trials (six evaluating gabexate mesylate and 11 evaluating aprotinin) found that intravenous protease inhibitors did not reduce mortality in AP (pooled risk difference −0.02; 95% CI −0.05 to 0.01). Subgroup analyses in moderate to severe AP also showed no meaningful benefit (pooled risk difference −0.03; 95% CI −0.07 to 0.01) (Seta et al., 2014). Moreover, a large analysis using the Japanese Diagnosis Procedure Combination database found that gabexate mesylate did not reduce mortality or hospital length of stay in either nonsevere or severe AP (Yasunaga et al., 2013). Overall, the evidence does not support the routine use of gabexate mesilate for preventing death in AP.

11.4.4 Nafamostat mesilate

Nafamostat mesilate is a synthetic serine protease inhibitor which is structurally similar to gabexate mesilate, but which additionally exerts potent and sustained anticoagulant effects through inhibition of thrombin, plasmin, and other serine proteases (Ino et al., 1986). When plasma concentrations are maintained at sufficiently high levels, this anticoagulant activity is hypothesized to mitigate the development of pancreatic necrosis by restoring pancreatic microcirculation through dissolution of thrombi formed in regional vessels injured by trypsin and DAMPs (Takeda, 2007).

A single-center randomized trial involving 78 patients compared continuous regional arterial infusion (CRAI), a technique that delivers high local concentrations of nafamostat directly into the arteries supplying the pancreas, with standard intravenous administration, and reported lower mortality (5.1% vs. 23.1%, $P=.02$) and reduced need for emergency surgery (10.3% vs. 33.3%, $P=.01$) in the CRAI group (Piaścik et al., 2010). Additionally, a multicenter, open-label randomized controlled trial enrolling 39 patients with predicted severe AP found no significant difference in the incidence of pancreatic necrosis between CRAI and intravenous nafamostat, although CRAI was associated with reduced analgesic requirements as well as two infusion-related adverse events (bleeding and splenic infarction) (Hirota et al., 2020). These trials share important methodological limitations, including small sample sizes, open-label designs, and heterogeneous patient selection, however, and the current evidence therefore remains insufficient to support the routine use of nafamostat mesilate administered via CRAI to reduce mortality in AP.

11.5 Other therapeutic agents

11.5.1 Heparin

Low-molecular-weight heparin (LMWH) possesses both antithrombin and anti-inflammatory activities(Salas et al., 2000). It improves pancreatic microcirculation by inhibiting the formation of microthrombi within the pancreas (Tozlu et al., 2019). In addition, heparin suppresses TNF-α–induced leukocyte rolling (Salas et al., 2000), and inhibits cathepsin G–induced platelet activation by blocking protease activity (Evangelista et al., 1992).

A randomized trial in 100 patients with moderate and severe AP which compared standard care alone with standard care plus LMWH (1 mg/kg twice daily for 7 days), showed that LMWH significantly reduced pancreatic necrosis (6.1% vs. 22.9%, $P < .05$) and other complications ($P < .05$) without increasing bleeding risk, without significantly affecting mortality ($P = .056$) (Tozlu et al., 2019). Moreover, a systematic review and meta-analysis of 16 randomized controlled trials in 1625 patients with severe AP further demonstrated that LMWH was associated with a shorter hospital stay ($P < .01$), significant reductions in mortality (RR 0.33, $P < .01$), multiple organ failure (RR 0.34, $P < .01$), pancreatic pseudocyst (RR 0.49, $P = .02$), and the need for operative intervention (RR 0.39, $P < .01$) (Qiu et al., 2019). Although LMWH has shown promise in reducing the severity of AP, evidence remains insufficient, and LMWH has not been recommended as routine therapy for AP in clinical practice guidelines (Párniczky et al., 2025). On the other hand, short-term (<3 days) use of LMWH, as well as insulin, may be considered in patients with hypertriglyceridemia-associated AP, albeit that neither heparin nor LMWH should be used in the presence of contraindications such as bleeding disorders or renal failure (Párniczky et al., 2025).

11.5.2 Somatostatin and octreotide

AP results in injury to the pancreas and peripancreatic fat. This injury is partly facilitated by autodigestive enzymes. Excessive stimulation of the exocrine pancreas exacerbates the symptoms of AP, which in turn raises the possibility that anti-secretory agents, which reduce pancreatic enzyme secretion, have potential utility as therapeutic options (Kambhampati et al., 2014). Somatostatin is a peptide hormone that inhibits the secretion of circulating hormones such as cholecystokinin (CCK) and secretin, both of which stimulate pancreatic exocrine secretion, and also suppresses gastrointestinal motility. Against this, however, the clinical use of somatostatin is limited by its extremely short half-life of only 2–3 min (Heintges et al., 2004; Katz and Erstad, 1989). In contrast, octreotide, a synthetic analogue of somatostatin with the comparatively longer half-life of approximately 1.7–1.9 h (Katz and Erstad, 1989), has been investigated as a potential therapeutic option for AP. However, a randomized, placebo-controlled trial in 302 patients with moderate to severe AP compared placebo ($n = 103$) with octreotide 100 μg ($n = 98$) or 200 μg ($n = 101$) subcutaneously three times daily for 7 days. Results showed no significant difference among groups in mortality, new complications, pain duration, surgical interventions, or hospital stay, and thus demonstrated no benefit of octreotide in AP (Uhl et al., 1999). Accordingly, octreotide has not yet been demonstrated to have any effect on the course of the disease (Moggia et al., 2017), and is not recommended as routine therapy for AP (Párniczky et al., 2025).

11.5.3 Corticosteroids

Corticosteroids have been shown to suppress the release of inflammatory cytokines in AP (Zanker et al., 1990). A meta-analysis of six randomized controlled trials demonstrated that corticosteroid therapy in patients with severe AP was associated with a shorter hospital stay, reduced need for surgical intervention, and lower mortality (Dong et al., 2015). Nevertheless, corticosteroids are not recommended as routine therapy for AP in major international guidelines due to insufficient high-quality evidence demonstrating improvement in mortality or major clinical outcomes and need further confirmation (Párniczky et al., 2025). A multicenter, double-blind, randomized, placebo-controlled clinical trial (NCT05160506) to compare the effects of corticosteroids versus placebo on organ injury in patients with severe AP is currently underway, with completion expected in April 2027 (Vine et al., 2024).

11.5.4 Probiotics

Probiotics were originally considered potentially beneficial in AP because their ability to modulate the intestinal microbiota and reduce bacterial translocation was thought to have the potential to prevent infectious complications. However, in a multicenter RCT of 296 patients with predicted severe AP (152 receiving probiotics, 144 receiving placebo), enteral prophylaxis with a multispecies probiotic preparation did not reduce infectious complications (30% vs. 28%, RR 1.06, 95% CI 0.75–1.51). Probiotic use was associated with higher mortality (16% vs. 6%, RR 2.53, 95% CI 1.22–5.25) and nine cases of bowel ischemia ($P = .004$), indicating that such prophylaxis should not be administered in this patient population (Besselink et al., 2008). Moreover, a meta-analysis of 13 RCTs in 950 patients which evaluated the effects of pre-, pro-, and synbiotics supplemental to standard enteral nutrition in predicted severe AP found that supplementation significantly reduced hospital stay in Chinese severe AP cohorts, but saw no consistent benefit in other clinical outcomes and noted the need for further confirmation (Tian et al., 2018). Accordingly, contemporary clinical guidelines do not recommend the use of probiotics to prevent infectious complications in patients with AP (Párniczky et al., 2025).

References

Afghani, E., Pandol, S.J., Shimosegawa, T., Sutton, R., Wu, B.U., Vege, S.S., Gorelick, F., Hirota, M., Windsor, J., Lo, S.K., Freeman, M.L., Lerch, M.M., Tsuji, Y., Melmed, G. Y., Wassef, W., Mayerle, J., 2015. Acute pancreatitis-progress and challenges a report on an international symposium. Pancreas 44 (8), 1195–1210. https://doi.org/10.1097/MPA.0000000000000500.

Bandyopadhyay, S., Samrat Samajdar, S., Das, S., 2025. Ulinastatin for the treatment of severe acute pancreatitis: a systematic review and meta-analysis. BMC Gastroenterology 25 (1). https://doi.org/10.1186/s12876-025-04239-6.

Banks, P.A., 1997. Practice guidelines in acute pancreatitis. American Journal of Gastroenterology 92 (3), 377–386.

Banks, P.A., Freeman, M.L., Fass, R., Baroni, D.S., Mutlu, E.A., Bernstein, D.E., Parkman, H. P., Bharucha, A.E., Prather, C., Brugge, W.R., Pratt, D.S., Chang, L., Roach, A.C., Chey, W., Sampliner, R.E., Cohen, M.E., Sridhar, S., Cunningham, J.T., Vakil, N., Edmundowicz, S.A., Valdovinos, M.A., Inadomi, J.M., Wong, B.C.Y., Koch, T.R., Zfass, A.M., 2006. Practice guidelines in acute pancreatitis. American Journal of Gastroenterology 101 (10), 2379–2400. https://doi.org/10.1111/j.1572-0241.2006.00856.x.

Banks, P.A., Bollen, T.L., Dervenis, C., Gooszen, H.G., Johnson, C.D., Sarr, M.G., Tsiotos, G. G., Vege, S.S., Windsor, J.A., Horvath, K.D., Mortele, K.J., Gardner, T.B., Van Santvoort, H., Pelaez-Luna, M., Yadav, D., Stefanidis, G., Delakidis, S., Morgan, D.E., Thoeni, R.F.L., Acosta, J.M., Amann, S.T., Andren-Sandberg, A., Aranha, G.V., Asciutti, S., Barauskas, G., Baron, T.H., Bassi, C., Behrman, S., Behrns, K.E., Belliappa, V., Berzin, T.M., Besselink, M.G.H., Bhasin, D.K., Biankin, A., Bishop, M. D., Bonini, C.J., Bradley, E.L., Buechler, M., Carter, R., Cavestro, G.M., Chari, S.T., Chavez-Rodriguez, J.J., Da Cunha, J.E., D'Agostino, H., De Campos, T., De-Madaria, E., Deprez, P.H., DiSario, J.A., Doria, C., Falconi, M., Fernandez-del Castillo, C., Freeny, P.C., Frey, C.F., Friess, H., Frossard, J.L., Fuchshuber, P., Gallagher, S.F., Garg, P.K., Ghattas, G., Glasgow, R., Gonzalez, J.A., Gress, T.M., Gumbs, A.A., Hallibruton, C., Helton, S., Hill, M.C., Hoyos, S., Imrie, C.W., Isenmann, R., Izbicki, J.R., Karagiannis, J.A., Klar, E., Kolokythas, O., Lau, J., Litvin, A.A., Longnecker, D.S., Lowenfels, A.B., Mackey, R., Mah'Moud, M., Malangoni, M., McFadden, D.W., Mishra, G., Moody, F.G., Morinville, V., Neoptolemos, J.P., Nordback, I., Pap, A., Papachristou, G.I., Parks, R., Pedrazolli, S., Pezzilli, R., Pitt, H.A., Prosanto, C., Ramesh, H., Ramirez, F.C., Raper, S.E., Rasheed, A., Reed, D.N., Romangnuolo, J., Rossaak, J., Sanabria, J., Schaefer, C., Schmidt, J., Schmidt, P.N., Serrablo, A., Senkowski, C.K., Sharma, M., Sigman, K.M., Singh, P., Steinberg, W., Steiner, J., Strasberg, S., Strum, W., Takada, T., Tanaka, M., Vaccaro, M., Villavicencio, R.L., Vrochides, D., Wagner, M., Warshaw, A.L., Wilcox, C.M., Wysocki, P., Zenilman, M.E., Zyromski, N.J., 2013. Classification of acute pancreatitis—2012: revision of the Atlanta classification and definitions by international consensus. Gut 62 (1), 102–111. https://doi.org/10.1136/gutjnl-2012-302779UnitedStates.

Baron, T.H., DiMaio, C.J., Wang, A.Y., Morgan, K.A., 2020. American gastroenterological association clinical practice update: management of pancreatic necrosis. Gastroenterology 158 (1), 67. https://doi.org/10.1053/j.gastro.2019.07.064.

Basurto Ona, X., Rigau Comas, D., Urrútia, G., 2013. Opioids for acute pancreatitis pain. Cochrane Database of Systematic Reviews 2013 (7), Cd009179.

Besselink, M.G., van Santvoort, H.C., Buskens, E., Boermeester, M.A., van Goor, H., Timmerman, H.M., Nieuwenhuijs, V.B., Bollen, T.L., van Ramshorst, B., Witteman, B. J., Rosman, C., Ploeg, R.J., Brink, M.A., Schaapherder, A.F., Dejong, C.H., Wahab, P. J., van Laarhoven, C.J., van der Harst, E., van Eijck, C.H., Cuesta, M.A., Akkermans, L.M., Gooszen, H.G., 2008. Probiotic prophylaxis in predicted severe acute pancreatitis: a randomised, double-blind, placebo-controlled trial. The Lancet 371 (9613), 651–659. https://doi.org/10.1016/S0140-6736(08)60207-X.

Besselink, M.G., van Santvoort, H.C., Boermeester, M.A., Nieuwenhuijs, V.B., van Goor, H., Dejong, C.H.C., Schaapherder, A.F., Gooszen, H.G., 2009. Timing and impact of infections in acute pancreatitis. British Journal of Surgery 96 (3), 267–273. https://doi.org/10.1002/bjs.6447.

Boxhoorn, L., Voermans, R.P., Bouwense, S.A., Bruno, M.J., Verdonk, R.C., Boermeester, M. A., van Santvoort, H.C., Besselink, M.G., 2020. Acute pancreatitis. The Lancet 396 (10252), 726–734. https://doi.org/10.1016/S0140-6736(20)31310-6.

Chen, H.M., Hwang, T.L., Chen, M.F., 1996. The effect of gabexate mesilate on pancreatic and hepatic microcirculation in acute experimental pancreatitis in rats. Journal of Surgical Research 66 (2), 147–153. https://doi.org/10.1006/jsre.1996.0387.

Chen, C., Wang, S., Tsay, S., Lee, F., Lu, R., Chang, F., Lee, S., 2006. Effects of gabexate mesilate on serum inflammatory cytokines in rats with acute necrotizing pancreatitis. Cytokine 33 (2), 95–99. https://doi.org/10.1016/j.cyto.2005.12.004.

Chen, H.-Z., Ji, L., Li, Le, Wang, G., Bai, X.-W., Cheng, C.-D., Sun, B., 2017. Early prediction of infected pancreatic necrosis secondary to necrotizing pancreatitis. Medicine 96 (30), e7487. https://doi.org/10.1097/md.0000000000007487.

Crockett, S.D., Wani, S., Gardner, T.B., Falck-Ytter, Y., Barkun, A.N., Feuerstein, J., Flamm, S., Gellad, Z., Gerson, L., Gupta, S., Hirano, I., Inadomi, J., Nguyen, G.C., Rubenstein, J.H., Singh, S., Smalley, W.E., Stollman, N., Street, S., Sultan, S., Vege, S. S., Wani, S.B., Weinberg, D., 2018. American gastroenterological association Institute guideline on initial management of acute pancreatitis. Gastroenterology 154 (4), 1096–1101. https://doi.org/10.1053/j.gastro.2018.01.032.

Dawson, A., Karunakaran, M., Sharma, Z.D., Ullah, S., Barreto, S.G., 2023. Fluid resuscitation in the early management of acute pancreatitis—evidence from a systematic review and meta-analysis. HPB 25 (12), 1451–1465. https://doi.org/10.1016/j.hpb.2023.08.013.

de-Madaria, E., Soler-Sala, G., Sánchez-Payá, J., Lopez-Font, I., Martínez, J., Gómez-Escolar, L., Sempere, L., Sánchez-Fortún, C., Pérez-Mateo, M., 2011. Influence of fluid therapy on the prognosis of acute pancreatitis: a prospective cohort study. American Journal of Gastroenterology 106 (10), 1843–1850. https://doi.org/10.1038/ajg.2011.236.

de-Madaria, E., Buxbaum, J.L., Maisonneuve, P., de Paredes, A.G.G., Zapater, P., Guilabert, L., Vaillo-Rocamora, A., Rodríguez-Gandía, M.Á., Donate-Ortega, J., Lozada-Hernández, E.E., Moreno, A.J.R.C., Lira-Aguilar, A., Llovet, L.P., Mehta, R., Tandel, R., Navarro, P., Sánchez-Pardo, A.M., Sánchez-Marin, C., Cobreros, M., Fernández-Cabrera, I., Casals-Seoane, F., Deza, D.C., Lauret-Braña, E., Martí-Marqués, E., Camacho-Montaño, L.M., Ubieto, V., Ganuza, M., Bolado, F., 2022. Aggressive or moderate fluid resuscitation in acute pancreatitis. New England Journal of Medicine 387 (11), 989–1000. https://doi.org/10.1056/nejmoa2202884.

Di Martino, M., Van Laarhoven, S., Ielpo, B., Ramia, J.M., Manuel-Vázquez, A., Martínez-Pérez, A., Pavel, M., Miranda, P.B., Orti-Rodríguez, R., de la Serna, S., Ortega Rabbione, G.J., Sanz-Garcia, A., Martín-Pérez, E., 2021. Systematic review and meta-analysis of fluid therapy protocols in acute pancreatitis: type, rate and route. HPB: The Official Journal of the International Hepato Pancreato Biliary Association 23 (11), 1629–1638.

Ding, X., Chen, B., 2023. Effect of aggressive intravenous fluid resuscitation versus nonaggressive fluid resuscitation in the treatment of acute pancreatitis: a systematic review and meta-analysis. Pancreas 52 (2), E89. https://doi.org/10.1097/MPA.0000000000002230.

Ding, L., Yu, C., Deng, F., He, W.H., Xia, L., Zhou, M., Lan, G.L., Huang, X., Lei, Y.P., Zhou, X.J., Zhu, Y., Lu, N.H., 2019. New risk factors for infected pancreatic necrosis secondary to severe acute pancreatitis: the role of initial contrast-enhanced computed tomography. Digestive Diseases and Sciences 64 (2), 553–560. https://doi.org/10.1007/s10620-018-5359-y.

Dong, L.H., Liu, Z.M., Wang, S.J., Zhao, S.J., Zhang, D., Chen, Y., Wang, Y.S., 2015. Corticosteroid therapy for severe acute pancreatitis: a meta-analysis of randomized, controlled trials. International Journal of Clinical and Experimental Pathology 8 (7), 7654–7660.

Dumnicka, P., Maduzia, D., Ceranowicz, P., Olszanecki, R., Drożdż, R., Kuśnierz-Cabala, B., 2017. The interplay between inflammation, coagulation and endothelial injury in the early phase of acute pancreatitis: clinical implications. International Journal of Molecular Sciences 18 (2), 354. https://doi.org/10.3390/ijms18020354.

Evangelista, V., Piccardoni, P., Maugeri, N., De Gaetano, G., Cerletti, C., 1992. Inhibition by heparin of platelet activation induced by neutrophil-derived cathepsin G. European Journal of Pharmacology 216 (3), 401–405. https://doi.org/10.1016/0014-2999(92)90437-9.

Feng, X., Ma, W., Chen, J., Jiao, W., Wang, Y., 2022. Ulinastatin alleviates early brain injury after traumatic brain injury by inhibiting oxidative stress and apoptosis. Acta Cirurgica Brasileira 37 (1). https://doi.org/10.1590/acb370108.

Fish, D.N., Teitelbaum, I., Abraham, E., 2005. Pharmacokinetics and pharmacodynamics of imipenem during continuous renal replacement therapy in critically ill patients. Antimicrobial Agents and Chemotherapy 49 (6), 2421–2428. https://doi.org/10.1128/AAC.49.6.2421-2428.2005.

Földi, M., Gede, N., Kiss, S., Vincze, Á., Bajor, J., Szabó, I., Szepes, Z., Izbéki, F., Gervain, J., Hamvas, J., Vitális, Z., Fehér, E., Crai, S., Sallinen, V., Maldonado, E.R., Meczker, Á., Varjú, P., Poropat, G., Stimac, D., Faluhelyi, N., Miseta, A., Nagy, T., Márton, Z., Vereczkei, A., Hegyi, P.J., Párniczky, A., Hegyi, P., Szentesi, A., 2022. The characteristics and prognostic role of acute abdominal on-admission pain in acute pancreatitis: a prospective cohort analysis of 1432 cases. European Journal of Pain 26 (3), 610–623. https://doi.org/10.1002/ejp.1885.

Frossard, J.L., Steer, M.L., Pastor, C.M., 2008. Acute pancreatitis. The Lancet 371 (9607), 143–152. https://doi.org/10.1016/S0140-6736(08)60107-5.

Gilbert, D.N., Chambers, H.F., Saag, M.S., Pavia, A., Boucher, H.W., 2025. The Sanford Guide to Antimicrobial Therapy 2025. 55th edition. Antimicrobial Therapy, Inc., Sperryville, VA, USA.

Gülen, B., Dur, A., Serinken, M., Karcioğlu, Ö., Sönmez, E., 2016. Pain treatment in patients with acute pancreatitis: a randomized controlled trial. Turkish Journal of Gastroenterology 27 (2), 192–196. https://doi.org/10.5152/tjg.2015.150398.

He, H.W., Zhang, H., 2020. The efficacy of different doses of ulinastatin in the treatment of severe acute pancreatitis. Annals of Palliative Medicine 9 (3), 730–737. https://doi.org/10.21037/apm.2020.04.19.

He, W.H., Zhu, Y., Zhu, Y., Jin, Q., Xu, H.R., Xion, Z.J., Yu, M., Xia, L., Liu, P., Lu, N.H., 2017. Comparison of multifactor scoring systems and single serum markers for the early prediction of the severity of acute pancreatitis. Journal of Gastroenterology and Hepatology 32 (11), 1895–1901. https://doi.org/10.1111/jgh.13803.

Heintges, T., Lüthen, R., Niederau, C., 2004. Inhibition of exocrine pancreatic secretion by somatostatin and its analogues. Digestion 55 (1), 1–9. https://doi.org/10.1159/000201179.

Helm, J.F., Venu, R.P., Geenen, J.E., Hogan, W.J., Dodds, W.J., Toouli, J., Arndorfer, R.C., 1988. Effects of morphine on the human sphincter of Oddi. Gut 29 (10), 1402–1407. https://doi.org/10.1136/gut.29.10.1402.

Herranz Prinz, L., de Castro, I., de Cesar Pontes Azevedo, L., Mauro Vieira, J., 2022. Uncontrolled pain in critically ill patients and acute kidney injury: a hypothesis-generating cohort study. BMC Nephrology 23 (1). https://doi.org/10.1186/s12882-022-02810-x. Available from: http://www.biomedcentral.com/bmcnephrol/.

Hirota, M., Shimosegawa, T., Kitamura, K., Takeda, K., Takeyama, Y., Mayumi, T., Ito, T., Takenaka, M., Iwasaki, E., Sawano, H., Ishida, E., Miura, S., Masamune, A., Nakai, Y., Mitoro, A., Maguchi, H., Kimura, K., Sanuki, T., Ito, T., Haradome, H., Kozaka, K., Gabata, T., Kataoka, K., Hirota, M., Isaji, S., Nakamura, R., Yamagiwa, K., Kayaba, C., Ikeda, K., 2020. Continuous regional arterial infusion versus intravenous administration of the protease inhibitor nafamostat mesilate for predicted severe acute pancreatitis: a multicenter, randomized, open-label, phase 2 trial. Journal of Gastroenterology 55 (3), 342–352. https://doi.org/10.1007/s00535-019-01644-z.

Hong, J., Li, Q., Wang, Y., Xiang, L., Zhou, Y., Fan, M., Lin, R., 2024. Comparison of fluid resuscitation with lactate ringer's versus normal saline in acute pancreatitis: an updated meta-analysis. Digestive Diseases and Sciences 69 (1), 262–274. https://doi.org/10.1007/s10620-023-08187-7.

Honore, P.M., De Bels, D., Gutierrez, L.B., Redant, S., Attou, R., Gallerani, A., Spapen, H.D., 2018. Optimizing micafungin dosing in critically ill patients: what about extracorporeal therapies? Critical Care 22 (1). https://doi.org/10.1186/s13054-018-2231-6. Available from: http://ccforum.com/content/17.

Horváth, I.L., Bunduc, S., Fehérvári, P., Váncsa, S., Nagy, R., Garmaa, G., Kleiner, D., Hegyi, P., Erőss, B., Csupor, D., 2022. The combination of ulinastatin and somatostatin reduces complication rates in acute pancreatitis: a systematic review and meta-analysis of randomized controlled trials. Scientific Reports 12 (1). https://doi.org/10.1038/s41598-022-22341-7. Available from: https://www.nature.com/srep/.

Howell, M.D., Davis, A.M., 2017. Management of sepsis and septic shock. JAMA 317 (8), 847. https://doi.org/10.1001/jama.2017.0131.

Ino, Y., Suzuki, K., Sato, T., Masahiro, I.W.A.K.I., 1986. Comparative studies of nafamostat mesilate and various serine protease inhibitors in vitro. Folia Pharmacologica Japonica 88 (6), 449–455. https://doi.org/10.1254/fpj.88.449.

Jakobs, R., Adamek, M.U., Von Bubnoff, A.C., Riemann, J.F., 2000. Buprenorphine or procaine for pain relief in acute pancreatitis. A prospective randomized study. Scandinavian Journal of Gastroenterology 35 (12), 1319–1323. https://doi.org/10.1080/003655200453692.

Ji, L., Lv, J.C., Song, Z.F., Jiang, M.T., Li, L., Sun, B., 2016. Risk factors of infected pancreatic necrosis secondary to severe acute pancreatitis. Hepatobiliary and Pancreatic Diseases International 15 (4), 428–433. https://doi.org/10.1016/S1499-3872(15)60043-1.

Juvonen, P.O., Tenhunen, J.J., Heino, A.A., Merasto, M., Paajanen, H.E.K., Alhava, E.M., Takala, J.A., 1999. Splanchnic tissue perfusion in acute experimental pancreatitis. Scandinavian Journal of Gastroenterology 34 (3), 308–314. https://doi.org/10.1080/00365529950173744.

Kahl, S., Zimmermann, S., Pross, M., Schulz, H.U., Schmidt, U., Malfertheiner, P., 2004. Procaine hydrochloride fails to relieve pain in patients with acute pancreatitis. Digestion 69 (1), 5–9. https://doi.org/10.1159/000076541.

Kambhampati, S., Park, W., Habtezion, A., 2014. Pharmacologic therapy for acute pancreatitis. World Journal of Gastroenterology 20 (45), 16868–16880. https://doi.org/10.3748/wjg.v20.i45.16868.

Kang, R., Lotze, M.T., Zeh, H.J., Billiar, T.R., Tang, D., 2014. Cell death and DAMPs in acute pancreatitis. Molecular Medicine 20, 466–477. https://doi.org/10.2119/molmed.2014.00117.

Kanjhan, R., 1995. Opioids and pain. Clinical and Experimental Pharmacology and Physiology 22 (6–7), 397–403.

Katz, M.D., Erstad, B.L., 1989. Octreotide, a new somatostatin analogue. Clinical Pharmacy 8 (4), 255–273.

Kinnala, P.J., Kuttila, K.T., Grönroos, J.M., Havia, T.V., Nevalainen, T.J., Niinikoski, J.H.A., 2002. Splanchnic and pancreatic tissue perfusion in experimental acute pancreatitis. Scandinavian Journal of Gastroenterology 37 (7), 845–849. https://doi.org/10.1080/713786530.

Kusterer, K., Enghofer, M., Zendler, S., Blochle, C., Usadel, K.H., 1991. Microcirculatory changes in sodium taurocholate-induced pancreatitis in rats. American Journal of Physiology—Gastrointestinal and Liver Physiology 260 (2), G346. https://doi.org/10.1152/ajpgi.1991.260.2.g346.

Mahapatra, S.J., Jain, S., Bopanna, S., Gupta, S., Singh, P., Trikha, A., Sreenivas, V., Shalimar, Garg, P.K., 2019. Pentazocine, a kappa-opioid agonist, is better than diclofenac for analgesia in acute pancreatitis: a randomized controlled trial. American Journal of Gastroenterology 114 (5), 813–821. https://doi.org/10.14309/ajg.0000000000000224.

Mayerle, J., Hlouschek, V., Lerch, M.M., 2005. Current management of acute pancreatitis. Nature Clinical Practice Gastroenterology & Hepatology 2 (10), 473–483. https://doi.org/10.1038/ncpgasthep0293.

Meng, W., Yuan, J., Zhang, C., Bai, Z., Zhou, W., Yan, J., Li, X., 2013. Parenteral analgesics for pain relief in acute pancreatitis: a systematic review. Pancreatology 13 (3), 201–206. https://doi.org/10.1016/j.pan.2013.02.003.

Mofidi, R., Duff, M.D., Wigmore, S.J., Madhavan, K.K., Garden, O.J., Parks, R.W., 2006. Association between early systemic inflammatory response, severity of multiorgan dysfunction and death in acute pancreatitis. British Journal of Surgery 93 (6), 738–744. https://doi.org/10.1002/bjs.5290.

Moggia, E., Koti, R., Belgaumkar, A.P., Fazio, F., Pereira, S.P., Davidson, B.R., Gurusamy, K. S., 2017. Pharmacological interventions for acute pancreatitis. Cochrane Database of Systematic Reviews 2017 (4). https://doi.org/10.1002/14651858.CD011384.pub2. Available from: http://as.wiley.com/WileyCDA/Brand/id-6.html.

Moran, R.A., Jalaly, N.Y., Kamal, A., Rao, S., Klapheke, R., James, T.W., Thiruvengadam, S. K., Makary, M.A., Hirose, K., Kumbhari, V., Stein, E.M., Khashab, M.A., Lennon, A.M., Kalloo, A.N., Zaheer, A., Hernaez, R., Singh, V.K., 2016. Ileus is a predictor of local infection in patients with acute necrotizing pancreatitis. Pancreatology 16 (6), 966–972. https://doi.org/10.1016/j.pan.2016.10.002.

Nelson, A.D., Lugo-Fagundo, N.S., Mahapatra, S.J., Cheungpastiporn, W., Thongprayoon, C., Wijarnpreecha, K., Muktesh, G., Kochhar, R., Garg, P., Wallace, M., Bi, Y., 2022. A systematic review and meta-analysis of opioids vs nonopioids in acute pancreatitis. Gastro Hep Advances 1 (1), 83–92. https://doi.org/10.1016/j.gastha.2021.09.006.

Ohashi, N., Kohno, T., 2020. Analgesic effect of acetaminophen: a review of known and novel mechanisms of action. Frontiers in Pharmacology 11. https://doi.org/10.3389/fphar.2020.580289.

Pandanaboyana, S., Knoph, C.S., Olesen, S.S., Jones, M., Lucocq, J., Samanta, J., Talukdar, R., Capurso, G., de-Madaria, E., Yadav, D., Siriwardena, A.K., Windsor, J., Drewes, A.M., Nayar, M., 2024. Opioid analgesia and severity of acute pancreatitis: an international multicentre cohort study on pain management in acute pancreatitis. United European Gastroenterology Journal 12 (3), 326–338. https://doi.org/10.1002/ueg2.12542.

Pappas, P.G., Kauffman, C.A., Andes, D.R., Clancy, C.J., Marr, K.A., Ostrosky-Zeichner, L., Reboli, A.C., Schuster, M.G., Vazquez, J.A., Walsh, T.J., Zaoutis, T.E., Sobel, J.D., 2016. Clinical practice guideline for the management of candidiasis: 2016 update by the infectious diseases society of America. Clinical Infectious Diseases 62 (4), e1. https://doi.org/10.1093/cid/civ933.

Párniczky, A., Mikó, A., Uc, A., Singh, A.N., Elhence, A., Saluja, A., Masamune, A., Abu Dayyeh, B.K., Davidson, B., Wilcox, C.M., Forsmark, C., Whitcomb, D., Gunjan, D., Yadav, D., Cahen, D., Nageshwar Reddy, D., Adler, D., Pleskow, D., Mao, E.-Q., de-Madaria, E., Zerem, E., Bolado, F., Gorelick, F., Papachristou, G., Cote, G., Trikudanathan, G., Alguel, H., Santvoort, H., Buxbaum, J., Werner, J., Bang, J.Y., Neoptolemos, J., Windsor, J., Mayerle, J., Choi, J.-H., Overbeek, K., Kikuta, K., Kitamura, K., Frulloni, L., Madhusudan, K., Besselink, M., Bruno, M., Lowe, M., Lerch, M., Freeman, M., Löhr, M., Hirota, M., Khashab, M., Zyromski, N., Guda, N., Singh, N., Baneerjee, N., Bakker, O.J., Mentula, P., Banks, P., Sahni, P., Hegyi, P., Szatmary, P., Garg, P., Gupta, R., Sharma, R., Kochhar, R., Sutton, R., Cartin-Ceba, R., Talukdar, R., Agarwal, S., Husain, S., Vege, S., Isaji, S., Varadarajulu, S., Gamanagatti, S., Jagannath, S., Rana, S., Bouwense, S., Pandol, S., Kumar, S., Chowdhury, S.D., Lakhtakia, S., Chari, S., Itoi, T., Hackert, T., Bollen, T., Gress, T., Gardner, T., Singh, V., Dudeja, V., Singh, V., Rebours, V., Park, W., Huang, W., He, W.-H., Kang, Y.-N., Tsuji, Y., Takeyama, Y., 2025. International association of pancreatology revised guidelines on acute pancreatitis 2025: supported and endorsed by the American pancreatic association, European pancreatic club, Indian pancreas club, and Japan pancreas society. Pancreatology 25 (6), 770–814. https://doi.org/10.1016/j.pan.2025.04.020.

Peiró, A.M., Martínez, J., Martínez, E., De Madaria, E., Llorens, P., Horga, J.F., Pérez-Mateo, M., 2008. Efficacy and tolerance of metamizole versus morphine for acute pancreatitis pain. Spain Pancreatology 8 (1), 25–29. https://doi.org/10.1159/000114852.

Piaścik, M., Rydzewska, G., Milewski, J., Olszewski, S., Furmanek, M., Walecki, J., Gabryelewicz, A., 2010. The results of severe acute pancreatitis treatment with continuous regional arterial infusion of protease inhibitor and antibiotic: a randomized controlled study. Pancreas 39 (6), 863–867. https://doi.org/10.1097/MPA.0b013e3181d37239.

Qiu, Q., Li, G.J., Tang, L., Guo, Y., Wen, L.Z., Wang, B., Chen, D.F., Liu, K.J., 2019. The efficacy of low molecular weight heparin in severe acute pancreatitis: a systematic review and meta-analysis of randomized controlled trials. Journal of Digestive Diseases 20 (10), 512–522. https://doi.org/10.1111/1751-2980.12815. Available from: http://onlinelibrary.wiley.com/journal/10.1111/(ISSN)1751-2980.

Rau, B., Pralle, U., Mayer, J.M., Beger, H.G., 1998. Role of ultrasonographically guided fine-needle aspiration cytology in the diagnosis of infected pancreatic necrosis. British Journal of Surgery 85 (2), 179–184. https://doi.org/10.1046/j.1365-2168.1998.00707.x.

Roberts, J.A., Lipman, J., 2009. Pharmacokinetic issues for antibiotics in the critically ill patient. Critical Care Medicine 37 (3), 840–851. https://doi.org/10.1097/ccm.0b013e3181961bff.

Sahu, S.K., Giri, S., Das, S., Deepak Patro, C., Praharaj, D.L., Mallick, B., Nath, P., Chandra Panigrahi, S., Anand, A.C., 2025. Approach to the diagnosis and management of infected pancreatic necrosis: a narrative review. Cureus 17 (4). https://doi.org/10.7759/cureus.83020.

Saini, M., Samanta, J., Kumar, A., Choudhury, A., Dhar, J., Jafra, A., Chauhan, R., Muktesh, G., Gupta, P., Gupta, V., Yadav, T.D., Kochhar, R., Capurso, G., De-Madaria, E., Facciorusso, A., 2024. Buprenorphine versus diclofenac for pain relief in acute pancreatitis: a double-blinded randomized controlled trial. Clinical Gastroenterology and Hepatology 22 (3), 532. https://doi.org/10.1016/j.cgh.2023.10.021.

Salas, A., Sans, M., Soriano, A., Reverter, J.C., Anderson, D.C., Piqué, J.M., Panés, J., 2000. Heparin attenuates TNF-α induced inflammatory response through a CD11b dependent mechanism. Gut 47 (1), 88–96. https://doi.org/10.1136/gut.47.1.88.

Seta, T., Noguchi, Y., Shikata, S., Nakayama, T., 2014. Treatment of acute pancreatitis with protease inhibitors administered through intravenous infusion: an updated systematic review and meta-analysis. BMC Gastroenterology 14 (1). https://doi.org/10.1186/1471-230x-14-102.

Singh, P., Garg, P.K., 2016. Pathophysiological mechanisms in acute pancreatitis: current understanding. Indian Journal of Gastroenterology 35 (3), 153–166. https://doi.org/10.1007/s12664-016-0647-y.

Stasi, A., Franzin, R., Caggiano, G., Losapio, R., Fiorentino, M., Alfieri, C., Gesualdo, L., Stallone, G., Castellano, G., 2023. New frontiers in sepsis-induced acute kidney injury and blood purification therapies: the role of polymethylmethacrylate membrane hemofilter. Blood Purification 52 (1), 71–84. https://doi.org/10.1159/000528685.

Stevens, M., Esler, R., Asher, G., 2002. Transdermal fentanyl for the management of acute pancreatitis pain. Applied Nursing Research 15 (2), 102–110. https://doi.org/10.1053/apnr.2002.29532.

Stigliano, S., Sternby, H., de Madaria, E., Capurso, G., Petrov, M.S., 2017. Early management of acute pancreatitis: a review of the best evidence. Digestive and Liver Disease 49 (6), 585–594. https://doi.org/10.1016/j.dld.2017.01.168.

Takada, T., Isaji, S., Mayumi, T., Yoshida, M., Takeyama, Y., Itoi, T., Sano, K., Iizawa, Y., Masamune, A., Hirota, M., Okamoto, K., Inoue, D., Kitamura, N., Mori, Y., Mukai, S., Kiriyama, S., Shirai, K., Tsuchiya, A., Higuchi, R., Hirashita, T., 2022. JPN clinical practice guidelines 2021 with easy-to-understand explanations for the management of acute pancreatitis. Journal of Hepato-Biliary-Pancreatic Sciences 29 (10), 1057–1083. https://doi.org/10.1002/jhbp.1146.

Takeda, K., 2007. Antiproteases in the treatment of acute necrotizing pancreatitis: continuous regional arterial infusion. Journal of the Pancreas 8 (4), 526–532.

Takeda, K., Mikami, Y., Fukuyama, S., Egawa, S., Sunamura, M., Ishibashi, T., Sato, A., Masamune, A., Matsuno, S., 2005. Pancreatic ischemia associated with vasospasm in the early phase of human acute necrotizing pancreatitis. Pancreas 30 (1), 40–49.

Tenner, S., Vege, S.S., Sheth, S.G., Sauer, B., Yang, A., Conwell, D.L., Yadlapati, R.H., Gardner, T.B., 2024. American college of gastroenterology guidelines: management of acute pancreatitis. American Journal of Gastroenterology 119 (3), 419–437. https://doi.org/10.14309/ajg.0000000000002645.

Thavanesan, N., White, S., Lee, S., Ratnayake, B., Oppong, K.W., Nayar, M.K., Sharp, L., Drewes, A.M., Capurso, G., De-Madaria, E., Siriwardena, A.K., Windsor, J.A., Pandanaboyana, S., 2022. Analgesia in the initial management of acute pancreatitis: a systematic review and meta-analysis of randomised controlled trials. World Journal of Surgery 46 (4), 878–890. https://doi.org/10.1007/s00268-021-06420-w.

Thompson, D., 2001. Narcotic analgesic effects on the sphincter of Oddi: a review of the data and therapeutic implications in treating pancreatitis. American Journal of Gastroenterology 96 (4), 1266–1272. https://doi.org/10.1016/s0002-9270(00)02329-7.

Tian, X., Pi, Y.P., Liu, X.L., Chen, H., Chen, W.Q., 2018. Supplemented use of pre-, pro-, and synbiotics in severe acute pancreatitis: an updated systematic review and meta-analysis of 13 randomized controlled trials. Frontiers in Pharmacology 9. https://doi.org/10.3389/fphar.2018.00690. Available from: http://www.frontiersin.org/Pharmacology.

Tozlu, M., Kayar, Y., Ince, A.T., Baysal, B., Senturk, H., 2019. Low molecular weight heparin treatment of acute moderate and severe pancreatitis: a randomized, controlled, open-label study. Turkish Journal of Gastroenterology 30 (1), 81–87. https://doi.org/10.5152/tjg.2018.18583.

Trikudanathan, G., Navaneethan, U., Vege, S.S., 2012. Current controversies in fluid resuscitation in acute pancreatitis: a systematic review. Pancreas 41 (6), 827–834. https://doi.org/10.1097/MPA.0b013e31824c1598.

Uhl, W., Büchler, M.W., Malfertheiner, P., Beger, H.G., Adler, G., Gaus, W., 1999. A randomised, double blind, multicentre trial of octreotide in moderate to severe acute pancreatitis. Gut 45 (1), 97–104. https://doi.org/10.1136/gut.45.1.97.

Van Baal, M.C., Bollen, T.L., Bakker, O.J., Van Goor, H., Boermeester, M.A., Dejong, C.H., Gooszen, H.G., Van Der Harst, E., Van Eijck, C.H., Van Santvoort, H.C., Besselink, M. G., 2014. The role of routine fine-needle aspiration in the diagnosis of infected necrotizing pancreatitis. Surgery 155 (3), 442–448. https://doi.org/10.1016/j.surg.2013.10.001.

Vane, J.R., Botting, R.M., 1995. New insights into the mode of action of anti-inflammatory drugs. Inflammation Research 44 (1), 1–10. https://doi.org/10.1007/BF01630479.

Vine, J., Berlin, N., Moskowitz, A., Berg, K.M., Liu, X., Balaji, L., Donnino, M.W., Grossestreuer, A.V., 2024. Corticosteroids to reduce inflammation in severe pancreatitis (CRISP) protocol and statistical analysis plan: a prospective, multicentre, double-blind, randomized, placebo-controlled clinical trial. Contemporary Clinical Trials 139. https://doi.org/10.1016/j.cct.2024.107486. Available from: https://www.sciencedirect.com/science/journal/15517144.

Wolbrink, D.R.J., Kolwijck, E., Ten Oever, J., Horvath, K.D., Bouwense, S.A.W., Schouten, J. A., 2020. Management of infected pancreatic necrosis in the intensive care unit: a narrative review. Clinical Microbiology and Infection 26 (1), 18–25. https://doi.org/10.1016/j.cmi.2019.06.017.

Wu, D., Bai, X., Lee, P., Yang, Y., Windsor, J., Qian, J., 2020. A systematic review of NSAIDs treatment for acute pancreatitis in animal studies and clinical trials. Clinics and Research in Hepatology and Gastroenterology 44, 100002. https://doi.org/10.1016/j.clirex.2019.100002.

Wu, Y., Han, C., Luo, R., Cai, W., Xia, Q., Jiang, R., Ferdek, P.E., Liu, T., Huang, W., 2023. Molecular mechanisms of pain in acute pancreatitis: recent basic research advances and therapeutic implications. Frontiers in Molecular Neuroscience 16. https://doi.org/10.3389/fnmol.2023.1331438. Available from: https://www.frontiersin.org/journals/molecular-neuroscience.

Yasunaga, H., Horiguchi, H., Hashimoto, H., Matsuda, S., Fushimi, K., 2013. Effect and cost of treatment for acute pancreatitis with or without gabexate mesylate: a propensity score analysis using a nationwide administrative database. Pancreas 42 (2), 260–264. https://doi.org/10.1097/MPA.0b013e31826495a0.

Yegneswaran, B., Kostis, J.B., Pitchumoni, C.S., 2011. Cardiovascular manifestations of acute pancreatitis. Journal of Critical Care 26 (2), 225.e11. https://doi.org/10.1016/j.jcrc.2010.10.013.

Zanker, B., Walz, G., Wieder, K.J., Strom, T.B., 1990. Evidence that glucocorticosteroids block expression of the human interleukin-6 gene by accessory cells. Transplantation 49 (1), 183–185. https://doi.org/10.1097/00007890-199001000-00040.

Zhao, L., Ma, W., 2022. Ulinastatin combined with glutamine improves liver function and inflammatory response in patients with severe acute pancreatitis. American Journal of Translational Research 14 (2), 918–926.

Zhu, L.X., Chen, Y., Chen, X.F., Sheng, N., Feng, P.F., 2025. Systematic review and meta-analysis of the clinical efficacy of octreotide in combination with ulinastatin in the treatment of acute pancreatitis. Drugs in R & D 25 (3), 195–207. https://doi.org/10.1007/s40268-025-00518-5.

CHAPTER

12 Management of pancreatitis in the intensive care unit

Hiroyuki Tomita

Department of Tumor Pathology, Gifu University Graduate School of Medicine, Gifu, Japan

12.1 Scope and ICU objectives

Most episodes of AP are mild, but a subset develops persistent organ failure and/or necrotizing disease requiring intensive care. In the ICU, management emphasizes rapid stabilization of organ dysfunction, prevention of secondary injury (fluid overload, unnecessary antibiotics, premature intervention), and multidisciplinary planning for time-sensitive procedures such as ERCP or drainage when indicated (Banks et al., 2013; Tenner et al., 2013; Leppäniemi et al., 2019; Petrov et al., 2010) (Fig. 12.1).

12.2 Indications for ICU admission and early triage

ICU admission is generally indicated for shock requiring vasopressors, hypoxemic respiratory failure requiring advanced support, severe metabolic derangements, or acute kidney injury with impending need for renal replacement therapy. Patients with early high-risk trajectories (persistent SIRS, rising blood urea nitrogen, escalating oxygen requirement) should be monitored in a high-acuity setting because severity often evolves over the first 24–72 h (Banks et al., 2013; Wu et al., 2008; Knaus et al., 1985; Vincent et al., 1996).

Etiology-directed actions should occur in parallel with stabilization. In biliary AP with acute cholangitis, urgent ERCP with biliary decompression is recommended. In the absence of cholangitis or clear ongoing obstruction, routine early ERCP is not universally beneficial and should be individualized (Fan et al., 1993; Fölsch et al., 1997; Neoptolemos et al., 1988; Oría et al., 2007).

12.3 Severity framework, monitoring, and imaging strategy

Severity classification should follow the Revised Atlanta framework, which defines severe AP primarily by persistent organ failure and describes the evolution of pancreatic/peripancreatic collections. Because infection of necrosis synergizes

Advanced Insights into Acute Pancreatitis. https://doi.org/10.1016/B978-0-443-44736-5.00008-5

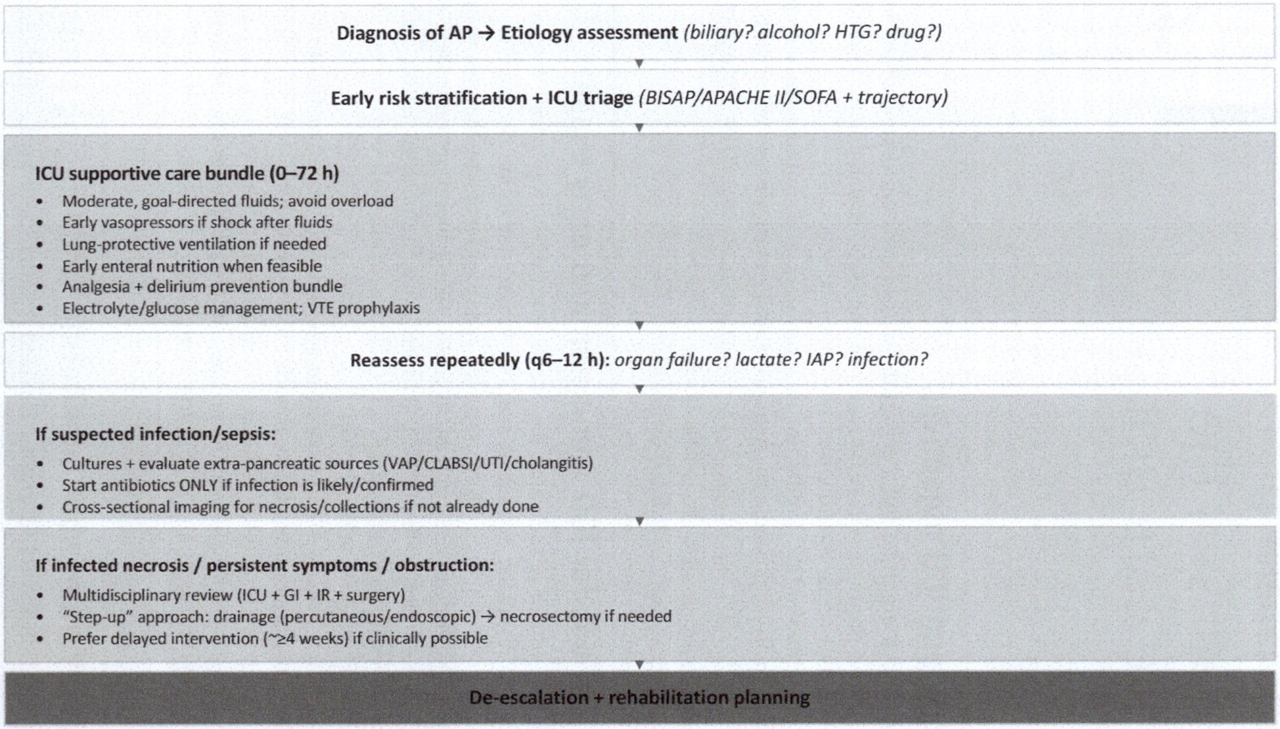

FIGURE 12.1 ICU care pathway for severe acute pancreatitis.

ICU pathway for severe acute pancreatitis.

with persistent organ failure to drive mortality, secondary deterioration later in the course should trigger systematic evaluation for infection and other ICU-acquired sources (Banks et al., 2013; Bradley, 1993; Petrov et al., 2010; Besselink et al., 2009).

Admission and serial severity scores support communication and trending but should not replace clinical trajectories. BISAP is useful early in the emergency setting, while APACHE II and SOFA are practical in the ICU for trending organ dysfunction over time (Wu et al., 2008; Knaus et al., 1985; Vincent et al., 1996; Marshall et al., 1995).

Contrast-enhanced CT (or MRI) is used to define necrosis, collections, and complications; it is typically most informative after 72 h unless the diagnosis is uncertain or the patient deteriorates unexpectedly. Gas in a necrotic collection is suggestive of infection but is not universally present, so diagnosis relies on the overall clinical context and microbiology when drainage occurs (Banks et al., 2013; Leppäniemi et al., 2019; van Grinsven et al., 2018).

12.4 Hemodynamic resuscitation and shock management

Early restoration of perfusion is critical, but excessive fluid loading increases pulmonary edema and intra-abdominal hypertension. A moderate, reassessment-driven approach is recommended, guided by perfusion markers, bedside echocardiography, urine output, and signs of overload (de-Madaria et al., 2022; de-Madaria et al., 2018; Tenner et al., 2013).

Balanced crystalloids are commonly preferred; lactated Ringer's has been associated with reduced systemic inflammation compared with normal saline in AP, and balanced crystalloids are favored in broader ICU populations to reduce kidney-related adverse events (Wu et al., 2011; de-Madaria et al., 2018; Semler et al., 2018; Self et al., 2018).

If hypotension persists after appropriate fluid challenges, early norepinephrine is recommended rather than continued large-volume crystalloids. When pancreatitis physiology overlaps with sepsis physiology, ICU shock management should follow standard sepsis guidance, including timely antimicrobial therapy and source control when infection is likely (Evans et al., 2021; Rivers et al., 2001; ProCESS Investigators, 2014; ARISE Investigators, 2014).

12.5 Respiratory support and pulmonary complications

Pulmonary complications (atelectasis, pleural effusions, pneumonia, and ARDS) are common reasons for ICU admission in AP. Respiratory failure is often multifactorial, reflecting systemic inflammation, capillary leak, aspiration, infection, and fluid overload (Banks et al., 2013; Petrov et al., 2010; Evans et al., 2021).

When invasive mechanical ventilation is required, lung-protective ventilation (low tidal volume and limitation of plateau pressure) is standard and improves outcomes in ARDS. In severe ARDS physiology, prone positioning and, in selected patients, short-course neuromuscular blockade can be considered as evidence-based adjuncts (ARDS Network, 2000; Guérin et al., 2013; Papazian et al., 2010).

12.6 Renal, metabolic, and endocrine management

Acute kidney injury in severe AP may result from hypoperfusion, systemic inflammation, intra-abdominal hypertension, nephrotoxins, and sepsis. Prevention focuses on optimized perfusion, avoidance of nephrotoxins, and recognition of abdominal compartment physiology (KDIGO, 2012; Kirkpatrick et al., 2013; Evans et al., 2021).

Renal replacement therapy should follow standard ICU indications (refractory acidosis, hyperkalemia, volume overload, uremic complications) with individualized timing in the absence of emergent criteria, consistent with randomized trial evidence (STARRT-AKI Investigators, 2020; Gaudry et al., 2016; Zarbock et al., 2016; Barbar et al., 2018).

Glycemic management should avoid tight control that increases hypoglycemia risk; moderate ICU targets are supported by large randomized trials. Electrolyte abnormalities (calcium, magnesium, phosphate) should be corrected, with attention to refeeding risk during nutrition escalation (Finfer et al., 2009; Van den Berghe et al., 2001; Arvanitakis et al., 2020).

12.7 Analgesia, sedation, delirium prevention, and early mobilization

Pain control is a core ICU intervention in AP because severe pain can worsen tachypnea, atelectasis, and delirium. ICU sedation should follow best practices emphasizing analgesia-first strategies, light sedation when possible, routine delirium monitoring, sleep promotion, and early mobilization (Devlin et al., 2018; Ely et al., 2004; Schweickert et al., 2009).

12.8 Nutrition strategy in severe acute pancreatitis

Enteral nutrition is preferred over parenteral nutrition in severe AP, supported by randomized trials and guideline recommendations. Early tube feeding can be provided on-demand when oral intake fails, with nasogastric feeding feasible in many patients; post-pyloric feeding may be used for intolerance or high aspiration risk (Kalfarentzos et al., 1997; Bakker et al., 2014; Eatock et al., 2005; Arvanitakis et al., 2020).

Probiotics should not be used routinely in predicted severe AP because randomized trial evidence demonstrated harm, including increased mortality and bowel ischemia signals (Besselink et al., 2008; Arvanitakis et al., 2020).

Tables 12.1 and 12.2 summarizing ICU admission triggers and evidence-based interventions.

Table 12.1 ICU admission triggers and monitoring targets.

Domain	ICU trigger (examples)	Monitoring focus (examples)	Key references
Circulation	Shock requiring vasopressors; rising lactate; persistent oliguria despite initial resuscitation	MAP; lactate trend; urine output; bedside echocardiography; cumulative fluid balance and signs of overload	Evans et al. (2021), de-Madaria et al. (2022), Vincent et al. (1996).
Respiration	Escalating oxygen requirement; need for non-invasive or invasive ventilation; ARDS physiology	SpO_2 and PaO_2/FiO_2; ventilator pressures (plateau/DP); imaging for effusions/ atelectasis; aspiration risk	ARDS Network (2000), Guérin et al. (2013), Banks et al. (2013)
Renal/ Metabolic	Progressive AKI; refractory hyperkalemia/acidosis; fluid overload limiting ventilation	KDIGO stage; electrolytes and acid-base status; fluid balance; RRT indications and response	KDIGO (2012), STARRT-AKI Investigators (2020), Kirkpatrick et al. (2013).
Inflammation/ Infection	Secondary clinical deterioration; suspected infected necrosis; suspected ICU-acquired infection (VAP/CLABSI/UTI/ cholangitis)	Cultures before antibiotics when feasible; imaging when indicated; de-escalation based on microbiology and response	Leppäniemi et al. (2019), Evans et al. (2021), Isenmann et al. (2004)
Abdomen	Progressive distension with worsening ventilation/renal function; suspected intra-abdominal hypertension/ACS	Bladder pressure (IAP) when suspected; response to decompression and drainage; perfusion and organ dysfunction	Kirkpatrick et al. (2013), Zarnescu et al. (2022)

Table 12.2 Evidence-based ICU interventions in severe AP.

Intervention domain	ICU recommendation	Avoid/Pitfalls	Key references
IV fluids and perfusion	Moderate, reassessment-driven resuscitation; prefer balanced crystalloids; early norepinephrine if shock persists	Fixed aggressive bolus strategies leading to fluid overload; delayed vasopressors with ongoing hypotension	de-Madaria et al. (2022), Wu et al. (2011), Semler et al. (2018), Evans et al. (2021)
Ventilation	Lung-protective ventilation; prone positioning for severe ARDS when appropriate	High tidal volumes and injurious pressures; ignoring fluid overload contribution to hypoxemia	ARDS Network (2000), Guérin et al. (2013), Papazian et al. (2010)
Nutrition	Oral feeding as tolerated; enteral nutrition when oral intake fails; NG route acceptable in many patients	Unnecessary delay in nutrition escalation; routine probiotics in predicted severe AP	Kalfarentzos et al. (1997), Bakker et al. (2014), Eatock et al. (2005), Besselink et al. (2008)
Antimicrobials	Treat proven or strongly suspected infection; obtain cultures; tailor and de-escalate based on results	Prophylactic antibiotics for sterile necrosis; prolonged broad-spectrum therapy without reassessment	Isenmann et al. (2004), Dellinger et al. (2007), Lim et al. (2015), Leppäniemi et al. (2019)
Necrotizing pancreatitis (source control)	Step-up approach: Percutaneous/endoscopic drainage → minimally invasive necrosectomy if needed; delay interventions until collections mature when feasible	Early open necrosectomy as first-line; drainage without multidisciplinary planning	van Santvoort et al. (2010), van Brunschot et al. (2018), Boxhoorn et al. (2021)
Intra-abdominal hypertension/ACS	Measure IAP when suspected; staged medical measures and drainage; decompressive laparotomy only if refractory	Missing ACS diagnosis in fluid-overloaded patients; late decompression after irreversible organ injury	Kirkpatrick et al. (2013), Zarnescu et al. (2022)
Sedation, delirium, mobilization	Analgesia-first; light sedation targets; delirium screening; early mobilization when feasible	Deep sedation without indication; lack of delirium surveillance; prolonged immobility	Devlin et al. (2018), Ely et al. (2004), Schweickert et al. (2009)
VTE prophylaxis and GI prophylaxis	Pharmacologic VTE prophylaxis unless contraindicated; individualized stress ulcer prophylaxis for high-risk patients	Withholding prophylaxis without clear bleeding risk; routine stress ulcer prophylaxis in low-risk patients	PROTECT Investigators (2011), Arabi et al. (2019), Krag et al. (2018)

12.9 Infection, diagnostics, and antimicrobial stewardship

Early in AP, fever and leukocytosis often reflect sterile inflammation. A key ICU task is to avoid anchoring on pancreatic infection and to evaluate for extrapancreatic and ICU-acquired sources (pneumonia, catheter-related bloodstream infection, urinary tract infection, cholangitis), particularly when secondary deterioration occurs (Besselink et al., 2009; De Waele et al., 2004; Evans et al., 2021).

Routine prophylactic antibiotics for sterile necrosis are not recommended because randomized trials and meta-analyses have not shown reduction in infected necrosis or mortality. Excess antimicrobial exposure increases selection of resistant organisms and is associated with fungal infection in severe AP (Isenmann et al., 2004; Dellinger et al., 2007; Xue et al., 2009; Jafri et al., 2009; Wittau et al., 2011; Lim et al., 2015; Kochhar et al., 2011; Lee et al., 2014).

When infection is proven or strongly suspected, antibiotics should be initiated promptly after cultures, tailored to local epidemiology, and de-escalated based on results and clinical response. Fine-needle aspiration is no longer routinely recommended because results rarely change management and may delay intervention (Evans et al., 2021; Leppäniemi et al., 2019; Banks et al., 1995; Gerzof et al., 1987).

12.10 Source control and procedural management of necrotizing pancreatitis

Infected necrosis and symptomatic sterile necrosis not resolving with supportive care are managed using a step-up approach, typically starting with percutaneous or endoscopic drainage and escalating to minimally invasive necrosectomy only when necessary. Randomized trials support step-up strategies over primary open necrosectomy in appropriate candidates (van Santvoort et al., 2010, 2011; van Brunschot et al., 2018; Mortelé et al., 2009).

Timing should be individualized. Delaying intervention until collections mature (often around 4 weeks) improves technical feasibility and outcomes, but earlier drainage may be required when sepsis cannot be controlled. Trials comparing immediate versus postponed intervention support selective postponement when physiology permits (van Santvoort et al., 2011; Boxhoorn et al., 2021; Onnekink et al., 2022; Leppäniemi et al., 2019).

12.11 Intra-abdominal hypertension, abdominal compartment syndrome, and gastrointestinal prophylaxis

Severe AP predisposes to intra-abdominal hypertension and abdominal compartment syndrome due to visceral edema, ileus, ascites, and iatrogenic fluid loading.

ICU teams should measure intra-abdominal pressure when suspected and apply staged management (optimize sedation, evacuate intraluminal contents, cautious diuresis/ultrafiltration, percutaneous drainage) before considering decompressive laparotomy in refractory cases (Kirkpatrick et al., 2013; Zarnescu et al., 2022).

Stress ulcer prophylaxis should be individualized to high-risk ICU patients. In a large randomized ICU trial, pantoprazole reduced clinically important gastrointestinal bleeding but did not improve mortality, supporting targeted use with attention to infectious risks (Krag et al., 2018; Alhazzani et al., 2018).

12.12 Venous thromboembolism prevention and ICU disposition

Critically ill patients with AP should receive standard pharmacologic venous thromboembolism prophylaxis unless contraindicated, coordinated with procedural plans and bleeding risk. Large ICU trials inform prophylaxis selection and demonstrate the incremental benefit of adjunctive mechanical prophylaxis in selected settings (PROTECT Investigators, 2011; Arabi et al., 2019).

Transition out of ICU should occur once organ failures resolve or stabilize with minimal support, nutrition is established, and an explicit plan exists for ongoing management of collections and definitive etiologic prevention (e.g., cholecystectomy timing in gallstone disease) (Banks et al., 2013; da Costa et al., 2015; Tenner et al., 2013).

References

Alhazzani, W., Alshamsi, F., Belley-Cote, E., et al., 2018. Efficacy and safety of stress ulcer prophylaxis in critically ill patients: a network meta-analysis of randomized trials. Intensive Care Medicine 44 (1), 1–11. https://doi.org/10.1007/s00134-017-5005-8.

Arabi, Y.M., Al-Hameed, F., Burns, K.E.A., Mehta, S., Alsolamy, S.J., Alshahrani, M.S., Mandourah, Y., Almekhlafi, G.A., Almaani, M., Ali, Al B., Simon, F., Arshad, Z., Khalid, I., Mehta, Y., Gaur, A., Hawa, H., Buscher, H., Lababidi, H., Abdulsalam, Al A., Abdukahil, S.A.I., Jose, J., Afesh, L.Y., Al-Dawood, A., 2019. Adjunctive intermittent pneumatic compression for venous thromboprophylaxis. New England Journal of Medicine 380 (14), 1305–1315. https://doi.org/10.1056/nejmoa1816150.

ARDS Network, 2000. Ventilation with lower tidal volumes as compared with traditional tidal volumes for acute lung injury and the acute respiratory distress syndrome. New England Journal of Medicine 342 (18), 1301–1308. https://doi.org/10.1056/NEJM200005043421801.

Arvanitakis, M., Ockenga, J., Bezmarevic, M., Gianotti, L., Krznarić, Ž., Lobo, D.N., Löser, C., Madl, C., Meier, R., Phillips, M., Rasmussen, H.H., Van Hooft, J.E., Bischoff, S.C., 2020. ESPEN guideline on clinical nutrition in acute and chronic pancreatitis. Clinical Nutrition 39 (3), 612–631. https://doi.org/10.1016/j.clnu.2020.01.004.

Bakker, O.J., Van Brunschot, S., Van Santvoort, H.C., Besselink, M.G., Bollen, T.L., Boermeester, M.A., Dejong, C.H., Van Goor, H., Bosscha, K., Ahmed Ali, U., Bouwense, S., Van Grevenstein, W.M., Heisterkamp, J., Houdijk, A.P., Jansen, J.M., Karsten, T.M., Manusama, E.R., Nieuwenhuijs, V.B., Schaapherder, A.F., Van Der Schelling, G.P., Schwartz, M.P., Spanier, B.W.M., Tan, A., Vecht, J., Weusten, B.L., Witteman, B.J., Akkermans, L.M., Bruno, M.J., Dijkgraaf, M.G., Van Ramshorst, B., Gooszen, H.G., 2014. Early versus on-demand nasoenteric tube feeding in acute pancreatitis. New England Journal of Medicine 371 (21), 1983–1993. https://doi.org/10.1056/NEJMoa1404393.

Banks, P.A., Gerzof, S.G., Langevin, R.E., Silverman, S.G., Sica, G.T., Hughes, M.D., 1995. CT-guided aspiration of suspected pancreatic infection—bacteriology and clinical outcome. International Journal of Pancreatology 18 (3), 265–270. https://doi.org/10.1007/BF02784951.

Banks, P.A., Bollen, T.L., Dervenis, C., Gooszen, H.G., Johnson, C.D., Sarr, M.G., Tsiotos, G. G., Vege, S.S., Windsor, J.A., Horvath, K.D., Mortele, K.J., Gardner, T.B., Van Santvoort, H., Pelaez-Luna, M., Yadav, D., Stefanidis, G., Delakidis, S., Morgan, D.E., Thoeni, R.F.L., Acosta, J.M., Amann, S.T., Andren-Sandberg, A., Aranha, G.V., Asciutti, S., Barauskas, G., Baron, T.H., Bassi, C., Behrman, S., Behrns, K.E., Belliappa, V., Berzin, T.M., Besselink, M.G.H., Bhasin, D.K., Biankin, A., Bishop, M. D., Bonini, C.J., Bradley, E.L., Buechler, M., Carter, R., Cavestro, G.M., Chari, S.T., Chavez-Rodriguez, J.J., Da Cunha, J.E., D'Agostino, H., De Campos, T., De-Madaria, E., Deprez, P.H., DiSario, J.A., Doria, C., Falconi, M., Fernandez-del Castillo, C., Freeny, P.C., Frey, C.F., Friess, H., Frossard, J.L., Fuchshuber, P., Gallagher, S.F., Garg, P.K., Ghattas, G., Glasgow, R., Gonzalez, J.A., Gress, T.M., Gumbs, A.A., Hallibruton, C., Helton, S., Hill, M.C., Hoyos, S., Imrie, C.W., Isenmann, R., Izbicki, J.R., Karagiannis, J.A., Klar, E., Kolokythas, O., Lau, J., Litvin, A.A., Longnecker, D.S., Lowenfels, A.B., Mackey, R., Mah'Moud, M., Malangoni, M., McFadden, D.W., Mishra, G., Moody, F.G., Morinville, V., Neoptolemos, J.P., Nordback, I., Pap, A., Papachristou, G.I., Parks, R., Pedrazolli, S., Pezzilli, R., Pitt, H.A., Prosanto, C., Ramesh, H., Ramirez, F.C., Raper, S.E., Rasheed, A., Reed, D.N., Romangnuolo, J., Rossaak, J., Sanabria, J., Schaefer, C., Schmidt, J., Schmidt, P.N., Serrablo, A., Senkowski, C.K., Sharma, M., Sigman, K.M., Singh, P., Steinberg, W., Steiner, J., Strasberg, S., Strum, W., Takada, T., Tanaka, M., Vaccaro, M., Villavicencio, R.L., Vrochides, D., Wagner, M., Warshaw, A.L., Wilcox, C.M., Wysocki, P., Zenilman, M.E., Zyromski, N.J., 2013. Classification of acute pancreatitis—2012: revision of the Atlanta classification and definitions by international consensus. Gut 62 (1), 102–111. https://doi.org/10.1136/gutjnl-2012-302779.

Barbar, S.D., Clere-Jehl, R., Bourredjem, A., Hernu, R., Montini, F., Bruyère, R., Lebert, C., Bohé, J., Badie, J., Eraldi, J.P., Rigaud, J.P., Levy, B., Siami, S., Louis, G., Bouadma, L., Constantin, J.M., Mercier, E., Klouche, K., Du Cheyron, D., Piton, G., Annane, D., Jaber, S., Van Der Linden, T., Blasco, G., Mira, J.P., Schwebel, C., Chimot, L., Guiot, P., Nay, M.A., Meziani, F., Helms, J., Roger, C., Louart, B., Trusson, R., Dargent, A., Binquet, C., Quenot, J.P., 2018. Timing of renal-replacement therapy in patients with acute kidney injury and sepsis. New England Journal of Medicine 379 (15), 1431–1442. https://doi.org/10.1056/NEJMoa1803213.

Besselink, M.G., van Santvoort, H.C., Buskens, E., Boermeester, M.A., van Goor, H., Timmerman, H.M., Nieuwenhuijs, V.B., Bollen, T.L., van Ramshorst, B., Witteman, B. J., Rosman, C., Ploeg, R.J., Brink, M.A., Schaapherder, A.F., Dejong, C.H., Wahab, P.

J., van Laarhoven, C.J., van der Harst, E., van Eijck, C.H., Cuesta, M.A., Akkermans, L. M., Gooszen, H.G., 2008. Probiotic prophylaxis in predicted severe acute pancreatitis: a randomised, double-blind, placebo-controlled trial. The Lancet 371 (9613), 651–659. https://doi.org/10.1016/S0140-6736(08)60207-X.

Besselink, M.G., van Santvoort, H.C., Boermeester, M.A., Nieuwenhuijs, V.B., van Goor, H., Dejong, C.H.C., Schaapherder, A.F., Gooszen, H.G., 2009. Timing and impact of infections in acute pancreatitis. British Journal of Surgery 96 (3), 267–273. https://doi.org/10.1002/bjs.6447.

Boxhoorn, L., van Dijk, S.M., van Grinsven, J., Verdonk, R.C., Boermeester, M.A., Bollen, T. L., Bouwense, S.A.W., Bruno, M.J., Cappendijk, V.C., Dejong, C.H.C., van Duijvendijk, P., van Eijck, C.H.J., Fockens, P., Francken, M.F.G., van Goor, H., Hadithi, M., Hallensleben, N.D.L., Willem Haveman, J., Jacobs, M.A.J.M., Jansen, J. M., Kop, M.P.M., van Lienden, K.P., Manusama, E.R., Mieog, J.S.D., Quintus Molenaar, I., Nieuwenhuijs, V.B., Poen, A.C., Jan-Werner, P., Poll, M.van de, Quispel, R., Römkens, T.E.H., Schwartz, M.P., Seerden, T.C., Stommel, M.W.J., Straathof, J.W.A., Timmerhuis, H.C., Venneman, N.G., Voermans, R.P., Vrie, W.van de, Witteman, B.J., Dijkgraaf, M.G.W., van Santvoort, H.C., Besselink, M.G., 2021. Immediate versus postponed intervention for infected necrotizing pancreatitis. New England Journal of Medicine 385 (15), 1372–1381. https://doi.org/10.1056/nejmoa2100826.

Bradley, E.L., 1993. A clinically based classification system for acute pancreatitis. Archives of Surgery 128 (5), 586–590. https://doi.org/10.1001/archsurg.1993.01420170122019.

Da Costa, D.W., Bouwense, S.A., Schepers, N.J., Besselink, M.G., Van Santvoort, H.C., Van Brunschot, S., Bakker, O.J., Bollen, T.L., Dejong, C.H., Van Goor, H., Boermeester, M.A., Bruno, M.J., Van Eijck, C.H., Timmer, R., Weusten, B.L., Consten, E.C., Brink, M.A., Spanier, B.W.M., Bilgen, E.J.S., Nieuwenhuijs, V.B., Hofker, H.S., Rosman, C., Voorburg, A.M., Bosscha, K., Van Duijvendijk, P., Gerritsen, J.J., Heisterkamp, J., De Hingh, I.H., Witteman, B.J., Kruyt, P.M., Scheepers, J.J., Molenaar, I.Q., Schaapherder, A.F., Manusama, E.R., Van Der Waaij, L.A., Van Unen, J., Dijkgraaf, M. G., Van Ramshorst, B., Gooszen, H.G., Boerma, D., 2015. Same-admission versus interval cholecystectomy for mild gallstone pancreatitis (PONCHO): a multicentre randomised controlled trial. The Lancet 386 (10000), 1261–1268. https://doi.org/10.1016/S0140-6736(15)00274-3.

De Waele, J.J., Vogelaers, D., Hoste, E., Blot, S., Colardyn, F., 2004. Emergence of antibiotic resistance in infected pancreatic necrosis. Archives of Surgery 139 (12), 1371–1375. https://doi.org/10.1001/archsurg.139.12.1371.

de-Madaria, E., Herrera-Marante, I., González-Camacho, V., Bonjoch, L., Quesada-Vázquez, N., Almenta-Saavedra, I., Miralles-Maciá, C., Acevedo-Piedra, N.G., Roger-Ibáñez, M., Sánchez-Marin, C., Osuna-Ligero, R., Gracia, Á., Llorens, P., Zapater, P., Singh, V.K., Moreu-Martín, R., Closa, D., 2018. Fluid resuscitation with lactated Ringer's solution vs normal saline in acute pancreatitis: a triple-blind, randomized, controlled trial. United European Gastroenterology Journal 6 (1), 63–72. https://doi.org/10.1177/2050640617707864.

de-Madaria, E., Buxbaum, J.L., Maisonneuve, P., Paredes, A.G.G.de, Zapater, P., Guilabert, L., Vaillo-Rocamora, A., Rodríguez-Gandía, M.Á., Donate-Ortega, J., Lozada-Hernández, E.E., Moreno, A.J.R.C., Lira-Aguilar, A., Llovet, L.P., Mehta, R., Tandel, R., Navarro, P., Sánchez-Pardo, A.M., Sánchez-Marin, C., Cobreros, M., Fernández-Cabrera, I., Casals-Seoane, F., Deza, D.C., Lauret-Braña, E., Martí-Marqués, E., Camacho-Montaño, L.M., Ubieto, V., Ganuza, M., Bolado, F., 2022. Aggressive or moderate fluid resuscitation in acute pancreatitis. New England Journal of Medicine 387 (11), 989–1000. https://doi.org/10.1056/nejmoa2202884.

Dellinger, E.P., Tellado, J.M., Soto, N.E., Ashley, S.W., Barie, P.S., Dugernier, T., Imrie, C.W., Johnson, C.D., Knaebel, H.P., Laterre, P.F., Maravi-Poma, E., Kissler, J.J.O., Sanchez-Garcia, M., Utzolino, S., 2007. Early antibiotic treatment for severe acute necrotizing pancreatitis: a randomized, double-blind, placebo-controlled study. Annals of Surgery 245 (5), 674–683. https://doi.org/10.1097/01.sla.0000250414.09255.84.

Devlin, J.W., Skrobik, Y., Gélinas, C., Needham, D.M., Slooter, A.J.C., Pandharipande, P.P., Watson, P.L., Weinhouse, G.L., Nunnally, M.E., Rochwerg, B., Balas, M.C., Van Den Boogaard, M., Bosma, K.J., Brummel, N.E., Chanques, G., Denehy, L., Drouot, X., Fraser, G.L., Harris, J.E., Joffe, A.M., Kho, M.E., Kress, J.P., Lanphere, J.A., McKinley, S., Neufeld, K.J., Pisani, M.A., Payen, J.F., Pun, B.T., Puntillo, K.A., Riker, R.R., Robinson, B.R.H., Shehabi, Y., Szumita, P.M., Winkelman, C., Centofanti, J.E., Price, C., Nikayin, S., Misak, C.J., Flood, P.D., Kiedrowski, K., Alhazzani, W., 2018. Clinical practice guidelines for the prevention and management of pain, agitation/sedation, delirium, immobility, and sleep disruption in adult patients in the ICU. Critical Care Medicine 46 (9), e825–e873. https://doi.org/10.1097/CCM.0000000000003299.

Eatock, F.C., Chong, P., Menezes, N., Murray, L., McKay, C.J., Carter, C.R., Imrie, C.W., 2005. A randomized study of early nasogastric versus nasojejunal feeding in severe acute pancreatitis. American Journal of Gastroenterology 100 (2), 432–439. https://doi.org/10.1111/j.1572-0241.2005.40587.x.

Ely, E.W., Shintani, A., Truman, B., et al., 2004. Delirium as a predictor of mortality in mechanically ventilated patients in the intensive care unit. JAMA 291 (14), 1753–1762. https://doi.org/10.1001/jama.291.14.1753.

Evans, L., Rhodes, A., Alhazzani, W., Antonelli, M., Coopersmith, C.M., Craig, F., Machado, F.R., Mcintyre, L., Ostermann, M., Prescott, H.C., Schorr, C., Simpson, S., Joost Wiersinga, W., Alshamsi, F., Angus, D.C., Arabi, Y., Azevedo, L., Beale, R., Beilman, G., Belley-Cote, E., Burry, L., Cecconi, M., Centofanti, J., Yataco, A.C., De Waele, J., Phillip Dellinger, R., Doi, K., Du, B., Estenssoro, E., Ferrer, R., Gomersall, C., Hodgson, C., Møller, M.H., Iwashyna, T., Jacob, S., Kleinpell, R., Klompas, M., Koh, Y., Kumar, A., Kwizera, A., Lobo, S., Masur, H., McGloughlin, S., Mehta, S., Mehta, Y., Mer, M., Nunnally, M., Oczkowski, S., Osborn, T., Papathanassoglou, E., Perner, A., Puskarich, M., Roberts, J., Schweickert, W., Seckel, M., Sevransky, J., Sprung, C.L., Welte, T., Zimmerman, J., Levy, M., 2021. Surviving sepsis campaign: international guidelines for management of sepsis and septic shock 2021. Intensive Care Medicine 47 (11), 1181–1247. https://doi.org/10.1007/s00134-021-06506-y.

Fan, S.T., Lai, E., Mok, F., lo, C.M., Zheng, S.S., Wong, J., 1993. Early treatment of acute biliary pancreatitis by endoscopic papillotomy. New England Journal of Medicine 328 (4), 228–232. https://doi.org/10.1056/NEJM199301283280402.

Finfer, S., Bellomo, R., Blair, D., Su, S.Y.S., Foster, D., Dhingra, V., Cook, D., Dodek, P., Henderson, W.R., Hébert, P.C., Heyland, D.K., McArthur, C., McDonald, E., Mitchell, I., Myburgh, J.A., Nor-ton, R., Potte, J., Robinson, B.G., NICE-SUGAR Study Investigators, 2009. Intensive versus conventional glucose control in critically Ill patients. New England Journal of Medicine 360 (13), 1283–1297. https://doi.org/10.1056/NEJMoa0810625.

Fölsch, U.R., Nitsche, R., Lüdtke, R., Hilgers, R.A., Creutzfeldt, W., 1997. Early ERCP and papillotomy compared with conservative treatment for acute biliary pancreatitis. New

England Journal of Medicine 336 (4), 237–242. https://doi.org/10.1056/NEJM199701233360401.

Gaudry, S., Hajage, D., Schortgen, F., Martin-Lefevre, L., Pons, B., Boulet, E., Boyer, A., Chevrel, G., Lerolle, N., Carpentier, D., Prost, N.de, Lautrette, A., Bretagnol, A., Mayaux, J., Nseir, S., Megarbane, B., Thirion, M., Forel, J.-M., Maizel, J., Yonis, H., Markowicz, P., Thiery, G., Tubach, F., Ricard, J.-D., Dreyfuss, D., 2016. Initiation strategies for renal-replacement therapy in the intensive care unit. New England Journal of Medicine 375 (2), 122–133. https://doi.org/10.1056/NEJMoa1603017.

Gerzof, S.G., Banks, P.A., Robbins, A.H., Johnson, W.C., Spechler, S.J., Wetzner, S.M., Snider, J.M., Langevin, R.E., Jay, M.E., 1987. Early diagnosis of pancreatic infection by computed tomography-guided aspiration. Gastroenterology 93 (6), 1315–1320. https://doi.org/10.1016/0016-5085(87)90261-7.

Guérin, C., Reignier, J., Richard, J.-C., Beuret, P., Gacouin, A., Boulain, T., Mercier, E., Badet, M., Mercat, A., Baudin, O., Clavel, M., Chatellier, D., Jaber, S., Rosselli, S., Mancebo, J., Sirodot, M., Hilbert, G., Bengler, C., Richecoeur, J., Gainnier, M., Bayle, F., Bourdin, G., Leray, V., Girard, R., Baboi, L., Ayzac, L., 2013. Prone positioning in severe acute respiratory distress syndrome. New England Journal of Medicine 368 (23), 2159–2168. https://doi.org/10.1056/NEJMoa1214103.

Isenmann, R., Rünzi, M., Kron, M., Kahl, S., Kraus, D., Jung, N., Maier, L., Malfertheiner, P., Goebell, H., Beger, H.G., 2004. Prophylactic antibiotic treatment in patients with predicted severe acute pancreatitis: a placebo-controlled, double-blind trial1. Gastroenterology 126 (4), 997–1004. https://doi.org/10.1053/j.gastro.2003.12.050.

Jafri, N.S., Mahid, S.S., Idstein, S.R., Hornung, C.A., Galandiuk, S., 2009. Antibiotic prophylaxis is not protective in severe acute pancreatitis: a systematic review and meta-analysis. The American Journal of Surgery 197 (6), 806–813. https://doi.org/10.1016/j.amjsurg.2008.08.016.

Kalfarentzos, F., Kehagias, J., Mead, N., Kokkinis, K., Gogos, C.A., 1997. Enteral nutrition is superior to parenteral nutrition in severe acute pancreatitis: results of a randomized prospective trial. British Journal of Surgery 84 (12), 1665–1669. https://doi.org/10.1002/bjs.1800841207.

KDIGO, 2012. KDIGO clinical practice guideline for acute kidney injury. Kidney International Supplements 2 (1), 1–138.

Kirkpatrick, A.W., Roberts, D.J., De Waele, J., Jaeschke, R., Malbrain, M.L.N.G., De Keulenaer, B., Duchesne, J., Bjorck, M., Leppaniemi, A., Ejike, J.C., Sugrue, M., Cheatham, M., Ivatury, R., Ball, C.G., Reintam Blaser, A., Regli, A., Balogh, Z.J., D'Amours, S., Debergh, D., Kaplan, M., Kimball, E., Olvera, C., 2013. Intra-abdominal hypertension and the abdominal compartment syndrome: updated consensus definitions and clinical practice guidelines from the World Society of the Abdominal Compartment Syndrome. Intensive Care Medicine 39 (7), 1190–1206. https://doi.org/10.1007/s00134-013-2906-z.

Knaus, W.A., Draper, E.A., Wagner, D.P., Zimmerman, J.E., 1985. Apache II: a severity of disease classification system. Critical Care Medicine 13 (10), 818–829. https://doi.org/10.1097/00003246-198510000-00009.

Kochhar, R., Noor, M.T., Wig, J., 2011. Fungal infections in severe acute pancreatitis. Journal of Gastroenterology and Hepatology 26 (6), 952–959. https://doi.org/10.1111/j.1440-1746.2011.06685.x.

Krag, M., Marker, S., Perner, A., Wetterslev, J., Wise, M.P., Schefold, J.C., Keus, F., Guttormsen, A.B., Bendel, S., Borthwick, M., Lange, T., Rasmussen, B.S.,

Siegemund, M., Bundgaard, H., Elkmann, T., Jensen, J.V., Nielsen, R.D., Liboriussen, L., Bestle, M.H., Elkjær, J.M., Palmqvist, D.F., Bäcklund, M., Laake, J.H., Bådstøløkken, P. M., Grönlund, J., Breum, O., Walli, A., Winding, R., Iversen, S., Jarnvig, I.-L., White, J. O., Brand, B., Madsen, M.B., Quist, L., Thornberg, K.J., Møller, A., Wiis, J., Granholm, A., Anthon, C.T., Meyhoff, T.S., Hjortrup, P.B., Aagaard, S.R., Andreasen, J.B., Sørensen, C.A., Haure, P., Jacob, H., Hollinger, A., Scheuzger, J., Tuchscherer, D., Vuilliomenet, T., Takala, J., Jakob, S.M., Vang, M.L., Pælestik, K.B., Andersen, K.L.D., van der Horst, I.C.C., Dieperink, W., Fjølner, J., Kjer, C.K.W., Sølling, C., Sølling, C.G., Karttunen, J., Morgan, M.P.G., Sjøbø, B., Engstrøm, J., Agerholm-Larsen, B., Møller, M.H., 2018. Pantoprazole in patients at risk for gastrointestinal bleeding in the ICU. New England Journal of Medicine 379 (23), 2199–2208. https://doi.org/10.1056/nejmoa1714919.

Lee, H.S., Lee, S.K., Park, D.H., Lee, S.S., Seo, D.W., Kim, M.H., Chong, Y.P., 2014. Emergence of multidrug resistant infection in patients with severe acute pancreatitis. Pancreatology 14 (6), 450–453. https://doi.org/10.1016/j.pan.2014.10.003.

Leppäniemi, A., Tolonen, M., Tarasconi, A., Segovia-Lohse, H., Gamberini, E., Kirkpatrick, A.W., Ball, C.G., Parry, N., Sartelli, M., Wolbrink, D., Goor, H.van, Baiocchi, G., Ansaloni, L., Biffl, W., Coccolini, F., Saverio, S.Di, Kluger, Y., Moore, E., Catena, F., 2019. WSES guidelines for the management of severe acute pancreatitis. World Journal of Emergency Surgery 14 (1). https://doi.org/10.1186/s13017-019-0247-0.

Lim, C.L.L., Lee, W., Liew, Y.X., Tang, S.S.L., Chlebicki, M.P., Kwa, A.L.H., 2015. Role of antibiotic prophylaxis in necrotizing pancreatitis: a meta-analysis. Journal of Gastrointestinal Surgery 19 (3), 480–491. https://doi.org/10.1007/s11605-014-2662-6.

Marshall, J.C., Cook, D.J., Christou, N.V., Bernard, G.R., Sprung, C.L., Sibbald, W.J., 1995. Multiple organ dysfunction score: a reliable descriptor of a complex clinical outcome. Critical Care Medicine 23 (10), 1638–1652. https://doi.org/10.1097/00003246-199510000-00007.

Mortelé, K.J., et al., 2009. CT-guided percutaneous catheter drainage of acute necrotizing pancreatitis: clinical experience and observations. American Journal of Roentgenology 192 (1), 110–116. https://doi.org/10.2214/AJR.08.1116.

Neoptolemos, J.P., London, N.J., James, D., Carr-Locke, D.L., Bailey, I.A., Fossard, D.P., 1988. Controlled trial of urgent endoscopic retrograde cholangiopancreatography and endoscopic sphincterotomy versus conservative treatment for acute pancreatitis due to gallstones. The Lancet 332 (8618), 979–983. https://doi.org/10.1016/s0140-6736(88)90740-4.

Oláh, A., Pardavi, G., Belágyi, T., Nagy, A., Issekutz, Á.A., Mohamed, G.E., 2002. Early nasojejunal feeding in acute pancreatitis is associated with a lower complication rate. Nutrition 18 (3), 259–262. https://doi.org/10.1016/s0899-9007(01)00755-9.

Onnekink, A.M., et al., 2022. Endoscopic versus surgical step-up approach for infected necrotizing pancreatitis: long-term follow-up. Gastroenterology 162 (4), 1203–1215.e3. https://doi.org/10.1053/j.gastro.2022.01.012.

Oría, A., Cimmino, D., Ocampo, C., Silva, W., Kohan, G., Zandalazini, H., Szelagowski, C., Chiappetta, L., 2007. Early endoscopic intervention versus early conservative management in patients with acute gallstone pancreatitis and biliopancreatic obstruction. Annals of Surgery 245 (1), 10–17. https://doi.org/10.1097/01.sla.0000232539.88254.80.

Papazian, L., Forel, J.-M., Gacouin, A., Penot-Ragon, C., Perrin, G., Loundou, A., Jaber, S., Arnal, J.-M., Perez, D., Seghboyan, J.-M., Constantin, J.-M., Courant, P., Lefrant, J.-Y.,

Guérin, C., Prat, G., Morange, S., Roch, A., Blockers, N., 2010. Neuromuscular blockers in early acute respiratory distress syndrome. New England Journal of Medicine 363 (12), 1107–1116. https://doi.org/10.1056/nejmoa1005372.

Petrov, M.S., Shanbhag, S., Chakraborty, M., Phillips, A.R.J., Windsor, J.A., 2010. Organ failure and infection of pancreatic necrosis as determinants of mortality in patients with acute pancreatitis. Gastroenterology 139 (3), 813–820. https://doi.org/10.1053/j.gastro.2010.06.010.

Rivers, E., Nguyen, B., Havstad, S., Ressler, J., Muzzin, A., Knoblich, B., Peterson, E., Tomlanovich, M., 2001. Early goal-directed therapy in the treatment of severe sepsis and septic shock. New England Journal of Medicine 345 (19), 1368–1377. https://doi.org/10.1056/NEJMoa010307.

Schweickert, W.D., Pohlman, M.C., Pohlman, A.S., Nigos, C., Pawlik, A.J., Esbrook, C.L., Spears, L., Miller, M., Franczyk, M., Deprizio, D., Schmidt, G.A., Bowman, A., Barr, R., McCallister, K.E., Hall, J.B., Kress, J.P., 2009. Early physical and occupational therapy in mechanically ventilated, critically ill patients: a randomised controlled trial. The Lancet 373 (9678), 1874–1882. https://doi.org/10.1016/S0140-6736(09)60658-9.

Self, W.H., Semler, M.W., Wanderer, J.P., Wang, L., Byrne, D.W., Collins, S.P., Slovis, C.M., Lindsell, C.J., Ehrenfeld, J.M., Siew, E.D., Shaw, A.D., Bernard, G.R., Rice, T.W., 2018. Balanced crystalloids versus saline in Noncritically Ill adults. New England Journal of Medicine 378 (9), 819–828. https://doi.org/10.1056/NEJMoa1711586.

Semler, M.W., Self, W.H., Wanderer, J.P., Ehrenfeld, J.M., Wang, L., Byrne, D.W., Stollings, J.L., Kumar, A.B., Hughes, C.G., Hernandez, A., Guillamondegui, O.D., May, A.K., Weavind, L., Casey, J.D., Siew, E.D., Shaw, A.D., Bernard, G.R., Rice, T. W., 2018. Balanced crystalloids versus saline in Critically Ill adults. New England Journal of Medicine 378 (9), 829–839. https://doi.org/10.1056/NEJMoa1711584.

STARRT-AKI Investigators, 2020. Timing of initiation of renal-replacement therapy in acute kidney injury. New England Journal of Medicine 383 (3), 240–251. https://doi.org/10.1056/nejmoa2000741.

Tenner, S., Baillie, J., DeWitt, J., Swaroop Vege, S., 2013. American college of gastroenterology guideline: management of acute pancreatitis. American Journal of Gastroenterology 108 (9), 1400–1415. https://doi.org/10.1038/ajg.2013.218.

The ARISE Investigators and the ANZICS Clinical Trials Group, 2014. Goal-directed resuscitation for patients with early septic shock. New England Journal of Medicine 371 (16), 1496–1506. https://doi.org/10.1056/NEJMoa1404380.

The ProCESS Investigators, 2014. A randomized trial of protocol-based care for early septic shock. New England Journal of Medicine 370 (18), 1683–1693. https://doi.org/10.1056/NEJMoa1401602.

The PROTECT Investigators for the Canadian Critical Care Trials Group and the Australian and New Zealand Intensive Care Society Clinical Trials Group, 2011. Dalteparin versus unfractionated Heparin in critically ill patients. New England Journal of Medicine 364 (14), 1305–1314. https://doi.org/10.1056/nejmoa1014475.

van Brunschot, S., van Grinsven, J., van Santvoort, H.C., et al., 2018. Endoscopic or surgical step-up approach for infected necrotising pancreatitis: a multicentre randomised trial. Lancet 391 (10115), 51–58. https://doi.org/10.1016/S0140-6736(17)32404-2.

Van den Berghe, G., Wouters, P., Weekers, F., Verwaest, C., Bruyninckx, F., Schetz, M., Vlasselaers, D., Ferdinande, P., Lauwers, P., Bouillon, R., 2001. Intensive insulin therapy in critically ill patients. New England Journal of Medicine 345 (19), 1359–1367. https://doi.org/10.1056/nejmoa011300.

van Grinsven, J., van Brunschot, S., van Baal, M.C., Besselink, M.G., Fockens, P., Goor, H. van, van Santvoort, H.C., Bollen, T.L., 2018. Natural history of gas configurations and encapsulation in necrotic collections during necrotizing pancreatitis. Journal of Gastrointestinal Surgery 22 (9), 1557–1564. https://doi.org/10.1007/s11605-018-3792-z.

Van Santvoort, H.C., Besselink, M.G., Bakker, O.J., Hofker, H.S., Boermeester, M.A., Dejong, C.H., Van Goor, H., Schaapherder, A.F., Van Eijck, C.H., Bollen, T.L., Van Ramshorst, B., Nieuwenhuijs, V.B., Timmer, R., Laméris, J.S., Kruyt, P.M., Manusama, E.R., Van Der Harst, E., Van Der Schelling, G.P., Karsten, T., Hesselink, E. J., Van Laarhoven, C.J., Rosman, C., Bosscha, K., De Wit, R.J., Houdijk, A.P., Van Leeuwen, M.S., Buskens, E., Gooszen, H.G., 2010. A step-up approach or open necrosectomy for necrotizing pancreatitis. New England Journal of Medicine 362 (16), 1491–1502. https://doi.org/10.1056/NEJMoa0908821.

Van Santvoort, H.C., Bakker, O.J., Bollen, T.L., Besselink, M.G., Ahmed Ali, U., Schrijver, A. M., Boermeester, M.A., Van Goor, H., Dejong, C.H., Van Eijck, C.H., Van Ramshorst, B., Schaapherder, A.F., Van Der Harst, E., Hofker, S., Nieuwenhuijs, V.B., Brink, M.A., Kruyt, P.M., Manusama, E.R., Van Der Schelling, G.P., Karsten, T., Hesselink, E.J., Van Laarhoven, C.J., Rosman, C., Bosscha, K., De Wit, R.J., Houdijk, A.P., Cuesta, M. A., Wahab, P.J., Gooszen, H.G., 2011. A conservative and minimally invasive approach to necrotizing pancreatitis improves outcome. Gastroenterology 141 (4), 1254–1263. https://doi.org/10.1053/j.gastro.2011.06.073.

Vincent, J.-L., Moreno, R., Takala, J., Willatts, S., De Mendonça, A., Bruining, H., Reinhart, C.K., Suter, P.M., Thijs, L.G., 1996. The SOFA (Sepsis-related Organ Failure Assessment) score to describe organ dysfunction/failure. Intensive Care Medicine 22 (7), 707–710. https://doi.org/10.1007/bf01709751.

Wittau, M., Mayer, B., Scheele, J., Henne-Bruns, D., Patchen Dellinger, E., Isenmann, R., 2011. Systematic review and meta-analysis of antibiotic prophylaxis in severe acute pancreatitis. Scandinavian Journal of Gastroenterology 46 (3), 261–270. https://doi.org/10.3109/00365521.2010.531486.

Wu, B.U., et al., 2008. The BISAP score: a bedside score to predict mortality in acute pancreatitis. Gut 57 (12), 1689–1692. https://doi.org/10.1136/gut.2008.152702.

Wu, B.U., Hwang, J.Q., Gardner, T.H., Repas, K., Delee, R., Song, Y., Smith, B., Banks, P.A., Conwell, D.L., 2011. Lactated ringer's solution reduces systemic inflammation compared with saline in patients with acute pancreatitis. Clinical Gastroenterology and Hepatology 9 (8), 710. https://doi.org/10.1016/j.cgh.2011.04.026.

Xue, P., Deng, L.-H., Zhang, Z.-D., Yang, X.-N., Wan, M.-H., Song, B., Xia, Q., 2009. Effect of antibiotic prophylaxis on acute necrotizing pancreatitis: results of a randomized controlled trial. Journal of Gastroenterology and Hepatology 24 (5), 736–742. https://doi.org/10.1111/j.1440-1746.2008.05758.x.

Zarbock, A., Kellum, J.A., Schmidt, C., Van Aken, H., Wempe, C., Pavenstädt, H., Boanta, A., Gerß, J., Meersch, M., 2016. Effect of early vs delayed initiation of renal replacement therapy on mortality in critically ill patients with acute kidney injury: the elain randomized clinical trial. JAMA, the Journal of the American Medical Association 315 (20), 2190–2199. https://doi.org/10.1001/jama.2016.5828.

Zarnescu, N.O., Dumitrascu, I., Zarnescu, E.C., Costea, R., 2022. Abdominal compartment syndrome in acute pancreatitis: a narrative review. Diagnostics 13 (1). https://doi.org/10.3390/diagnostics13010044.

CHAPTER

Endoscopic interventions in acute pancreatitis

13

Akinori Maruta, Takuya Koizumi, Yosuke Ohashi, Shota Iwata, Shinya Uemura, Masahito Shimizu

First Department of Internal Medicine, Gifu University Hospital, Gifu, Japan

ABBREVIATIONS AND ACRONYMS

AE	adverse event
EN	endoscopic necrosectomy
EPST	endoscopic pancreatic sphincterotomy
ERCP	endoscopic retrograde cholangiopancreatography
EST	endoscopic sphincterotomy
EUS	endoscopic ultrasonography
EUS-PD	endoscopic ultrasound-guided pancreatic duct drainage
EUS-TD	endoscopic ultrasound transgastrointestinal drainage
LAMS	lumen apposing metal stent
PFC	pancreatic fluid collection
POPS	peroral pancreatoscopy
SEMS	self-expandable metallic stents
WON	walled-off necrosis

13.1 Introduction

Although the prognosis for acute pancreatitis has improved, it remains a serious disease with a high mortality rate. According to an epidemiological survey in Japan, the mortality rate within 2 weeks of the onset of pancreatitis improved significantly from 6.6% in 2011 to 2.7% in 2016, but remained almost unchanged at 3.5%–3.4% after 2 weeks of onset (Masamune et al., 2020). Among patients with severe pancreatitis who died after two weeks of onset, 42.3% had walled-off necrosis (WON). Even among patients with acute pancreatitis overall, the mortality rate in those with WON was 6.7%, significantly higher than the 1.3% in those without WON. To further improve outcomes in the treatment of acute pancreatitis, it is important to manage late complications such as WON in addition to providing appropriate treatment during the acute phase.

Endoscopic treatments for pancreatitis are highly diverse, including endoscopic sphincterotomy (EST) and stone extraction for biliary pancreatitis, pancreatic duct stent placement for pancreatic duct strictures due to chronic pancreatitis, endoscopic pancreatic sphincterotomy (EPST) and peroral pancreatoscopy (POPS) for pancreatic duct

Advanced Insights into Acute Pancreatitis. https://doi.org/10.1016/B978-0-443-44736-5.00009-7

stones, and endoscopic drainage procedures for pancreatic fluid collections (PFCs), which are local complications following acute pancreatitis such as WON. Endoscopic treatment is widely performed as a minimally invasive and effective option, and as devices for each treatment are continually being improved, the importance of endoscopic treatment in the management of pancreatitis is expected to further increase in the future. This article explains endoscopic treatment for pancreatitis.

13.2 Biliary pancreatitis

Biliary pancreatitis is a type of pancreatitis caused by bile duct stones or bile sludge impacting the papilla, resulting in pancreatic duct obstruction and papillary edema. The severity of biliary pancreatitis has been reported to depend on the duration of pancreatic duct obstruction, so it is recommended to perform endoscopic treatment by endoscopic retrograde cholangiopancreatography (ERCP) as soon as possible after diagnosis (Senninger et al., 1986; Acosta et al., 1997).

However, in some cases of biliary pancreatitis, bile duct stones and biliary sludge become impacted in the papilla, causing pancreatitis and cholangitis, and are then spontaneously expelled into the duodenum. The usefulness of performing ERCP in such cases is unclear, and there is a risk of further worsening of the condition due to procedure-related complications such as post-ERCP pancreatitis. Therefore, the acute pancreatitis guidelines from Japan recommend early ERCP if cholangitis or bile stasis is present and an impacted stone is diagnosed using various imaging examinations.

13.2.1 Diagnosis of biliary pancreatitis

Biliary pancreatitis differs from other causes of pancreatitis in that the underlying cause can be treated. If the diagnosis is confirmed, ERCP should be performed as described above. However, performing ERCP in stone-negative cases or in cases after stone passage may worsen the condition due to complications such as post-ERCP pancreatitis. Therefore, it is extremely important to determine whether ERCP is appropriate in cases of suspected biliary pancreatitis.

In many cases, bile duct stones are diagnosed using computed tomography (CT) imaging. However, bile duct stones with low calcium content are difficult to visualize on CT, and the reported sensitivity of CT for biliary stones is only 40%–53% (Wang et al., 1988; Moon et al., 2005). CT is useful for assessing the severity of acute pancreatitis but not for ruling out biliary pancreatitis. Other diagnostic modalities include magnetic resonance imaging (MRI) and endoscopic ultrasonography (EUS). Magnetic resonance cholangiopancreatgraphy (MRCP) is noninvasive and can provide images of the bile duct and pancreatic duct without contrast agents, and may detect stones even when other tests are negative. The sensitivity of MRCP for bile duct stones has been reported to be 77%–100%, and the specificity 73%–99%. However, the detection rate is low when the bile duct diameter is large, when ascites is present, or when breath-holding is poor, and small stones less than 5 mm in size may be missed (Giljaca et al., 2015; Jendresen et al., 2002).

EUS can visualize the bile duct without influence from the abdominal wall or digestive tract and is considered superior to transabdominal ultrasonography in detecting common bile duct stones. The diagnostic accuracy of EUS is not affected by bile duct or stone diameter, with reported sensitivity of 75%–100% and specificity of 85%–100% (Giljaca et al., 2015). Biliary pancreatitis is often caused by small stones, and EUS may be useful even when other imaging tests such as MRCP do not detect them. In a meta-analysis of three randomized controlled trials, Petrov et al. reported that performing EUS first significantly reduced complications (13/153 versus 34/150, $P = .001$) (Petrov, 2009). In a review of seven studies, DeLisi et al. reported that performing EUS first in 545 cases avoided unnecessary ERCP in 71.2% of cases, and no EUS-related adverse events (AEs) were observed (De Lisi, Leandro and Buscarini, 2011).

These findings suggest in cases of suspected biliary pancreatitis, the rate of passed stones is not low, and it is considered useful and safe to perform EUS first to determine whether ERCP is appropriate (Fig. 13.1).

13.2.2 Endoscopic procedure for biliary pancreatitis

ERCP for biliary pancreatitis aims to eliminate the cause of pancreatitis by removing impacted stones, separating the bile duct and pancreatic duct through endoscopic sphincterotomy (EST), and performing biliary drainage if cholangitis is present (Fig. 13.2). During ERCP and EST, accidental insertion of contrast media and guidewires into the pancreatic duct should be avoided whenever possible to prevent pancreatic injury.

If bile duct stones or impacted stones are identified, it is advisable to perform EST before stone removal. When a small stone is impacted in the papilla, it may be easily removed by gently touching it with the catheter tip. In contrast, when a relatively large stone is impacted, it may be difficult to pass not only a catheter but also a guidewire, and precutting the papilla with a needle knife to facilitate stone removal can be useful.

Dubravcsik et al. conducted a comparative study in which spontaneously detachable pancreatic stents were placed in high-risk patients during early ERCP for acute biliary pancreatitis complicated by cholangitis, and reported that the group with pancreatic stent placement had lower rate of AEs and mortality than the group without stent placement (Dubravcsik et al., 2012). Based on these results, placement of a pancreatic duct stent should be considered if a guidewire is inadvertently inserted into the pancreatic duct during ERCP.

13.3 Chronic pancreatitis

Chronic pancreatitis is defined as a condition characterized by chronic changes within the pancreas, including irregular fibrosis, inflammatory cell infiltration,

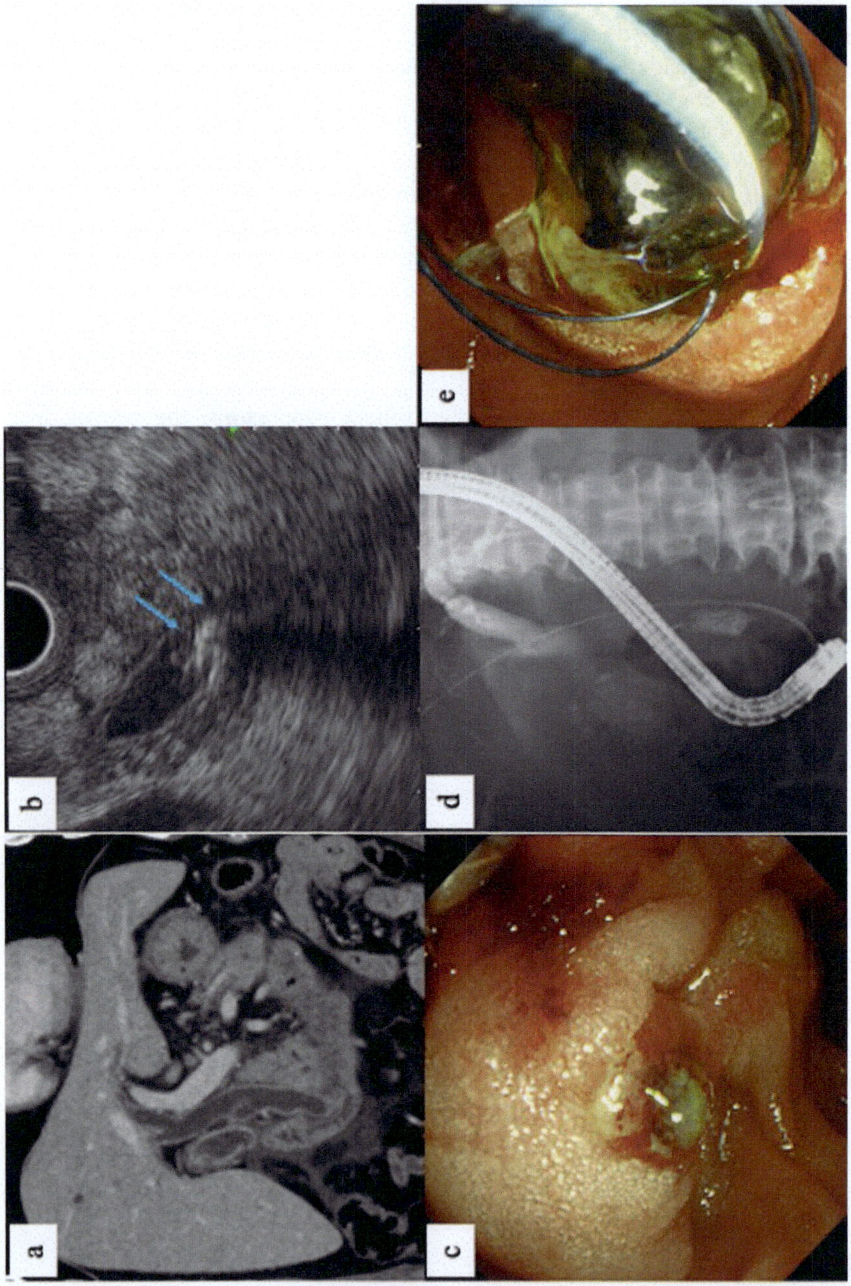

FIGURE 13.1 Diagnosis and treatment of gallstone pancreatitis.

(a) No obvious bile duct stones were found on the CT image. (b) EUS examination revealed a hyperechoic area near the papilla (blue arrow). (c) During ERCP, a stone impacted in the papilla was found. (d) After EST, the stone was extracted using a basket catheter (fluoroscopic image). (e) The stone was successfully removed (endoscopic image).

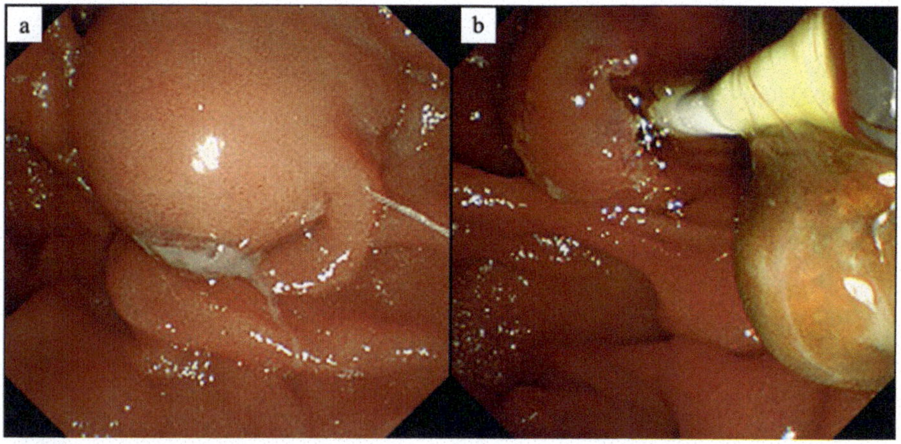

FIGURE 13.2 Emergency endoscopic treatment for biliary pancreatitis.

(a) Impaction of a bile duct stone in the papilla. (b) Endoscopic stone removal using a balloon catheter after EST.

parenchymal loss, granulation tissue formation, pancreatic stone formation, and irregular dilation of the pancreatic duct, which, as it progresses, is accompanied by a decline in both endocrine and exocrine pancreatic function. The disease progresses irreversibly from the compensated stage, when subjective symptoms such as pain appear, through a transitional stage, to the decompensated stage, which is accompanied by pancreatic endocrine and exocrine dysfunction.

As chronic pancreatitis progresses, pancreatic duct strictures and pancreatic stones may develop, causing pancreatic juice stasis and increased intraductal pressure, which can result in intractable abdominal pain, acute exacerbation of chronic pancreatitis, and pancreatic endocrine and exocrine insufficiency. Pancreatic duct decompression may relieve pain that does not improve with conservative treatment and may help prevent a decline in pancreatic endocrine and exocrine secretion. Treatment approaches are broadly divided into medical treatments (primarily endoscopy and lithotripsy) and surgical treatments. This article explains endoscopic treatment for pancreatic duct strictures and pancreatic stones associated with chronic pancreatitis.

13.3.1 Endoscopic treatment for pancreatic duct strictures

The first-choice endoscopic treatment for pancreatic duct strictures is pancreatic duct stent placement by ERCP (Fig. 13.3). Pancreatic duct stents were first placed by Fuji et al., in 1983, and reports from the 1990s described their usefulness for relieving pain in chronic pancreatitis and for treating pancreatic duct strictures. Because pancreatic duct strictures caused by chronic pancreatitis are typically severe, temporary balloon dilation or dilation with a bougie is insufficient, and continuous dilation by stent placement is necessary (Chandrasekhara et al., 2015).

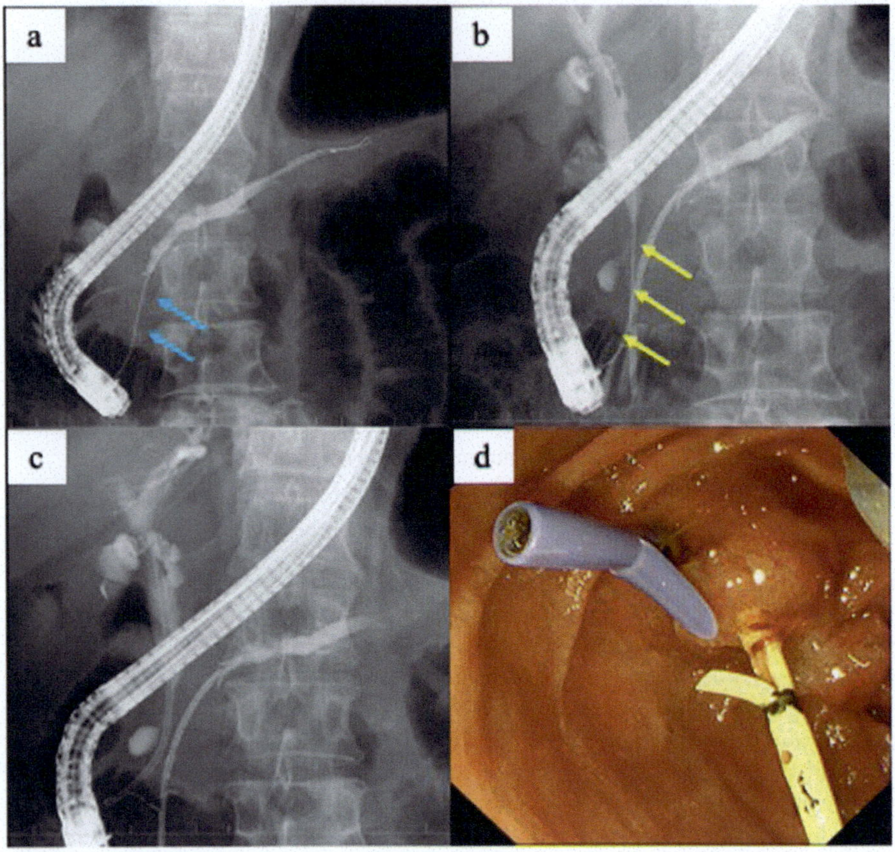

FIGURE 13.3 Endoscopic stent placement for pancreatic and bile duct strictures due to chronic pancreatitis.

(a) Pancreatic duct stricture due to chronic pancreatitis (blue arrow). (b) Bile duct stricture due to chronic pancreatitis (yellow arrow). (c) Placement of pancreatic and biliary stents (fluoroscopic image). (d) Placement of pancreatic and biliary stents (endoscopic image).

Depending on the degree and location of the stricture, the final success rate of pancreatic duct stent placement is approximately 90%, and the reported clinical success rate is 74%–94% (Binmoeller et al., 1995; Dumonceau et al., 2012). However, pancreatic duct stents have been reported to completely resolve strictures in fewwr than 10% of cases (Cremer et al., 1991), and the placement of multiple plastic stents is considered effective for refractory strictures.Costamagna et al. reported that in 19 cases of intractable pancreatic duct stricture, two to four 8.5–11.5 Fr stents were placed. After an average of 7 months of placement, the stricture improved in 18 cases (95%), allowing the stents to be removed. During an average observation period of 38 months after removal, 16 cases (84%) showed no recurrence of symptoms (Costamagna et al.,

2006). The placement of multiple plastic stents for pancreatic duct strictures has shown good results, and the guidelines of the European Society of Gastrointestinal Endoscopy (ESGE) and the American Society of Gastrointestinal Endoscopy (ASGE) recommend it as the first-choice treatment for refractory strictures. The effectiveness of large-diameter self-expandable metallic stents (SEMS) has also been reported compared with plastic stents. For benign strictures, uncovered SEMS, which may become irremovable, should be avoided, and removable fully covered SEMS (FCSEMS) are used. However, because the pancreatic duct has multiple branches, FCSEMS are known to carry a high risk of pancreatitis and ductitis due to obstruction of branch ducts, as well as a high risk of stent migration. In recent years, several studies have shown that FCSEMS can have therapeutic efficacy without serious AEs. Shen et al. reported that FCSEMS placement for refractory pancreatic duct strictures resulted in pain relief in 82.5% of cases, but stent migration occurred in 8.2% of cases (Shen et al., 2014). In addition, a dumbbell-shaped stent with terminal flanges to reduce the risk of stent migration and stent-induced pancreatic duct stricture has been developed. In a study of 25 cases, strictures were relieved in all cases, and after a median follow-up of 34 months, only two cases (8%) required stent reinsertion (Lee et al., 2020). FCSEMS may be a useful treatment option for pancreatic duct strictures, but further validation is needed.

Although the transpapillary approach is the first choice for stent placement for pancreatic duct strictures, endoscopic ultrasound-guided pancreatic duct drainage (EUS-PD) has been reported as a treatment for recurrent pancreatitis in patients with difficult ERCP or postoperative pancreaticojejunostomy strictures (Kurihara et al., 2013; Will et al., 2015). EUS-PD was first reported by Bataille et al., in 2002 (Bataille and Deprez, 2002), and can be divided into two methods.

1. EUS-rendezvous (EUS-RV): puncture is performed via the stomach, and a guidewire is passed antegrade through the pancreatic duct across the stricture, followed by ERCP using a duodenoscope or balloon-assisted endoscope.
2. EUS-guided pancreaticogastrostomy (EUS-PGS): an anastomosis is created between the pancreatic duct and the stomach or small intestine via the gastrointestinal tract, and a stent is placed antegrade (Fig. 13.4).

EUS-PGS provides excellent access to the main pancreatic duct and is a useful salvage method for cases in which ERCP is difficult. A review of 33 clinical studies involving 517 cases reported a technical success rate of 78.7% for EUS-PD, with 21.8% experiencing AEs, including pancreatitis, pancreatic leak, pancreatic pseudocyst, peritonitis, bleeding, and perforation (Nakai et al., 2019). Therefore, the procedure should be performed only at facilities equipped with the expertise and devices needed to manage serious AEs.

13.3.2 Endoscopic treatment for pancreatic stones

The presence of pancreatic stones in chronic pancreatitis can increase intraductal pressure, leading not only to abdominal and back pain but also to a decline in endocrine

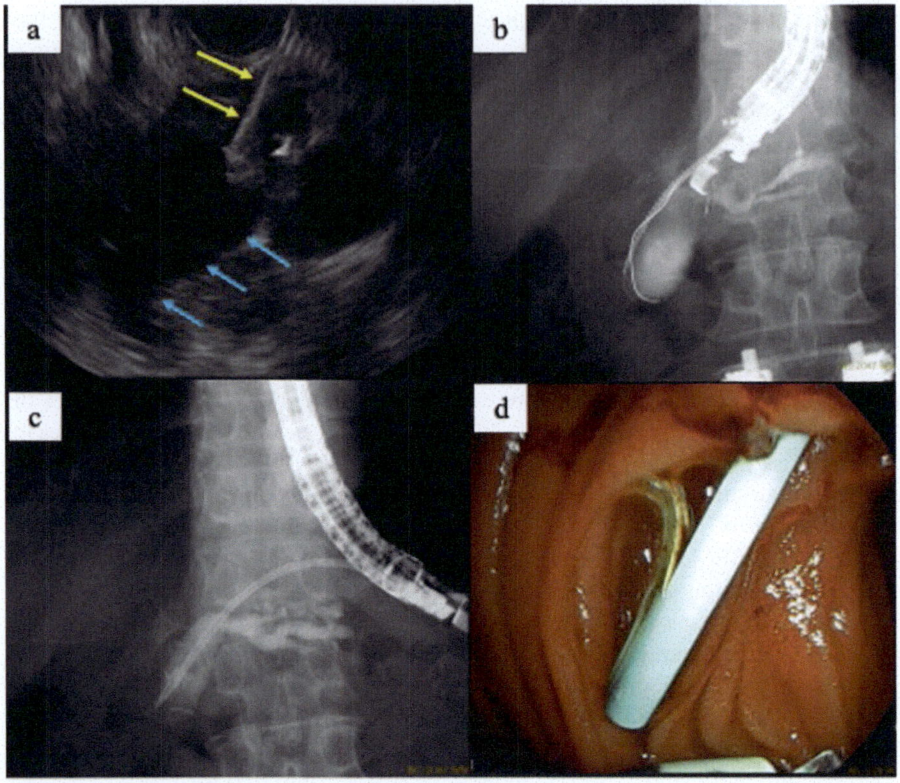

FIGURE 13.4 EUS-guided pancreatic duct drainage.

(a) After visualization of the pancreatic duct by EUS (blue arrow), the pancreatic duct was punctured (yellow arrow). (b) A guidewire was advanced into the pancreatic duct, and the fistula was dilated with a drill dilator. (c) A plastic stent was placed in the pancreatic duct (fluoroscopic image). (d) A plastic stent was placed in the pancreatic duct (endoscopic image).

and exocrine pancreatic function due to persistent chronic inflammation. In addition, rupture of the pancreatic duct caused by increased intraductal pressure can result in pseudocysts and pancreatic juice leakage. Therefore, treating pancreatic stones in chronic pancreatitis is clinically important as it can affect patient outcomes.

The pancreatic duct and pancreatic stones are evaluated by ERP, and after advancing a guidewire into the pancreatic duct, endoscopic pancreatic sphincterotomy (EPST) is performed using a papillotome, similar to standard EST. To safely remove pancreatic stones, it is generally recommended to crush them to a size of 5 mm or less (Dumonceau et al., 2012). Extracorporeal shock wave lithotripsy (ESWL) for pancreatic stones was first reported by Sauerbruch et al., in 1987 (Sauerbruch et al., 1987), and has been shown to be effective for small stones of 5–6 mm or less, floating stones, and X-ray-positive pancreatic stones. There are also reports that larger pancreatic stones can be removed after repeated ESWL sessions (Ito et al., 2023). Furthermore, even if the stone is not visible on X-ray, it can be fragmented by ESWL if focusing is possible using pancreatography through an endoscopic nasopancreatic drainage (ENPD).

In recent years, the usefulness of pancreatic stone fragmentation under peroral pancreatoscopy (POPS) has been reported for cases where fragmentation by ESWL is difficult (Fig. 13.5). Pancreatic stone fragmentation devices include holmium-YAG laser lithotripsy (LL) and electrohydraulic shock wave lithotripsy (EHL). Under direct visualization with POPS, the stones are crushed by applying the probe. Iwata et al. conducted a retrospective comparative study of POPS ($n = 19$) and ESWL ($n = 47$) for pancreatic duct stones. Although there were no significant difference in stone clearance, AEs, or recurrence rates, POPS required fewer sessions to achieve pancreatic stone clearance ($P < .01$) (Iwata et al., 2024). Conventional POPS scopes were relatively thick, and in cases with pancreatic duct stricture, insertion and manipulation within the duct were sometimes difficult. However, thinner scopes have been developed in recent years, and further advances in POPS-assisted lithotripsy are expected in the future.

13.4 Concept and classification of pancreatic fluid collections

Pancreatic fluid collections (PFCs) are fluid accumulations surrounding pancreas and often occur as local complications of acute pancreatitis. According to the revised Atlanta classification, PFCs are categorized as acute peripancreatic fluid collection (APFC), acute necrotic collection (ANC), pancreatic pseudocyst (PC), and walled-off necrosis (WON) based on the time elapsed after the onset of acute pancreatitis (<4 weeks versus ≥ 4 weeks) and the presence of necrosis. Infected pancreatic

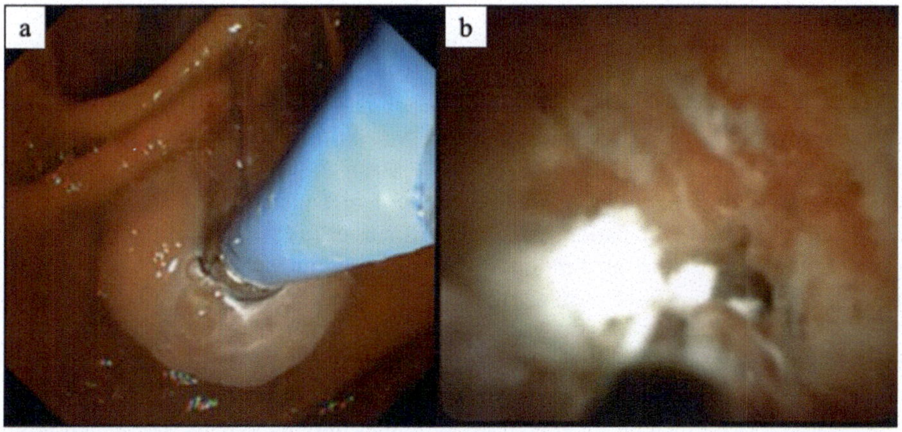

FIGURE 13.5 Treatment of pancreatic duct stones with POPS.

(a) Insertion of POPS into the pancreatic duct. (b) Pancreatic stone crushing by EHL.

necrosis refers to ANC or WON complicated by bacterial or fungal infection (Banks et al., 2013) (Fig. 13.6). For infected pancreatic necrosis, conservative treatment should be attempted first (Van Santvoort et al., 2011; Iwashita et al., 2023), and if the clinical course is poor, aggressive drainage treatment should be considered.

13.4.1 Clinical outcomes for PFCs by drainage approach

For infected pancreatic necrosis, it is recommended not to immediately remove necrotic material using laparotomy or other methods, but instead to follow a step-up approach, which involves minimally invasive drainage followed by gradual progression to necrosectomy as needed (IAP/APA evidence-based guidelines for the management of acute pancreatitis, 2013). The PANTER trial, conducted in the Netherlands, compared a step-up approach (gradually increasing the invasiveness of procedures such as percutaneous or endoscopic drainage) with open necrotomy. While no significant difference in mortality was observed, the step-up approach group had a significantly lower incidence of multiple organ failure (12% versus 40%) (van Brunschot et al., 2018). The step-up approach is divided into endoscopic approach (endoscopic ultrasound-guided transluminal drainage [EUS-TD] followed by direct endoscopic necrosectomy [EN]) and a surgical approach (percutaneous drainage followed by video-assisted retroperitoneal debridement). Brunschot et al. conducted a randomized trial comparing the endoscopic step-up approach ($n = 51$) and the surgical step-up approach ($n = 47$) for pancreatic necrosis (van Brunschot et al., 2018). Major AEs occurred in 43% (22/51) in the endoscopy group and in 45% (21/47) in the surgery group ($P = .88$), and mortality did not differ between groups (18% versus 13%, $P = .50$). However, the rates of pancreatic fistulas and the length of hospital stay were lower in the endoscopy group. Another randomized trial by Bang et al. compared the endoscopic step-up approach (transluminal drainage with or without necrosectomy, $n = 34$) with minimally invasive surgery (laparoscopic or video-assisted retroperitoneal debridement, depending on the location of the collection, $n = 32$) (Bang et al., 2019). Major AEs occurred in 11.8% of patients who underwent the endoscopic approach and in 40.6% of those who underwent minimally invasive surgery ($P = .007$). Mortality did not differ significantly (endoscopy 8.8% versus surgery 6.3%, $P = .999$). None of the patients treated with the endoscopic approach developed enteral or pancreatic-cutaneous fistulas, compared with 28.1% of patients treated with surgery ($P = .001$). Physical health scores for quality of life at 3 months were better with the endoscopic approach ($P = .039$), and mean total cost was lower ($75,830 versus $117,492, $P = .039$). Based on these results, an endoscopic step-up approach is recommended for the interventional treatment of infected pancreatic necrosis. However, for areas that are anatomically inaccessible by endoscopy, such as the pelvic cavity or right paracolic gutter, a combined surgical step-up approach may be considered. When both options are feasible, the endoscopic step-up approach is preferred because it can shorten hospital stays.

		Less than 4 weeks after onset	More than 4 weeks after onset
Acute intestinal edematous pancreatitis (without necrosis)	without infection	Acute peripancreatic fluid collection (APFC) →	Pancreatic pseudocyst (PC)
	with infection		
Acute necrotizing pancreatitis (with necrosis)	without infection	Acute necrotic collection (ANC) →	Walled off necrosis (WON)
	with infection		

FIGURE 13.6

The revised Atlanta classification of acute pancreatitis.

13.4.2 The optimal timing of interventional treatment for PFCs

The optimal timing of intervention for infected pancreatic necroses, particularly in relation to the time elapsed since the onset of acute pancreatitis (<4 weeks versus ≥4 weeks), remains debated. Several comparative studies have evaluated the timing of intervention, and delaying intervention until at least 4 weeks after the onset of acute pancreatitis has been associated with safer treatment outcomes (Nakai et al., 2023). However, if clinical improvement is poor, proactive intervention may be considered even within 4 weeks of onset (Ramai, 2023), and the optimal timing of intervention is currently controversial. Therefore, it was hypothesized that encapsulation status may be more strongly related to clinical outcomes than the time from disease onset, and this was examined in a retrospective study (Maruta et al., 2025). Patients who underwent endoscopic ultrasound-guided drainage for WON were divided into two groups based on encapsulation status at the time of drainage: the partial/no encapsulation group (n = 46) and the complete encapsulation group (n = 176). Complete encapsulation was associated with a higher clinical success rate (88.0% versus 52.1%, $P < .01$). When patients were classified by both time elapse and encapsulation status, encapsulation status was more strongly associated with clinical success than time elapsed. Encapsulation status was also associated with AE rates and mortality. In multivariable analysis, encapsulation status and the diameter of the WON were identified as significant factors associated with clinical success. These results suggest that the degree of encapsulation is more strongly associated with the clinical outcomes of EUS-guided drainage for WON than the time elapsed since disease onset, although larger randomized controlled trials are needed to comfirm this finding.

13.4.3 Endoscopic ultrasound-guided drainage for PFCs

13.4.3.1 Advance preparation

Before treatment, contrast-enhanced CT should be performed to evaluate the location and size of the PFCs and their relationship to surrounding organs, and to plan the appropriate puncture and drainage route. It is also important to check for the presence of large-caliber arteries within the cyst, pseudoaneurysms caused by pancreatitis, or varices due to portal hypertension (Mukai et al., 2015b). If a pseudoaneurysm is detected, vascular embolization using interventional radiology (IVR) should be performed before drainage treatment (Fig. 13.7).

13.4.3.2 Puncture, fistula dilation, and stent placement

First, a convex ultrasound endoscope is inserted to visualize the PFC. Lesions composed mainly of fluid, such as PCs, are easy to recognize, but in cases such as WON—where hyperechoic necrotic tissue is present within the encapsulated fluid—or when air is present, the lesion may not be clearly visualized by EUS. In such cases, the lesion should be identified carefully, taking into account the relative positions of the gastrointestinal tract and the lesion on CT, and the position of the scope under fluoroscopy.

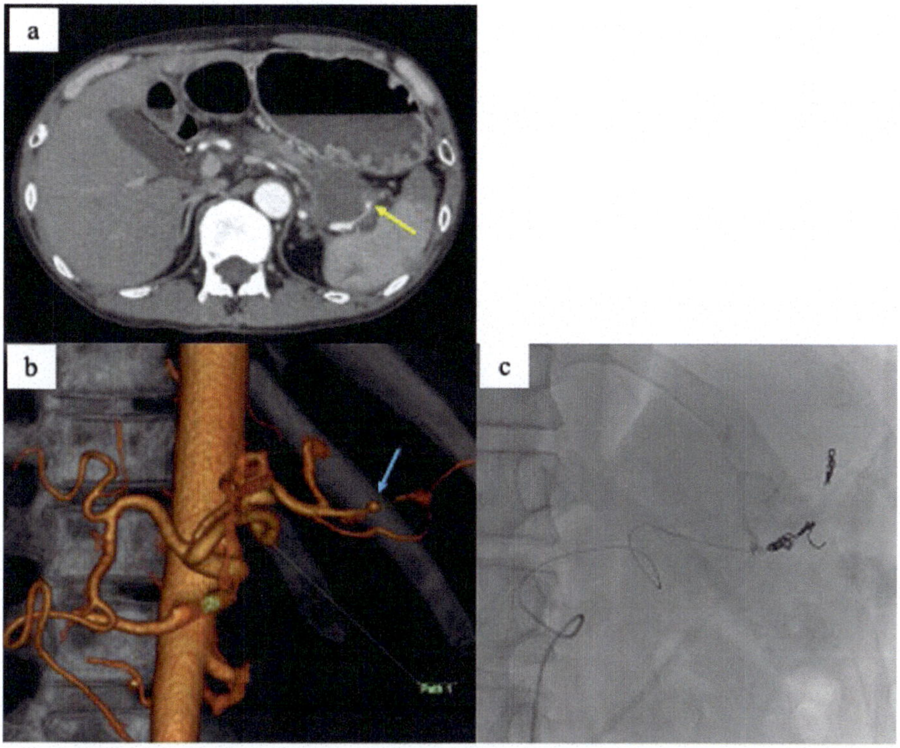

FIGURE 13.7 Diagnosis and treatment of pseudoaneurysms before EUS-guided drainage for PFC.

(a) Pseudoaneurysm (yellow arrow) formed in PFC (CT image). (b) Pseudoaneurysm (blue arrow) formed in PFC (3D-CT angiography image). (c) Coiling vascular embolization (fluoroscopic image).

After confirming with color Doppler that there are no blood vessels along the puncture route, the PFC is punctured under EUS guidance. After aspirating the cyst fluid and confirming intracystic imaging, a long guidewire is looped inside the cyst and left in place. The fistula is then dilated using a mechanical dilator or a 4–6 mm biliary dilation balloon. If the stomach wall is too stiff to dilate, an electrosurgical dilator is used. In Japan, many facilities perform internal and external drainage as the final form of drainage, which involves placing a 7-Fr double-ended pigtail-type plastic stent and a 5-Fr or 6-Fr pigtail-tipped nasal drainage tube (Fig. 13.8). Techniques to improve drainage efficiency include the multiple transluminal gateway technique (MTGT), which involves puncturing multiple sites (Varadarajulu et al., 2011), and the single transluminal gateway transcystic multiple drainages (SGTMD), which involves placing multiple drainage tubes in different directions from a single puncture cavity (Mukai et al., 2014). Both techniques are considered to be highly effective for multilocular WON.

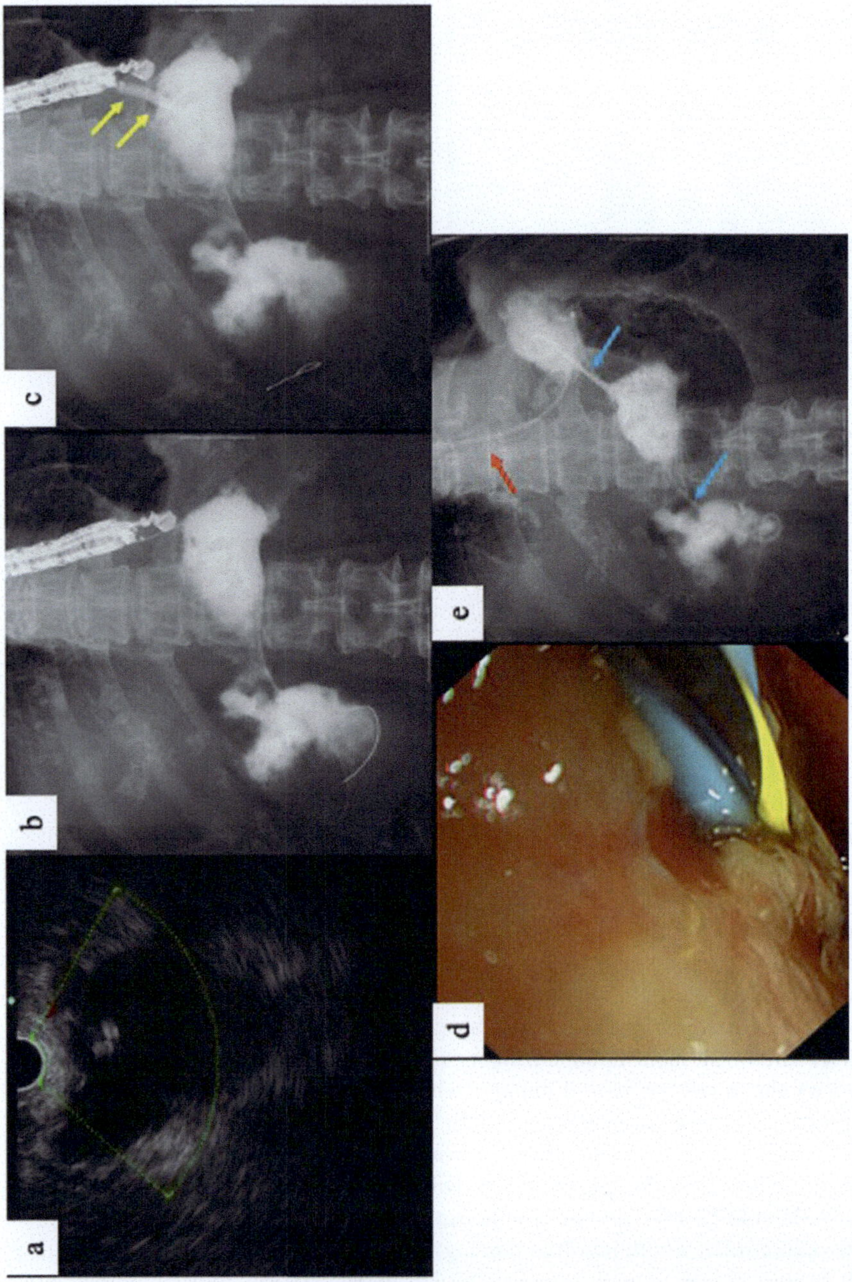

FIGURE 13.8 EUS-guided drainage for PFC.

(a) PFC visualized by EUS. (b) After puncturing the PFC, a guidewire was advanced into the cavity. (c) Balloon dilation of the fistula (red arrow). (d) Plastic stents placed in the PFC via the fistula. (e) Two internal fistula stents (blue arrow) and one external fistula drainage tube (yellow arrow) placed.

13.4.3.3 Endoscopic ultrasound-guided drainage using lumen-apposing metal stent

Previously, several studies reported placing fully covered metallic stents for the bile duct into PFCs in an attempt to achieve good drainage and early fistula formation (Weilert et al., 2012). Although this approach is simpler than placing multiple plastic stents and provide high drainage efficiency due to the large diameter, it carries risks such as misplacement, stent migration, and displacement. As the cyst shrinks during treatment, there is also concern about stent displacement and intracystic bleeding caused by the stent end contacting the cyst wall.

In recent years, lumen-apposing metal stents (LAMS), which are metal stents with anchors at both ends specifically designed for EUS-guided drainage, have been developed (Binmoeller and Shah, 2011; Itoi et al., 2012). LAMS differ from previous plastic stents and fully covered metallic stents for the bile duct. They are designed specifically for EUS-guided drainage, drawing the lumens together with flanges on both ends to create a fistula. The large diameter of LAMS allows efficient drainage, and once placed, the scope can be inserted and removed repeatedly, which make them very useful for performing endoscopic necrosectomy, as described later (Mukai et al., 2015a). LAMS delivery systems have an electrocautery-enhanced tip, allowing puncture, tract dilation, and stent placement in a single step (HOT-Axios, Boston Scientific, Marlborough, MA) (Rinninella et al., 2015) (Fig. 13.9). One method of placement is to insert the Hot-Axios delivery system after puncturing with an FNA needle and placing a guidewire to secure the route. However, the single-step method (freehand or freestyle method), in which the Hot-Axios is inserted while energizing it without prior puncture or guidewire placement, has recently become the mainstream approach. This method does not require multiple device exchanges and avoids separation between the target and gastrointestinal wall caused by guidewire manipulation, enabling quick and safe treatment. It is also useful in emergency situations when the patient's condition is unstable and the procedure must be completed rapidly.

13.4.3.4 Clinical outcomes using LAMS in the EUS-guided drainage for PFCs

Several comparative studies have reported on the clinical outcomes of different stents used for PFC drainage (Tian et al., 2024; Guzmán-Calderón et al., 2022; Kamal et al., 2024; Ardengh et al., 2026). A meta-analysis by Tian et al. reviewed nine studies involving 707 patients who underwent EUS-guided drainage for PFCs using either LAMS or traditional SEMS (Tian et al., 2024). The clinical success rate with LAMS tended to be higher than with SEMS. LAMS were associated with higher clinical success, significantly shorter procedure times, and fewer overall AEs ($P = .004$). Specifically, LAMS were associated with fewer cases of migration, occlusion, and infection. The recurrence rate was also lower in the LAMS than in the SEMS group ($P = .007$). Another meta-analysis by Guzman-Calderon et al. compared clinical outcomes of EUS-guided drainage for PFCs using LAMS ($n = 663$) and plastic stent (PS) ($n = 921$) from 13 studies (Guzmán-Calderón

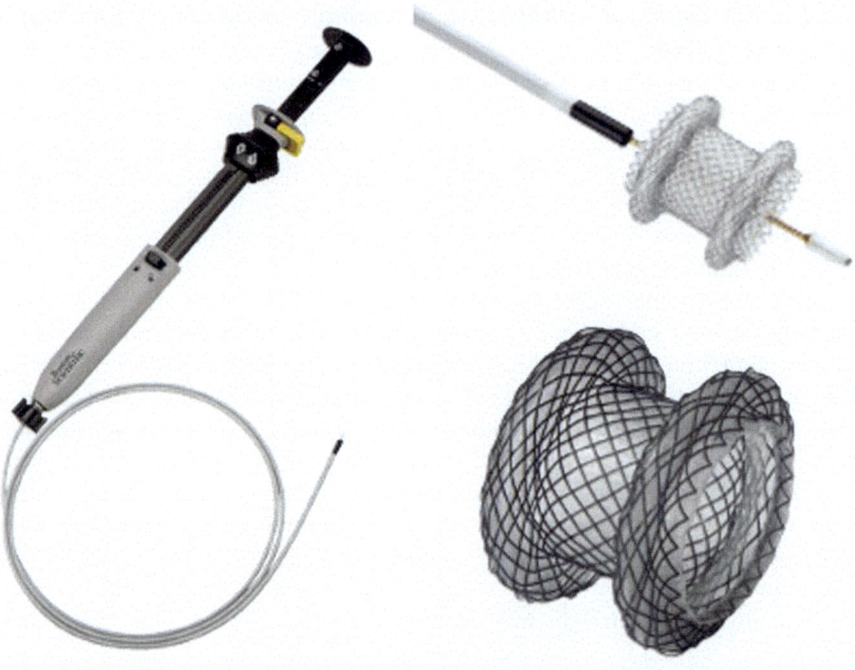

FIGURE 13.9

Lumen apposing metal stent (HOT Axios, system).

et al., 2022). Although there were no significant difference technical success, clinical success, or mortality between the two groups, the LAMS group had a significantly lower AE rate than the PS groups (16.0% versus 20.2%, $P = .009$). Bleeding was the most common AE in the LAMS group (5.0%), whereas infection was the most common AE in the PS group (6.1%). The stent migration rate was lower with LAMS than with PS (0.9% versus 2.2%, $P = .05$). These findings suggest that LAMS offer high technical and clinical success rates, shorter procedure times, fewer AEs, and lower recurrence rates in EUS-guided drainage of PFCs compared with conventional SEMS or PS. However, a concern with using LAMS instead of plastic stents is the potential increased risk of bleeding (Facciorusso et al., 2022; Holmes et al., 2022; Fugazza et al., 2020; Bang et al., 2020b). Because LAMS are self-expanding, they may damage surrounding tissues along the puncture tract, leading to mucosal or pseudoaneurysmal bleeding (Abdallah et al., 2022; Takasaki et al., 2023). Another concern is the potential for higher total inpatient costs (Chen et al., 2018a, 2018b). Hamada et al. conducted a nationwide retrospective study to evaluate the impact of LAMS (approved in Japan in 2018) on clinical outcomes in real-world settings (Hamada et al., 2025). In the pre-LAMS period, 3787 patients were treated with plastic stents, while among 2098 cases treated since 2018, the use of LAMS stabilized at approximately 50%. Compared to the pre-LAMS period, patients treated in the post-LAMS period had a higher risk of bleeding when LAMS were used (adjusted OR 1.81, 95% CI: 1.55–2.13), but not when plastic stents were used (adjusted OR 0.96, 95% CI: 0.74–1.26). The post-LAMS period was also associated with increased total costs (coefficient: USD 2813; 95% CI: 1503–4122). In analyses stratified by PFC types (WON versus PC), the association between the post-LAMS period and bleeding risk appeared stronger for PCs than for WON. Patients with WON had shorter hospital stays and lower total costs in the post-LAMS period. These findings do not support the routine use of LAMS for EUS-guided treatment of all PFCs. LAMS may need to be used selectively depending on the type of PFC and individual case characteristics, but further studies are needed to establish the most effective treatment strategy.

13.4.3.5 Endoscopic necrosectomy for PFCs

ANC/WON, which contain large amount of necrotic material, are often difficult to treat with EUS-TD alone. Therefore, endoscopic necrosectomy (EN) was developed, in which an endoscope is inserted into the necrotic cavity through a fistula created by EUS-TD to remove infected necrotic tissue.

In 1996, Baron et al. first reported the removal of necrotic tissue from WON using endoscopic insertion of a large-diameter catheter and irrigation (Baron et al., 1996), and in 2000, Seifert et al. reported direct EN via a retroperitoneal approach (Seifert et al., 2000). Subsequently, direct EN via the gastrointestinal tract was reported (Seewald et al., 2005; Schrover et al., 2008).

In 2013, Yasuda et al. summarized the outcomes of 57 cases of EN for infected WON performed at 16 Japanese institutions between 2005 and 2011, during the early period following the introduction of EN. They reported a clinical success rate of 75% (43/57), an AE rate of 33% (19/57), a mortality rate of 11% (6/57), and three recurrences (Yasuda et al., 2013). EN has been reported to be more effective and safer than conventional open surgical approaches and laparoscopic or robotic-assisted minimally invasive surgical approaches (Traverso and Kozarek, 2005; Bang et al., 2020a; Tang et al., 2023), and the endoscopic approach is now considered the first choice. Previously, EN was performed by dilating the fistula with a large-diameter balloon several days after EUS-TD. However, with the recent development of LAMS, which arrow easier insertion and removal of the endoscope into the necrotic cavity, EN can now be performed more efficiently than before.

13.4.3.6 Endoscopic necrosectomy technique

The endoscope is advanced into the necrotic cavity through the lumen of the LAMS, and the water jet function is used to wash the cavity with saline as needed, carefully removing the necrotic material to avoid vascular injury or cyst wall perforation. Various treatment tools can be used to remove the necrotic material, including basket catheters for stone removal, snares, biopsy forceps, retrieval nets, and tripod or pentapod forceps (Fig. 13.10). Because prolonged procedure times increase the risk of

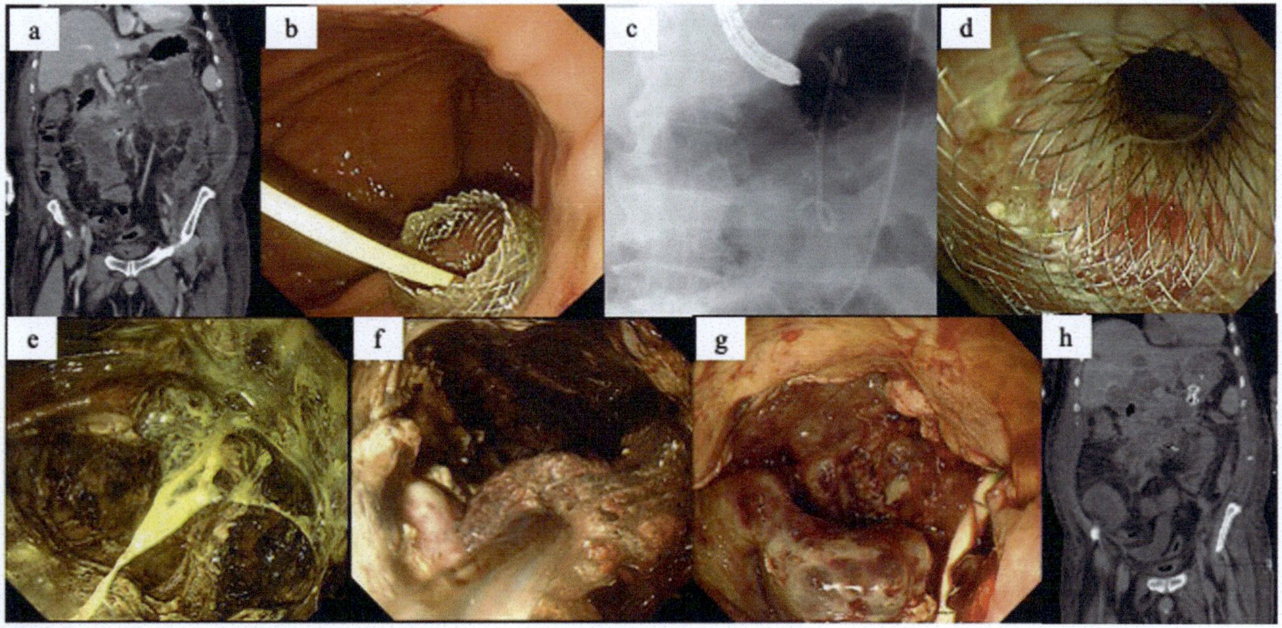

FIGURE 13.10 Endoscopic necrosectomy for WON.

(a) CT image before WON treatment. (b) LAMS placement for WON (endoscopic image). (c) LAMS placement for WON (fluoroscopic image). (d) Endoscope insertion into WON via LAMS. (e) Necrotic material in the WON. (f) Removal of necrotic material by endoscopic necrosectomy. (g) The WON after multiple endoscopic necrosectomy sessions. (h) CT image after WON treatment.

complications such as air embolism and aspiration pneumonia, it is reasonable to limit the duration of each session to about 1 hour and to perform treatment two to three times per week, taking into consideration the patient's general condition. It is also important to evaluate the treatment effectiveness with CT scans as needed and to adjust the approach to the necrotic cavity as treatment progresses. Although endoscopic treatment is effective for many cases of WON, when the necrotic cavity extends to the pelvic cavity or right paracolic gutter, treatment using a transgastrointestinal endoscopic approach alone can be difficult and has limitations. In such cases, a combination of percutaneous and surgical treatment should be considered.

EN is considered complete when most of the necrotic material inside the WON has been removed and new granulation tissue growth is confirmed. The placement period of LAMS is limited to 60 days to avoid the risk of bleeding associated with long-term placement. There is no clear consensus on whether to place a PS at the fistula site after LAMS removal.

A meta-analysis by Iwashita et al. reviewed nine studies involving 380 patients with long-term transmural PS and 289 patients without PS after EUS-guided treatment for PFCs (Iwashita et al., 2024). The rate of PFC recurrence was significantly lower in patients with transmural PS (pooled OR = 0.23, 95%CI, 0.08–0.65, $P = .005$). In a subgroup analysis limited to studies focusing on patients with disconnected pancreatic duct syndrome (DPDS), which has been reported to be a risk factor for PFC recurrence (Chen et al., 2019; Hamada et al., 2022), the OR was numerically lower than that for the entire cohort (OR = 0.14, 95%CI, 0.04–0.46). The rate of AEs was significantly higher with long-term transmural PS. These results suggest that long-term PS placement reduces the risk of PFC recurrence, especially in cases complicated with DPDS. However, AEs such as intestinal perforation and empyema have been reported with long-term stent placement, and further studies are needed to determine which cases require PS placement after LAMS removal and the optimal duration of placement.

13.4.3.7 Adverse events of endoscopic necrotomy and management

The most common AE of EN is bleeding, which has been reported in 12.6% (80/633) of cases. The severity of bleeding was reported as mild in 67 of 80 cases, moderate in 2cases, severe in 90 cases, and unspecified in 2 cases (Isayama et al., 2016). As noted above, if a pseudoaneurysm is detected on preoperative CT, transcatheter embolization should be considered. However, because necrosectomy is a prolonged treatment process, new pseudoaneurysms may develop due to ongoing inflammation. Therefore, it is advisable to repeat contrast-enhanced CT scans as appropriate during the treatment period.

For endoscopic hemostasis in intraoperative bleeding, local epinephrine injection and clipping are commonly used. In recent years, the usefulness of gel (Viscoclear; Otsuka Pharmaceutical Factory, Inc., Tokushima, Japan) for securing the visual field and self-assembling peptide hydrogel (3-D Matrix Europe SAS) for hemostasis has also been reported (Hama et al., 2024; Yamamoto et al., 2023). However, if the bleeding point cannot be identified or if hemostasis is not achievable with these methods, vascular embolization using interventional radiology should be considered.

Author contributions

Akinori Maruta wrote the manuscript. Akinori Maruta, Takuya Koizumi, Yosuke Ohashi, Shota Iwata, Shinya Uemura, Masahito Shimizu managed the patients. All authors read and approved the final manuscript.

Conflicts of interest

Authors declare no conflicts of interest for this article.

References

Abdallah, M., Vantanasiri, K., Young, S., Azeem, N., Amateau, S.K., Mallery, S., Freeman, M. L., Trikudanathan, G., 2022. Visceral artery pseudoaneurysms in necrotizing pancreatitis: risk of early bleeding with lumen-apposing metal stents. Gastrointestinal Endoscopy 95 (6), 1150–1157. https://doi.org/10.1016/j.gie.2021.11.030.

Acosta, J.M., Galli, O.M.R., Rossi, R., Chinellato, A.V., Pellegrini, C.A., 1997. Effect of duration of ampullary gallstone obstruction on severity of lesions of acute pancreatitis. Journal of the American College of Surgeons 184 (5), 499–504.

Ardengh, A.O., Visconti, T.A.d.C., Dall'Agnol, M.K., Fernandes, Y.R., Veras, M.d.O., Valentim dos Santos, E.S., Lera do Santos, M.E., Matuguma, S.E., Ardengh, J.C., Bernardo, W. M., Hourneaux de Moura, E.G., 2026. Endoscopic ultrasound-guided drainage with lumen-apposing metal stent versus plastic stent for the treatment of pancreatic pseudocyst: a systematic review and meta-analysis. DEN Open 6 (1). https://doi.org/10.1002/deo2.70165. Available from: https://onlinelibrary.wiley.com/loi/26924609/year/2022.

Bang, J.Y., Arnoletti, J.P., Holt, B.A., Sutton, B., Hasan, M.K., Navaneethan, U., Feranec, N., Wilcox, C.M., Tharian, B., Hawes, R.H., Varadarajulu, S., 2019. An endoscopic transluminal approach, compared with minimally invasive surgery, reduces complications and costs for patients with necrotizing pancreatitis. Gastroenterology 156 (4), 1027. https://doi.org/10.1053/j.gastro.2018.11.031.

Bang, J.Y., Wilcox, C.M., Arnoletti, J.P., Varadarajulu, S., 2020a. Superiority of endoscopic interventions over minimally invasive surgery for infected necrotizing pancreatitis: meta-analysis of randomized trials. Digestive Endoscopy 32 (3), 298–308. https://doi.org/10.1111/den.13470. Available from: http://onlinelibrary.wiley.com/journal/10.1111/(ISSN)1443-1661.

Bang, J.Y., Hawes, R.H., Varadarajulu, S., 2020b. Lumen-apposing metal stent placement for drainage of pancreatic fluid collections: predictors of adverse events. Gut 69 (8), 1379–1381. https://doi.org/10.1136/gutjnl-2019-320539.

Banks, P.A., Bollen, T.L., Dervenis, C., Gooszen, H.G., Johnson, C.D., Sarr, M.G., Tsiotos, G. G., Vege, S.S., Windsor, J.A., Horvath, K.D., Mortele, K.J., Gardner, T.B., Van Santvoort, H., Pelaez-Luna, M., Yadav, D., Stefanidis, G., Delakidis, S., Morgan, D.E., Thoeni, R.F.L., Acosta, J.M., Amann, S.T., Andren-Sandberg, A., Aranha, G.V., Asciutti, S., Barauskas, G., Baron, T.H., Bassi, C., Behrman, S., Behrns, K.E., Belliappa, V., Berzin, T.M., Besselink, M.G.H., Bhasin, D.K., Biankin, A., Bishop, M. D., Bonini, C.J., Bradley, E.L., Buechler, M., Carter, R., Cavestro, G.M., Chari, S.T.,

Chavez-Rodriguez, J.J., Da Cunha, J.E., D'Agostino, H., De Campos, T., De-Madaria, E., Deprez, P.H., DiSario, J.A., Doria, C., Falconi, M., Fernandez-del Castillo, C., Freeny, P. C., Frey, C.F., Friess, H., Frossard, J.L., Fuchshuber, P., Gallagher, S.F., Garg, P.K., Ghattas, G., Glasgow, R., Gonzalez, J.A., Gress, T.M., Gumbs, A.A., Hallibruton, C., Helton, S., Hill, M.C., Hoyos, S., Imrie, C.W., Isenmann, R., Izbicki, J.R., Karagiannis, J.A., Klar, E., Kolokythas, O., Lau, J., Litvin, A.A., Longnecker, D.S., Lowenfels, A.B., Mackey, R., Mah'Moud, M., Malangoni, M., McFadden, D.W., Mishra, G., Moody, F.G., Morinville, V., Neoptolemos, J.P., Nordback, I., Pap, A., Papachristou, G.I., Parks, R., Pedrazolli, S., Pezzilli, R., Pitt, H.A., Prosanto, C., Ramesh, H., Ramirez, F.C., Raper, S.E., Rasheed, A., Reed, D.N., Romangnuolo, J., Rossaak, J., Sanabria, J., Schaefer, C., Schmidt, J., Schmidt, P.N., Serrablo, A., Senkowski, C.K., Sharma, M., Sigman, K.M., Singh, P., Steinberg, W., Steiner, J., Strasberg, S., Strum, W., Takada, T., Tanaka, M., Vaccaro, M., Villavicencio, R.L., Vrochides, D., Wagner, M., Warshaw, A.L., Wilcox, C.M., Wysocki, P., Zenilman, M. E., Zyromski, N.J., 2013. Classification of acute pancreatitis—2012: revision of the Atlanta classification and definitions by international consensus. Gut 62 (1), 102–111. https://doi.org/10.1136/gutjnl-2012-302779.

Baron, T.H., Thaggard, W.G., Morgan, D.E., Stanley, R.J., 1996. Endoscopic therapy for organized pancreatic necrosis. Gastroenterology 111 (3), 755–764. https://doi.org/10.1053/gast.1996.v111.pm8780582.

Bataille, L., Deprez, P., 2002. A new application for therapeutic EUS: main pancreatic duct drainage with a "pancreatic rendezvous technique". Gastrointestinal Endoscopy 55 (6), 740–743. https://doi.org/10.1067/mge.2002.123621.

Binmoeller, K.F., Shah, J., 2011. A novel lumen-apposing stent for transluminal drainage of nonadherent extraintestinal fluid collections. Endoscopy 43 (4), 337–342. https://doi.org/10.1055/s-0030-1256127.

Binmoeller, K.F., Jue, P., Seifert, H., Nam, W.C., Izbicki, J., Soehendra, N., 1995. Endoscopic pancreatic stent drainage in chronic pancreatitis and a dominant stricture: long-term results. Endoscopy 27 (9), 638–644. https://doi.org/10.1055/s-2007-1005780.

Chandrasekhara, V., Chathadi, K.V., Acosta, R.D., Anton Decker, G., Early, D.S., Eloubeidi, M.A., Evans, J.A., Faulx, A.L., Fanelli, R.D., Fisher, D.A., Foley, K., Fonkalsrud, L., Hwang, J.H., Jue, T.L., Khashab, M.A., Lightdale, J.R., Raman Muthusamy, V., Pasha, S.F., Saltzman, J.R., Sharaf, R., Shaukat, A., Shergill, A.K., Wang, A., Cash, B.D., DeWitt, J.M., 2015. The role of endoscopy in benign pancreatic disease. Gastrointestinal Endoscopy 82 (2), 203–214. https://doi.org/10.1016/j.gie.2015.04.022.

Chen, Y.-I., Khashab, M., Adam, V., Bai, G., Singh, V., Bukhari, M., Gutierrez, O.B., Elmunzer, B., Moran, R., Fayad, L., Zein, M.El, Kumbhari, V., Repici, A., Barkun, A., 2018a. Plastic stents are more cost-effective than lumen-apposing metal stents in management of pancreatic pseudocysts. Endoscopy International Open 06 (07), E780. https://doi.org/10.1055/a-0611-5082.

Chen, et al., 2018b. Cost-effectiveness analysis comparing lumen-apposing metal stents with plastic stents in the management of pancreatic walled-off necrosis. Gastrointestinal Endoscopy 88 (2).

Chen, Y., Jiang, Y., Qian, W., Yu, Q., Dong, Y., Zhu, H., Liu, F., Du, Y., Wang, D., Li, Z., 2019. Endoscopic transpapillary drainage in disconnected pancreatic duct syndrome after acute pancreatitis and trauma: long-term outcomes in 31 patients. BMC Gastroenterology 19 (1). https://doi.org/10.1186/s12876-019-0977-1.

Costamagna, G., Bulajic, M., Tringali, A., Pandolfi, M., Gabbrielli, A., Spada, C., Petruzziello, L., Familiari, P., Mutignani, M., 2006. Multiple stenting of refractory pancreatic duct strictures in severe chronic pancreatitis: long-term results. Endoscopy 38 (3), 254–259. https://doi.org/10.1055/s-2005-921069.

Cremer, M., Deviere, J., Delhaye, M., Baize, M., Vandermeeren, A., 1991. Stenting in severe chronic pancreatitis: results of medium-term follow-up in seventy-six patients. Endoscopy 23 (3), 171–176. https://doi.org/10.1055/s-2007-1010649.

De Lisi, S., Leandro, G., Buscarini, E., 2011. Endoscopic ultrasonography versus endoscopic retrograde cholangiopancreatography in acute biliary pancreatitis: a systematic review. European Journal of Gastroenterology and Hepatology 23 (5), 367–374. https://doi.org/10.1097/MEG.0b013e3283460129.

Dubravcsik, Z., Hritz, I., Fejes, R., Balogh, G., Virányi, Z., Hausinger, P., Székely, A., Szepes, A., Madácsy, L., 2012. Early ERCP and biliary sphincterotomy with or without small-caliber pancreatic stent insertion in patients with acute biliary pancreatitis: better overall outcome with adequate pancreatic drainage. Scandinavian Journal of Gastroenterology 47 (6), 729–736. https://doi.org/10.3109/00365521.2012.660702.

Dumonceau, J.-M., Delhaye, M., Tringali, A., Dominguez-Munoz, J., Poley, J.-W., Arvanitaki, M., Costamagna, G., Costea, F., Devière, J., Eisendrath, P., Lakhtakia, S., Reddy, N., Fockens, P., Ponchon, T., Bruno, M., 2012. Endoscopic treatment of chronic pancreatitis: European society of gastrointestinal endoscopy (ESGE) clinical guideline. Endoscopy 44 (08), 784–800. https://doi.org/10.1055/s-0032-1309840.

Facciorusso, A., Amato, A., Francesco Crinò, S., Sinagra, E., Maida, M., Fugazza, A., Binda, C., Repici, A., Tarantino, I., Anderloni, A., Fabbri, C., 2022. Nomogram for prediction of adverse events after lumen-apposing metal stent placement for drainage of pancreatic fluid collections. Digestive Endoscopy 34 (7), 1459–1470. https://doi.org/10.1111/den.14354.

Fugazza, A., Sethi, A., Trindade, A.J., Troncone, E., Devlin, J., Khashab, M.A., Vleggaar, F.P., Bogte, A., Tarantino, I., Deprez, P.H., Fabbri, C., Aparicio, J.R., Fockens, P., Voermans, R. P., Uwe, W., Vanbiervliet, G., Charachon, A., Packey, C.D., Benias, P.C., El-Sherif, Y., Paiji, C., Ligresti, D., Binda, C., Martínez, B., Correale, L., Adler, D.G., Repici, A., Anderloni, A., 2020. International multicenter comprehensive analysis of adverse events associated with lumen-apposing metal stent placement for pancreatic fluid collection drainage. Gastrointestinal Endoscopy 91 (3), 574–583. https://doi.org/10.1016/j.gie.2019.11.021.

Giljaca, V., Selvan Gurusamy, K., Takwoingi, Y., Higgie, D., Poropat, G., Štimac, D., Davidson, B.R., 2015. Endoscopic ultrasound versus magnetic resonance cholangiopancreatography for common bile duct stones. Cochrane Database of Systematic Reviews 2015 (2). https://doi.org/10.1002/14651858.cd011549.

Guzmán-Calderón, E., Chacaltana, A., Díaz, R., Li, B., Martinez-Moreno, B., Aparicio, J.R., 2022. Head-to-head comparison between endoscopic ultrasound guided lumen apposing metal stent and plastic stents for the treatment of pancreatic fluid collections: a systematic review and meta-analysis. Journal of Hepato-Biliary-Pancreatic Sciences 29 (2), 198–211. https://doi.org/10.1002/jhbp.1008. Available from: http://onlinelibrary.wiley.com/journal/10.1002/(ISSN)1868-6982.

Hama, K., Toyonaga, H., Ishii, T., Motoya, M., Kin, T., Hayashi, T., Katanuma, A., 2024. Enhancing visibility and hemostasis during necrosectomy for walled-off necrosis: the "dual-gel method". Endoscopy 56 (S 01), E597. https://doi.org/10.1055/a-2344-8503.

Hamada, T., Iwashita, T., Saito, T., Shiomi, H., Takenaka, M., Isayama, H., Yasuda, I., Nakai, Y., 2022. Disconnected pancreatic duct syndrome and outcomes of endoscopic ultrasound-guided treatment of pancreatic fluid collections: systematic review and meta-analysis. Digestive Endoscopy 34 (4), 676–686. https://doi.org/10.1111/den.14142.

Hamada, T., Michihata, N., Saito, T., Kimura, Y., Iwashita, T., Matsui, H., Shiomi, H., Takenaka, M., Iwata, K., Maruta, A., Masuda, A., Matsubara, S., Fushimi, K., Yasuda, I., Isayama, H., Yasunaga, H., Nakai, Y., 2025. Real-world impact of implementing lumen-apposing metal stents for pancreatic fluid collections: a nationwide Japanese study. Gut. https://doi.org/10.1136/gutjnl-2025-335067 gutjnl-2025-335067.

Holmes, I., Shinn, B., Mitsuhashi, S., Boortalary, T., Bashir, M., Kowalski, T., Loren, D., Kumar, A., Schlachterman, A., Chiang, A., 2022. Prediction and management of bleeding during endoscopic necrosectomy for pancreatic walled-off necrosis: results of a large retrospective cohort at a tertiary referral center. Gastrointestinal Endoscopy 95 (3), 482–488. https://doi.org/10.1016/j.gie.2021.10.015.

IAP/APA evidence-based guidelines for the management of acute pancreatitis. Pancreatology 13, 2013.

Isayama, H., Nakai, Y., Rerknimitr, R., Khor, C., Lau, J., Wang, H.-P., Seo, D.W., Ek, T.R., Lakhtakia, S., Ang, T.L., Ryozawa, S., Hayashi, T., Kawakami, H., Yamamoto, N., Iwashita, T., Itokawa, F., Kuwatani, M., Kitano, M., Hanada, K., Kogure, H., Hamada, T., Ponnudurai, R., Moon, J.H., Itoi, T., Yasuda, I., Irisawa, A., Maetani, I., 2016. Asian consensus statements on endoscopic management of walled-off necrosis. Part 2: endoscopic management. Journal of Gastroenterology and Hepatology 31 (9), 1555–1565. https://doi.org/10.1111/jgh.13398.

Ito, K., Okano, N., Takuma, K., Iwasaki, S., Watanabe, K., Kimura, Y., Yamada, Y., Yoshimoto, K., Hara, S., Kishimoto, Y., Matsuda, T., Igarashi, Y., 2023. Are newer extracorporeal shock wave lithotripsy models truly improving pancreatolithiasis lithotripsy performance? A Japanese single-center study using endoscopic adjunctive treatment. Gut and Liver 17 (4), 647–658. https://doi.org/10.5009/gnl220204.

Itoi, T., Binmoeller, K.F., Shah, J., Sofuni, A., Itokawa, F., Kurihara, T., Tsuchiya, T., Ishii, K., Tsuji, S., Ikeuchi, N., Moriyasu, F., 2012. Clinical evaluation of a novel lumen-apposing metal stent for endosonography-guided pancreatic pseudocyst and gallbladder drainage (with videos). Gastrointestinal Endoscopy 75 (4), 870–876. https://doi.org/10.1016/j.gie.2011.10.020.

Iwashita, T., Iwata, K., Hamada, T., Saito, T., Shiomi, H., Takenaka, M., Maruta, A., Uemura, S., Masuda, A., Matsubara, S., Mukai, T., Takahashi, S., Hayashi, N., Isayama, H., Yasuda, I., Nakai, Y., 2023. Supportive treatment during the periprocedural period of endoscopic treatment for pancreatic fluid collections: a critical review of current knowledge and future perspectives. Journal of Gastroenterology 58 (2), 98–111. https://doi.org/10.1007/s00535-022-01935-y.

Iwashita, T., Sato, T., Hamada, T., Saito, T., Iwata, K., Shiomi, H., Takenaka, M., Maruta, A., Uemura, S., Masuda, A., Matsubara, S., Mukai, T., Isayama, H., Yasuda, I., Nakai, Y., 2024. Risk of recurrence with or without plastic stent after EUS-guided treatment of peri-pancreatic fluid collections: a systematic review and meta-analysis. Endoscopy International Open 12 (02), E188. https://doi.org/10.1055/a-2226-1237.

Iwata, K., Iwashita, T., Mukai, T., Iwasa, Y., Okuno, M., Yoshida, K., Maruta, A., Uemura, S., Yasuda, I., Shimizu, M., 2024. Peroral pancreatoscopy-guided lithotripsy compared with extracorporeal shock wave lithotripsy in the management of pancreatic duct stones in chronic pancreatitis: a multicenter retrospective cohort study. Diagnostics 14 (9), 891. https://doi.org/10.3390/diagnostics14090891.

Jendresen, M.B., Thorbøll, J.E., Adamsen, S., Nielsen, H., Grønvall, S., Hart-Hansen, O., 2002. Preoperative routine magnetic resonance cholangiopancreatography before laparoscopic cholecystectomy: a prospective study. European Journal of Surgery 168 (12), 690–694. https://doi.org/10.1080/11024150201680024.

Kamal, F., Aziz, M., Lee-Smith, W., Sharma, S., Acharya, A., Ali Khan, M., Farooq, U., Tarar, Z.I., Gangwani, M.K., Hayat, U., Kumar, A., Schlacterman, A., Kowalski, T., Adler, D.G., 2024. Comparative efficacy and safety of different stent types for pancreatic fluid collections: a systematic review and network meta-analysis. Digestive Diseases and Sciences 69 (9), 3466–3480. https://doi.org/10.1007/s10620-024-08538-y.

Kurihara, T., Itoi, T., Sofuni, A., Itokawa, F., Moriyasu, F., 2013. Endoscopic ultrasonography-guided pancreatic duct drainage after failed endoscopic retrograde cholangiopancreatography in patients with malignant and benign pancreatic duct obstructions. Digestive Endoscopy 25 (2), 109–116. https://doi.org/10.1111/den.12100.

Lee, Y.N., Moon, J.H., Park, J.K., Jo, S.J., Lee, T.H., Cha, S.W., Cho, Y.D., Park, S.H., 2020. Preliminary study of a modified, nonflared, short, fully covered metal stent for refractory benign pancreatic duct strictures (with videos). Gastrointestinal Endoscopy 91 (4), 826–833. https://doi.org/10.1016/j.gie.2019.11.011.

Maruta, A., Iwashita, T., Saito, T., Gonoi, W., Yoshida, K., Omoto, S., Takenaka, M., Tsujimae, M., Masuda, A., Hamada, T., Uemura, S., Ota, S., Shiomi, H., Suda, K., Matsubara, S., Takahashi, S., Fujisawa, T., Iwasa, Y., Iwata, K., Hayashi, N., Mukai, T., Isayama, H., Yasuda, I., Nakai, Y., 2025. Encapsulation levels rather than elapsed time after the onset of acute pancreatitis are related to clinical outcomes of pancreatic walled-off necrosis. Endoscopy. https://doi.org/10.1055/a-2654-4317.

Masamune, A., Kikuta, K., Hamada, S., Tsuji, I., Takeyama, Y., Shimosegawa, T., Okazaki, K., Matsumoto, R., Kume, K., Miura, S., Takikawa, T., Hongo, S., Tanaka, Y., Sakai, T., Ueno, M., Unno, J., Toki, M., Nishimura, M., Kitamura, K., Kobashigawa, K., Ashida, K., Takahashi, K., Watanabe, T., Sato, Y., Satoh, A., Yoshida, N., Tanaka, K., Kondo, H., Suzuki, N., Sato, H., Mitoro, A., Minami, K., Iwasaki, E., Okuwaki, K., Mukai, T., Kudo, Y., Azuma, S., Takeda, Y., Setsuda, M., Tsuchida, K., Irisawa, A., Motoya, D., Shirahata, N., Iwata, K., Oza, N., Kawaguchi, Y., Hoshino, A., Hirano, A., Sakakihara, I., Watanabe, N., Funayama, H., Nakamura, Y., Yamamoto, T., Uchida, H.H., Iizawa, Y., Imamura, M., Sekine, M., Inui, K., Yamamoto, S., Sato, S., Ishii, S., Yakushijin, T., Yamamoto, K., Masaka, T., Yoshida, M., Arizumi, T., Goto, M., Takenaka, M., Miyakawa, H., Otsuka, Y., Nishi, M., Suwa, K., Sakagami, J., Satoh, M., Koga, H., Takagi, T., Nagahama, M., Hatayama, K., Kaneko, N., Mori, M., Nakai, Y., Iwashita, Y., Iwatsuka, K., Hayashi, K., Terai, S., Kaneto, H., Nishikawa, T., Uchi, M., Honjo, M., Takagi, T., Mizukami, K., Tamura, T., Gushima, T., Sato, A., Shimizu, K., Aruga, Y., Nonogaki, K., Kaino, M., Iwashita, T., Tsuruoka, M., Miura, M., Hasegawa, H., Takai, T., Yokota, H., Fujisawa, T., Uetsuki, K., Kobayashi, T., yamaji, N., Itoi, T., Muraki, T., Watanabe, T., Inatomi, O., Ushijima, T., Takehara, Y., Kawamura, H., Fukasawa, M., Hagiwara, H., Tamura, H., Moriya, T., Kuwatani, M., Kawaji, Y., Kamisawa, T., Abue, M., Fukunishi, M., Onoyama, H., Mochizuki, S., Sumimoto, K., Koyabu, M., Nishio, A., Sakao, M., Imamura, Y., Ohta, H., Nishimura, N., Shiihara, M., Anan, H., Motoya, M., Chiba, M., Deguchi, T., Kofunato, Y., Sawada, N., Katada, K., Matsumoto, S., Higashide, S., Okano, H., Ishida, M., Yamazaki, M.,

Tamura, N., Muraoka, T., Uza, N., Kodama, Y., Muro, S., Fujimori, N., Ikeda, T., Shimokawahara, N., Otsu, N., Yano, Y., Sugawara, S., Takaura, K., Ogura, Y., Furuuchi, T., Shibasaki, M., Kishimoto, M., Ishida, M., 2020. Clinical practice of acute pancreatitis in Japan: an analysis of nationwide epidemiological survey in 2016. Pancreatology 20 (4), 629–636. https://doi.org/10.1016/j.pan.2020.04.013.

Moon, J.H., Cho, Y.D., Cha, S.W., Cheon, Y.K., Ahn, H.C., Kim, Y.S., Kim, Y.S., Lee, J.S., Lee, M.S., Lee, H.K., Shim, C.S., Kim, B.S., 2005. The detection of bile duct stones in suspected biliary pancreatitis: comparison of MRCP, ERCP, and intraductal US. American Journal of Gastroenterology 100 (5), 1051–1057. https://doi.org/10.1111/j.1572-0241.2005.41057.x.

Mukai, S., Itoi, T., Sofuni, A., Itokawa, F., Kurihara, T., Tsuchiya, T., Ishii, K., Tsuji, S., Ikeuchi, N., Tanaka, R., Umeda, J., Tonozuka, R., Honjo, M., Moriyasu, F., 2014. Novel single transluminal gateway transcystic multiple drainages after EUS-guided drainage for complicated multilocular walled-off necrosis (with videos). Gastrointestinal Endoscopy 79 (3), 531–535. https://doi.org/10.1016/j.gie.2013.10.004.

Mukai, S., Itoi, T., Baron, T.H., Sofuni, A., Itokawa, F., Kurihara, T., Tsuchiya, T., Ishii, K., Tsuji, S., Ikeuchi, N., Tanaka, R., Umeda, J., Tonozuka, R., Honjo, M., Gotoda, T., Moriyasu, F., Yasuda, I., 2015a. Endoscopic ultrasound-guided placement of plastic vs. biflanged metal stents for therapy of walled-off necrosis: a retrospective single-center series. Endoscopy 47 (1), 47–55. https://doi.org/10.1055/s-0034-1377966.

Mukai, S., Itoi, T., Tsuchiya, T., Ishii, K., Tonozuka, R., 2015b. Pulsating pseudoaneurysm in a walled-off necrosis. Gastrointestinal Endoscopy 81 (5), 1262. https://doi.org/10.1016/j.gie.2014.08.031.

Nakai, Y., Kogure, H., Isayama, H., Koike, K., 2019. Endoscopic ultrasound-guided pancreatic duct drainage. Saudi Journal of Gastroenterology 25 (4), 210–217. https://doi.org/10.4103/sjg.SJG_474_18.

Nakai, Y., Hamada, T., Saito, T., Shiomi, H., Maruta, A., Iwashita, T., Iwata, K., Takenaka, M., Masuda, A., Matsubara, S., Sato, T., Mukai, T., Yasuda, I., Isayama, H., 2023. Time to think prime times for treatment of necrotizing pancreatitis: pendulum conundrum. Digestive Endoscopy 35 (6), 700–710. https://doi.org/10.1111/den.14598.

Petrov, M.S., 2009. Early use of ERCP in acute biliary pancreatitis with(out) jaundice: an unjaundiced view. Journal of the Pancreas 10 (1), 1–7.

Ramai, D., 2023. Early (ESCAPED_CHARACTER_LESS_THAN4 weeks) versus standard (>/=4 weeks) endoscopic drainage of pancreatic walled-off fluid collections: a systematic review and meta-analysis. Gastrointestinal Endoscopy 97 (3), 415–421.

Rinninella, E., Kunda, R., Dollhopf, M., Sanchez-Yague, A., Will, U., Tarantino, I., Gornals Soler, J., Ullrich, S., Meining, A., Esteban, J.M., Enz, T., Vanbiervliet, G., Vleggaar, F., Attili, F., Larghi, A., 2015. EUS-guided drainage of pancreatic fluid collections using a novel lumen-apposing metal stent on an electrocautery-enhanced delivery system: a large retrospective study (with video). Gastrointestinal Endoscopy 82 (6), 1039–1046. https://doi.org/10.1016/j.gie.2015.04.006.

Sauerbruch, T., Holl, J., Sackmann, M., Werner, R., Wotzka, R., Paumgartner, G., 1987. Disintegration of a pancreatic duct stone with extracorporeal shock waves in a patient with chronic pancreatitis. Endoscopy 19 (5), 207–208. https://doi.org/10.1055/s-2007-1018284.

Schrover, I.M., Weusten, B.L.A.M., Besselink, M.G.H., Bollen, T.L., Van Ramshorst, B., Timmer, R., 2008. EUS-guided endoscopic transgastric necrosectomy in patients with infected necrosis in acute pancreatitis. Pancreatology 8 (3), 271–276. https://doi.org/10.1159/000134275.

Seewald, S., Groth, S., Omar, S., Imazu, H., Seitz, U., Weerth, A.de, Soetikno, R., Zhong, Y., Sriram, P.V.J., Ponnudurai, R., Sikka, S., Thonke, F., Soehendra, N., 2005. Aggressive endoscopic therapy for pancreatic necrosis and pancreatic abscess: a new safe and effective treatment algorithm (videos). Gastrointestinal Endoscopy 62 (1), 92–100. https://doi.org/10.1016/s0016-5107(05)00541-9.

Seifert, H., Wehrmann, T., Schmitt, T., Zeuzem, S., Caspary, W.F., 2000. Retroperitoneal endoscopic debridement for infected peripancreatic necrosis. The Lancet 356 (9230), 653–655. https://doi.org/10.1016/s0140-6736(00)02611-8.

Senninger, N., Moody, F.G., Coelho, J.C.U., Van Buren, D.H., 1986. The role of biliary obstruction in the pathogenesis of acute pancreatitis in the opossum. Surgery 99 (6), 688–693.

Shen, Y., Liu, M., Chen, M., Li, Y., Lu, Y., Zou, X., 2014. Covered metal stent or multiple plastic stents for refractory pancreatic ductal strictures in chronic pancreatitis: a systematic review. Pancreatology 14 (2), 87–90. https://doi.org/10.1016/j.pan.2013.12.005.

Takasaki, Y., Isayama, H., Shin, K.S., Ishii, S., Fujisawa, T., Moon, J.P., Fukuma, T., Ushio, M., Takahashi, S., Suzuki, A., Ito, K., Tomishima, K., 2023. Measurement of the anchoring force of covered self-expandable and lumen-apposing metal stents for interventional endoscopic ultrasonography. Digestive Endoscopy 35 (1), 96–102. https://doi.org/10.1111/den.14406.

Tang, P., Ali, K., Khizar, H., Ni, Y., Cheng, Z., Xu, B., Qin, Z., Zhang, W., 2023. Endoscopic versus minimally invasive surgical approach for infected necrotizing pancreatitis: a systematic review and meta-analysis of randomized controlled trials. Annals of Medicine 55 (2). https://doi.org/10.1080/07853890.2023.2276816.

Tian, Y., Yin, C., Ma, Y., Fu, G., Liu, R., Ran, H., Pan, T., Xiao, Y., Wen, X., 2024. Lumen-apposing metal stents versus traditional self-expanding metal stents for endoscopic ultrasound-guided drainage of pancreatic fluid collections: a systematic review and meta-analysis. Surgical Endoscopy 38 (2), 586–596. https://doi.org/10.1007/s00464-023-10636-3.

Traverso, L.W., Kozarek, R.A., 2005. Pancreatic necrosectomy: definitions and technique. Journal of Gastrointestinal Surgery 9 (3), 436–439. https://doi.org/10.1016/j.gassur.2004.05.013.

van Brunschot, S., van Grinsven, J., van Santvoort, H.C., Bakker, O.J., Besselink, M.G., Boermeester, M.A., Bollen, T.L., Bosscha, K., Bouwense, S.A., Bruno, M.J., Cappendijk, V.C., Consten, E.C., Dejong, C.H., van Eijck, C.H., Erkelens, W.G., Goor, H.van, van Grevenstein, W.M.U., Haveman, J.-W., Hofker, S.H., Jansen, J.M., Laméris, J.S., Lienden, K.P van, Meijssen, M.A., Mulder, C.J., Nieuwenhuijs, V.B., Jan-Werner, P., Quispel, R., Ridder, R.J de, Römkens, T.E., Scheepers, J.J., Schepers, N.J., Schwartz, M.P., Seerden, T., Marcel Spanier, B.W., Straathof, J.W.A., Strijker, M., Timmer, R., Venneman, N.G., Vleggaar, F.P., Voermans, R.P., Witteman, B.J., Gooszen, H.G., Dijkgraaf, M.G., Paul, F., Manusama, E.R., Hadithi, M., Rosman, C., Schaapherder, A.F., Schoon, E.J., 2018. Endoscopic or surgical step-up approach for infected necrotising pancreatitis: a multicentre randomised trial. The Lancet 391 (10115), 51–58. https://doi.org/10.1016/S0140-6736(17)32404-2.

Van Santvoort, H.C., Bakker, O.J., Bollen, T.L., Besselink, M.G., Ahmed Ali, U., Schrijver, A.M., Boermeester, M.A., Van Goor, H., Dejong, C.H., Van Eijck, C.H., Van Ramshorst, B., Schaapherder, A.F., Van Der Harst, E., Hofker, S., Nieuwenhuijs, V.B., Brink, M.A.,

Kruyt, P.M., Manusama, E.R., Van Der Schelling, G.P., Karsten, T., Hesselink, E.J., Van Laarhoven, C.J., Rosman, C., Bosscha, K., De Wit, R.J., Houdijk, A.P., Cuesta, M.A., Wahab, P.J., Gooszen, H.G., 2011. A conservative and minimally invasive approach to necrotizing pancreatitis improves outcome. Gastroenterology 141 (4), 1254–1263. https://doi.org/10.1053/j.gastro.2011.06.073.

Varadarajulu, S., Phadnis, M.A., Christein, J.D., Mel Wilcox, C., 2011. Multiple transluminal gateway technique for EUS-guided drainage of symptomatic walled-off pancreatic necrosis. Gastrointestinal Endoscopy 74 (1), 74–80. https://doi.org/10.1016/j.gie.2011.03.1122.

Wang, S.S., Lin, X.Z., Tsai, Y.T., Lee, S.D., Pan, H.B., Chou, Y.H., Su, C.H., Lee, C.H., Shiesh, S.C., Lin, C.Y., Lin, H.C., 1988. Clinical significance of ultrasonography, computed tomography, and biochemical tests in the rapid diagnosis of gallstone-related pancreatitis: a prospective study. Pancreas 3 (2), 153–158. https://doi.org/10.1097/00006676-198804000-00007.

Weilert, F., Binmoeller, K.F., Shah, J.N., Bhat, Y.M., Kane, S., 2012. Endoscopic ultrasound-guided drainage of pancreatic fluid collections with indeterminate adherence using temporary covered metal stents. Endoscopy 44 (8), 780–783. https://doi.org/10.1055/s-0032-1309839.

Will, U., Reichel, A., Fueldner, F., Meyer, F., 2015. Endoscopic ultrasonography-guided drainage for patients with symptomatic obstruction and enlargement of the pancreatic duct. World Journal of Gastroenterology 21 (46), 13140–13151. https://doi.org/10.3748/wjg.v21.i46.13140. Available from: http://www.wjgnet.com/esps/DownLoadFile.aspx?Type=Digital&SubType=1&DOI=10.3748%2fwjg.v21.i46.13140&FilePath=Pub%5c10.3748%5cv21%5ci46%5cWJG-21-13140.pdf.

Yamamoto, K., Mukai, S., Tonozuka, R., Asai, Y., Matsunami, Y., Hirakawa, N., Itoi, T., 2023. A novel self-assembling hemostatic agent as an effective therapeutic tool for intracavitary bleeding from peripancreatic fluid collection after endoscopic ultrasonography-guided drainage using a lumen-apposing metal stent. Journal of Hepato-Biliary-Pancreatic Sciences 30 (6), e38. https://doi.org/10.1002/jhbp.1254. Available from: http://onlinelibrary.wiley.com/journal/10.1002/(ISSN)1868-6982.

Yasuda, I., Nakashima, M., Iwai, T., Isayama, H., Itoi, T., Hisai, H., Inoue, H., Kato, H., Kanno, A., Kubota, K., Irisawa, A., Igarashi, H., Okabe, Y., Kitano, M., Kawakami, H., Hayashi, T., Mukai, T., Sata, N., Kida, M., Shimosegawa, T., 2013. Japanese multicenter experience of endoscopic necrosectomy for infected walled-off pancreatic necrosis: the JENIPaN study. Endoscopy 45 (8), 627–634. https://doi.org/10.1055/s-0033-1344027.

CHAPTER

Minimally invasive surgical approaches in acute pancreatitis

14

Fuminori Yamaji[1], Tetsuya Fukuta[2]

[1]*Graduate School of Medical and Dental Sciences, Science Tokyo, Department of Emergency and Disaster Medicine, Tokyo, Japan;* [2]*Graduate School of Medicine, Gifu University, Department of Emergency and Disaster Medicine, Gifu, Japan*

14.1 Overview

Historically, the primary therapeutic strategy for infected pancreatic necrosis (IPN) was open surgical necrosectomy, namely a laparotomic procedure in which necrotic pancreatic and peripancreatic tissue is bluntly removed and large-bore drains are positioned for postoperative lavage. While this technique reliably allowed mechanical clearance of devitalized tissue, it was also associated with substantial surgical trauma, very high rates of postoperative complications, and considerable mortality. For many years, such unfavorable risk profiles prompted surgeons and intensivists to seek less invasive solutions that could control sepsis while minimizing iatrogenic injury.

A decisive change in clinical practice occurred with the publication of the PANTER trial in 2010. This landmark randomized study demonstrated that a step-up approach—initiating treatment with minimally invasive drainage and escalating to necrosectomy only if required—significantly reduced adverse outcomes, including mortality, compared with primary open necrosectomy (Van Santvoort et al., 2010). In this paradigm, percutaneous or transluminal drainage serves as the first interventional step to control infection, decompress collections, and stabilize organ function; only if clinical improvement is inadequate does the algorithm advance to minimally invasive debridement. Since PANTER, the step-up approach has become the dominant standard of care for IPN in many centers.

Subsequent evidence refined the choice of access route within the step-up strategy. In 2018, the TENSION trial compared surgical and endoscopic step-up approaches and found no significant difference in mortality between the two arms. However, the incidence of pancreatic fistula was significantly higher in the surgical group (van Brunschot et al., 2018). On the basis of these findings, the endoscopic step-up approach is widely regarded as the current first-line option when the anatomical relationship between the walled-off necrosis and gastric/duodenal wall permits safe transmural access. Consistently, multiple contemporary guidelines recommend beginning with transgastric drainage and proceeding to endoscopic necrosectomy

Advanced Insights into Acute Pancreatitis. https://doi.org/10.1016/B978-0-443-44736-5.00011-5

should drainage alone prove insufficient (Leppäniemi et al., 2019; Working Group of the International Association of Pancreatology, 2025; Takada et al., 2022).

Despite the broad adoption of endoscopic techniques, a meaningful subset of patients is not amenable to endoscopic access. Examples include collections located at a distance from the stomach or duodenum, such as necrosis confined to the pancreatic tail or within the left retroperitoneum, where the lack of apposition to the gastric or duodenal wall renders transluminal access technically unsafe or unfeasible. In such contexts, surgical approaches retain an essential role within the step-up framework.

Concurrently, the past 2 decades have witnessed rapid progress in minimally invasive surgery (MIS) across multiple domains, particularly laparoscopic and robot-assisted platforms. Building on the principles crystallized by PANTER and subsequent long-term observations, minimally invasive surgical necrosectomy has shown clear advantages over open surgery with respect to complication profiles and overall procedural invasiveness. The present chapter focuses specifically on surgical options within the step-up algorithm that leverage minimally invasive access. We first describe video-assisted retroperitoneal debridement (VARD)—the canonical surgical escalation after percutaneous drainage—and then discuss laparoscopic necrosectomy, including the retroperitoneal laparoscopic approach.

14.2 Video-assisted retroperitoneal debridement (VARD)

14.2.1 Definition, position in the step-up algorithm, and historical note

VARD represents the next step after percutaneous drainage in the surgical branch of the step-up strategy for infected pancreatic necrosis. Conceptually, VARD exploits the trajectory of the existing percutaneous drain tract to gain controlled retroperitoneal access to the necrotic cavity, thereby minimizing additional tissue trauma and preserving peritoneal integrity. The technique was first reported in 2007 by H. C. van Santvoort and colleagues and has since remained a central operative option in the surgical step-up repertoire (Van Santvoort et al., 2007).

14.2.2 Indications and anatomic considerations

VARD is particularly suited to patients with localized walled-off necrosis (WON) situated away from the stomach and duodenum, most commonly involving the pancreatic tail and/or left retroperitoneal space. Because transluminal endoscopic access depends on direct or near-direct apposition of the collection to the gastric/duodenal wall, lesions that lack such apposition are poor candidates for endoscopic drainage and debridement. In these scenarios, a retroperitoneal route can provide a more direct, safer corridor while avoiding entry into the peritoneal cavity.

Building on the same logic, a hybrid variant termed video-assisted retroperitoneal and/or transperitoneal debridement (VARTD) has been reported. In VARTD, a small laparotomy allows entry to the lesser sac (omental bursa) for transperitoneal assistance, which can be combined with retroperitoneal access to address complex or compartmentalized necrotic spaces (Wei et al., 2023). The choice between VARD and VARTD hinges on preoperative imaging, anticipated access angles, and the surgeon's experience.

14.2.3 Characteristic features, outcomes, and complications

A defining feature of VARD is the use of the pre-existing percutaneous drainage route as a guide and access path, which often simplifies the approach to the target cavity. Reported major complications of VARD include (Wei et al., 2023; Budkule et al., 2019):

Hemorrhage: 2.5%–18.2%

Gastrointestinal leakage/fistula: ~17.5%

Pancreatic fistula: ~18.2%–21%

When compared with endoscopic step-up strategies, the overall complication burden appears broadly similar; however, the rate of pancreatic fistula has been reported to be higher after VARD (van Brunschot et al., 2018). Mortality after VARD varies among series but is generally quoted in the range of 2.5%–18.2%, with postoperative hemorrhage, sepsis, and multiorgan failure being the leading causes of death. Typical operative times of approximately 90–150 min further underscore the minimally invasive nature of VARD relative to open necrosectomy.

14.3 Operative technique (modified based on reference Van Santvoort et al., 2007)

The following outline reflects a commonly adopted technique; institutional preferences and equipment may introduce minor variations, but the core steps remain consistent.

Preoperative access and imaging: Guided by contrast-enhanced CT, a 12–14 French percutaneous drainage catheter is inserted through the left lateral abdominal wall into the retroperitoneal necrotic cavity. This initial drainage often precedes VARD by hours to days and serves both diagnostic (microbiology) and therapeutic (source control) purposes.

Criteria for escalation: If clinical improvement (e.g., resolution of sepsis, decreasing inflammatory markers, improving organ function) is not achieved with drainage alone, the strategy escalates to VARD as the next step in the algorithm.

Positioning: The patient is placed supine with the left flank elevated by approximately 30–40 degrees to optimize the trajectory into the left retroperitoneum. Adequate padding and attention to pressure points are essential, given the often-prolonged critical illness preceding surgery.

Skin incision: Centered on the exit site of the percutaneous drain, a ~5-cm transverse incision is made at the level of the mid-axillary line. This positioning aligns the incision with the pre-formed tract and maximizes the likelihood of a direct corridor to the cavity.

Retroperitoneal entry: Using the drain as a guide, the surgeon performs blunt dissection through the abdominal wall musculature and fascia to enter the retroperitoneal space. Meticulous technique minimizes bleeding and avoids violation of the peritoneum.

Initial debridement under direct vision: Long grasping forceps and suction–irrigation devices are introduced to evacuate purulent material and remove loose necrotic debris. Copious irrigation facilitates visualization, reduces bacterial load, and helps mobilize slough adherent to the cavity walls.

Videoscopic assistance for deep debridement: When direct line-of-sight is insufficient to reach deeper recesses, a videoscope is inserted through the incision. Under videoscopic guidance, laparoscopic instruments are used to perform targeted debridement of residual necrosis. The surgeon proceeds gradually, avoiding overly aggressive manipulation that could injure adjacent vessels or viscera.

Drain placement and lavage planning: After thorough debridement and irrigation, two large-bore drains are placed—one into the deepest portion of the cavity and one more superficial—to permit continuous postoperative lavage and dependable egress of inflammatory effluent. Correct positioning is crucial for continued source control.

Closure: The fascia and skin are closed in standard fashion. Dressing choice should anticipate high-volume output and permit frequent inspection.

Intraoperative caveat: Profuse hemorrhage can occur if fragile vessels within or beneath the necrotic bed are injured. Retroperitoneal packing is the first-line temporizing measure. If hemostasis remains elusive or if a pseudoaneurysm is suspected, transcatheter arterial embolization (TAE) may be necessary as part of a coordinated surgical–interventional radiology strategy.

14.4 Laparoscopic necrosectomy

14.4.1 Background and early experience

The expanding capabilities of laparoscopic surgery since the early 2000s have encouraged its application to increasingly complex inflammatory pathologies, including necrotizing pancreatitis. In 2006, Parekh D reported 19 cases of hand-assisted laparoscopic necrosectomy, with a mean operative time of 153 ± 10 min, mean blood loss of 352.6 ± 103 mL, and no fatal complications (Parekh, 2006). These data suggested that a minimally invasive transperitoneal route could safely replicate the goals of open necrosectomy while attenuating operative trauma.

Subsequently, in 2014, Mathew MJ and colleagues described 28 cases of completely laparoscopic necrosectomy, encompassing transgastric and transcolonic approaches among others. The mortality rate was 3.6%, and although pancreatic fistula occurred in 28.6% of patients, all cases resolved with conservative management

(Mathew et al., 2014). Collectively, these series provided early evidence that laparoscopic debridement—whether hand-assisted or fully laparoscopic—could be delivered with acceptable morbidity and low mortality in experienced hands.

14.4.2 Laparoscopic retroperitoneal necrosectomy (LRN)

More recently, Jinjian Li and collaborators reported on laparoscopic retroperitoneal necrosectomy (LRN), a technique that merges the benefits of laparoscopy with the anatomic advantages of a retroperitoneal corridor (Li et al., 2025). Compared with VARD, LRN has been suggested to provide greater precision of debridement; lesions that might require multiple VARD sessions may be adequately cleared in a single laparoscopic procedure. Furthermore, mortality and overall complication rates have been reported lower with LRN than with VARD in selected cohorts, reinforcing its profile as a safe and minimally invasive modality.

Mechanistically, the retroperitoneal approach confers two major benefits:

It avoids peritoneal incision and exposure, thereby reducing contamination of the intraperitoneal cavity and limiting adhesiogenic trauma; and

It allows direct lateral or posterior access to necrosis in the pancreatic tail and left retroperitoneum without extensive visceral mobilization. These features can translate into fewer postoperative ileus episodes, potentially shorter hospital stays, and easier drain management—provided that the cavity is reachable via a safe retroperitoneal window.

14.4.3 Contemporary outcomes

Recent reports of laparoscopic necrosectomy document favorable operative metrics and postoperative courses. Representative figures include (Sivakumar et al., 2024):

Mean operative time: 56.40 ± 20.48 min.

Mean intraoperative blood loss: 120 ± 31.45 mL.

Mean postoperative length of stay: 11.19 ± 7.09 days.

Overall postoperative complication rate: 26%

Mortality: 4.9%

Postoperative pancreatic fistula rate: 13.9%

These data, while influenced by case-mix and institutional expertise, emphasize that laparoscopic necrosectomy—whether transperitoneal, retroperitoneal, or hybrid—can be conducted efficiently and safely within a structured step-up program.

14.4.4 Advantages, limitations, and patient selection

Advantages. Laparoscopy offers magnified visualization of complex anatomic planes, especially when coupled with pneumoperitoneum to create working space. This improved exposure facilitates careful debridement of necrotic tissue under direct vision, minimizes collateral trauma, and supports meticulous hemostasis.

Smaller incisions translate into reduced postoperative pain, lower wound complication rates, and faster rehabilitation. For retroperitoneal targets, LRN can obviate peritoneal violation while maintaining the ergonomic benefits of laparoscopic instrumentation.

Limitations. Laparoscopic necrosectomy is technically demanding. The inflammatory milieu, tissue friability, and distorted anatomy of necrotizing pancreatitis impose a steep learning curve, and conversion to open surgery may be required if visualization is poor, bleeding is uncontrolled, or necrosis is more extensive than imaging suggested. Access limitations arise when collections are deep, multifocal, or intimately adherent to critical structures. Optimal outcomes thus rely on rigorous preoperative imaging review, multiplanar trajectory planning, and the availability of advanced laparoscopic skill sets within a coordinated multidisciplinary team.

Patient selection. Selection integrates clinical (hemodynamic stability, ongoing sepsis, organ dysfunction), radiologic (size, location, degree of encapsulation, proximity to hollow viscera), and logistic (institutional experience, interventional radiology and endoscopy resources) factors. In the step-up setting, laparoscopy is typically reserved for patients with ongoing infection or clinical deterioration despite adequate drainage, or for those whose anatomy precludes endoscopic intervention but is favorable for a minimally invasive surgical approach.

14.4.4.1 Practical synthesis and clinical integration

Within a modern step-up algorithm for infected pancreatic necrosis:

First-line intervention is minimally invasive drainage (percutaneous or, when favorable apposition exists, transgastric endoscopic).

If the patient fails to improve, surgical escalation is considered. Here, VARD is a well-established option that leverages the existing drain tract to achieve controlled retroperitoneal debridement with limited additional access trauma.

When anatomy and expertise permit, laparoscopic necrosectomy—especially retroperitoneal laparoscopic access—can provide more comprehensive single-session debridement, potentially with lower complication rates relative to VARD in selected cases.

Throughout, open necrosectomy remains a salvage strategy for situations in which minimally invasive modalities are infeasible or unsuccessful.

The common denominators of success are timing (prefer debridement once collections are walled off unless emergent indications demand earlier action), multidisciplinary coordination, and meticulous postoperative care (drain management, nutritional optimization, antimicrobial stewardship, and vigilant surveillance for hemorrhage, fistula, or persistent sepsis). In this carefully orchestrated environment, minimally invasive surgical approaches—VARD and laparoscpic necrosectomy—have firmly established themselves as cornerstone techniques that complement endoscopic therapy and significantly reduce the historical burdens of morbidity and mortality that accompanied open necrosectomy (Fig. 14.1).

	Open necrosectomy (Conventional)	Laparoscopic Necrosectomy (LPN)	Video-Assisted Retroperitoneal Debridement (VARD)	Endoscopic Transluminal Necrosectomy (ETN)
Approach	midline abdominal incision	Through the omental bursa or mesocolon	Extension of a percutaneous drainage tract	Transluminal access via stomach, duodenum
Anatomical Target	Any location; indicated for diffuse, extensive, or complicated necrosis	Extensive necrosis, central or anterior lesions, pancreatic head involvement	Pancreatic body or tail, especially left retroperitoneal	Necrosis adjacent to the stomach or duodenum
Visualization and Operability	Full open field with direct manual debridement; maximum exposure	Wide operative field	Limited working space	Direct endoscopic visualization
Degree of Invasiveness	Most invasive	Moderate	Moderate	Least invasive
Bleeding Control	Excellent; direct control of major vessel bleeding possible	Feasible	Difficult	Challenging
Major Complications	High rates of wound infection, incisional hernia, postoperative organ failure	Pancreatic fistula, bleeding, reoperation	Bleeding, enteric fistula, residual necrosis	Fistula (gastric/colonic), bleeding
Reported Mortality	15–30%	5–15%	10–20%	5–10%
Reintervention Rate	<10% (though morbidity higher)	10–20%	20–40%	20–40%
Advantages	Complete debridement possible; suitable for uncontrolled sepsis	Allows extensive debridement	Minimizes peritoneal contamination	Minimally invasive
Limitations	Highly invasive; increased morbidity and mortality	Requires pneumoperitoneum	not suitable for lesions beyond retroperitoneum	Restricted to areas adjacent to the gastrointestinal tract

FIGURE 14.1

Comparison of open surgery, laparoscopic surgery, VARD, and endoscopic necrosectomy.

14.5 Robotic surgery for infected pancreatic necrosis

14.5.1 Background

Since the early 2020s, reports on robotic-assisted surgery for infected pancreatic necrosis (IPN) have been steadily increasing. This growing body of literature reflects a broader global trend toward minimally invasive techniques in the management of severe pancreatitis. Traditionally, open necrosectomy had been the definitive treatment for IPN; however, its substantial morbidity and mortality prompted the development of less invasive alternatives. Most contemporary reports describe robotic surgery being performed as a second- or third-line intervention—typically after endoscopic or percutaneous drainage has failed to achieve adequate clinical improvement.

Among the published studies, the report by Wang et al. stands out for its detailed clinical data and relatively large patient cohort. In their series of 57 patients, the median operative time was 187 min (range, 91–344 min), and the 90-day mortality rate was only 1.8% (1 of 57 cases), suggesting that robotic necrosectomy is both feasible and safe when performed by experienced surgeons (Wang et al., 2023). The rate of freedom from reintervention reached 93% (53 of 57 patients), indicating durable efficacy, while conversion to open surgery occurred in merely 3.5% of cases. Moreover, the median postoperative hospital stay was 4 days (range, 2–38 days), a finding that underscores the minimally invasive nature of this approach compared with traditional open procedures. Taken together, these data suggest that robotic surgery may offer excellent short-term outcomes with low morbidity and high procedural precision.

14.5.2 Technical and clinical advantages

The robotic platform provides several key advantages in the context of pancreatic necrosectomy. The three-dimensional, high-definition visualization markedly enhances depth perception, allowing surgeons to navigate safely through the complex and inflamed retroperitoneal space. The articulated robotic instruments, with their multiple degrees of freedom, enable superior dexterity and fine control compared with conventional laparoscopic instruments. This is particularly advantageous when dissecting necrotic tissue adjacent to major vascular structures such as the splenic artery, portal vein, or superior mesenteric vessels. Furthermore, the combination of robotic technology with intraoperative ultrasonography allows precise delineation of necrotic areas and their spatial relationships to vital structures (Mahendraraj et al., 2021). This integration improves the safety and completeness of debridement, reducing the risk of bleeding and injury to the surrounding pancreas or bowel.

14.5.3 Limitations and current challenges

Despite encouraging clinical outcomes, robotic surgery for IPN remains in an early phase of adoption, and several challenges must be addressed before it can be widely accepted. The number of published studies remains limited, and most reports are single-center retrospective analyses with small sample sizes. Significant heterogeneity exists in patient selection, timing of intervention, and surgical technique, making direct comparison between studies difficult. Additionally, robotic pancreatic surgery requires specialized equipment and multidisciplinary expertise, restricting its availability to high-volume tertiary centers. Operator experience plays a substantial role in determining outcomes, introducing a degree of institutional and surgeon-dependent bias. Cost considerations also represent a non-negligible barrier to widespread implementation, particularly in healthcare systems with constrained resources.

14.5.4 Future directions

To clarify the true role of robotic surgery in the management of infected pancreatic necrosis, prospective multicenter randomized controlled trials (RCTs) are warranted. Future research should aim to directly compare robotic, laparoscopic, and open necrosectomy in terms of mortality, morbidity, length of stay, postoperative recovery, and cost-effectiveness. Long-term outcomes such as pancreatic function, quality of life, and recurrence of infection should also be evaluated to fully assess the clinical value of robotic intervention. In parallel, ongoing technological advancements—including improved imaging integration, haptic feedback, and reduced system cost—may further enhance the accessibility and precision of robotic procedures.

AI disclosure

During the preparation of this work the author(s) used Chat GPT 5 in order to For English translation. After using this tool/service, the author(s) reviewed and edited the content as needed and take(s) full responsibility for the content of the publication.

References

Budkule, D., Desai, G., Pande, P., Narkhede, R., Wagle, P., Varty, P., 2019. An outcome analysis of videoscopic assisted retroperitoneal debridement in infected pancreatic necrosis: a single centre experience. Turkish Journal of Surgery 35 (3), 214–222. https://doi.org/10.5578/turkjsurg.4289.

Leppäniemi, A., Tolonen, M., Tarasconi, A., Segovia-Lohse, H., Gamberini, E., Kirkpatrick, A.W., Ball, C.G., Parry, N., Sartelli, M., Wolbrink, D., Goor, H.van, Baiocchi, G., Ansaloni, L., Biffl, W., Coccolini, F., Saverio, S.Di, Kluger, Y., Moore, E., Catena, F., 2019. WSES guidelines for the management of severe acute pancreatitis. World Journal of Emergency Surgery 14 (1). https://doi.org/10.1186/s13017-019-0247-0.

Li, J., Huang, Z., Shen, B., Yu, H., 2025. Laparoscopic retroperitoneal necrosectomy in patients with necrotizing pancreatitis: addressing the limitations of traditional treatment options. BMC Gastroenterology 25 (1). https://doi.org/10.1186/s12876-025-04001-y.

Mahendraraj, K., Kosari, K., Voidonikolas, G., Nissen, N., 2021. Robotic pancreatic necrosectomy for necrotizing pancreatitis following unsuccessful non-operative management: a novel and effective approach. HPB 23, S642. https://doi.org/10.1016/j.hpb.2021.06.364.

Mathew, M.J., Parmar, A.K., Sahu, D., Reddy, P.K., 2014. Laparoscopic necrosectomy in acute necrotizing pancreatitis: our experience. Journal of Minimal Access Surgery 10 (3), 126–131. https://doi.org/10.4103/0972-9941.134875.

Parekh, D., 2006. Laparoscopic-assisted pancreatic necrosectomy: a new surgical option for treatment of severe necrotizing pancreatitis. Archives of Surgery 141 (9), 895–902. https://doi.org/10.1001/archsurg.141.9.895.

Sivakumar, S.G., Sekaran, M., Muthukrishnan, S., Natesan, A.V., Nalankilli, V.P., Senthilnathan, P., Palanivelu, C., 2024. Laparoscopic necrosectomy for acute necrotising pancreatitis: retrospective analysis of a decade-long experience from a tertiary centre. Journal of Minimal Access Surgery 20 (2), 127–135. https://doi.org/10.4103/jmas.jmas_215_22.

Takada, T., Isaji, S., Mayumi, T., Yoshida, M., Takeyama, Y., Itoi, T., Sano, K., Iizawa, Y., Masamune, A., Hirota, M., Okamoto, K., Inoue, D., Kitamura, N., Mori, Y., Mukai, S., Kiriyama, S., Shirai, K., Tsuchiya, A., Higuchi, R., Hirashita, T., 2022. JPN clinical practice guidelines 2021 with easy-to-understand explanations for the management of acute pancreatitis. Journal of Hepato-Biliary-Pancreatic Sciences 29 (10), 1057–1083. https://doi.org/10.1002/jhbp.1146.

van Brunschot, S., van Grinsven, J., van Santvoort, H.C., Bakker, O.J., Besselink, M.G., Boermeester, M.A., Bollen, T.L., Bosscha, K., Bouwense, S.A., Bruno, M.J., Cappendijk, V.C., Consten, E.C., Dejong, C.H., van Eijck, C.H., Erkelens, W.G., van Goor, H., van Grevenstein, W.M.U., Haveman, J.W., Hofker, S.H., Jansen, J.M., Laméris, J.S., van Lienden, K.P., Meijssen, M.A., Mulder, C.J., Nieuwenhuijs, V.B., Poley, J.W., Quispel, R., de Ridder, R.J., Römkens, T.E., Scheepers, J.J., Schepers, N.J., Schwartz, M.P., Seerden, T., Spanier, B.W.M., Straathof, J.W.A., Strijker, M., Timmer, R., Venneman, N.G., Vleggaar, F.P., Voermans, R.P., Witteman, B.J., Gooszen, H.G., Dijkgraaf, M.G., Fockens, P., Manusama, E.R., Hadithi, M., Rosman, C., Schaapherder, A.F., Schoon, E.J., 2018. Endoscopic or surgical step-up approach for infected necrotising pancreatitis: a multicentre randomised trial. The Lancet 391 (10115), 51–58. https://doi.org/10.1016/S0140-6736(17)32404-2.

Van Santvoort, H.C., Besselink, M.G.H., Horvath, K.D., Sinanan, M.N., Bollen, T.L., Van Ramshorst, B., Gooszen, H.G., Dutch Acute Pancreatis Study Group, 2007. Videoscopic assisted retroperitoneal debridement in infected necrotizing pancreatitis. HPB 9 (2), 156–159. https://doi.org/10.1080/13651820701225688.

Van Santvoort, H.C., Besselink, M.G., Bakker, O.J., Hofker, H.S., Boermeester, M.A., Dejong, C.H., Van Goor, H., Schaapherder, A.F., Van Eijck, C.H., Bollen, T.L., Van Ramshorst, B., Nieuwenhuijs, V.B., Timmer, R., Laméris, J.S., Kruyt, P.M., Manusama, E.R., Van Der Harst, E., Van Der Schelling, G.P., Karsten, T., Hesselink, E. J., Van Laarhoven, C.J., Rosman, C., Bosscha, K., De Wit, R.J., Houdijk, A.P., Van Leeuwen, M.S., Buskens, E., Gooszen, H.G., 2010. A step-up approach or open necrosectomy for necrotizing pancreatitis. New England Journal of Medicine 362 (16), 1491–1502. https://doi.org/10.1056/NEJMoa0908821.

Wang, Y., Yoshino, O., Driedger, M.R., Beckman, M.J., Vrochides, D., Martinie, J.B., 2023. Robotic pancreatic necrosectomy and internal drainage for walled-off pancreatic necrosis. HPB 25 (7), 813–819. https://doi.org/10.1016/j.hpb.2023.03.013.

Wei, W., Tang, Y., Peng, Z., Xie, J., Deng, Z., Yuan, T., Tang, C., Pi, R., Wang, S., Zhao, S., Wang, L., Li, C., Wang, Y., Zhang, P., Wu, Z., Wan, Y., Ma, Y., Wen, T., Liang, X., Liu, K., Wang, W., Liang, X., Zeng, D., Shan, L., Liu, H., 2023. Minimal-access video-assisted retroperitoneal and/or transperitoneal debridement (VARTD) in the management of infected walled-off pancreatic necrosis with deep extension: initial experience from a prospective single-arm study. European Journal of Medical Research 28 (1). https://doi.org/10.1186/s40001-023-01030-9.

Working Group of the International Association of Pancreatology (IAP), 2025. Revised guidelines on acute pancreatitis 2025: supported and endorsed by the American pancreatic association, European pancreatic club, Indian pancreas club, and Japan pancreas society. Pancreatology 25, 770–814.

CHAPTER

Open surgical interventions in acute pancreatitis

15

Masashi Kuno

Gifu University, Graduate School of Medicine, Department of Gastroenterological Surgery, Gifu, Japan

15.1 Introduction

Acute pancreatitis (AP) is a common gastrointestinal emergency with a variable clinical course. While most patients recover with conservative management, up to 20% develop severe necrotizing pancreatitis, associated with organ failure and high mortality (Leppäniemi et al., 2019; Petrov et al., 2010). Historically, early open necrosectomy was considered the standard of care; however, this strategy was associated with morbidity rates of up to 80% and mortality exceeding 30% (van Brunschot et al., 2018). Over the last 2 decades, international consensus has shifted toward delayed, minimally invasive, and step-up approaches, supported by randomized controlled trials (RCTs) and updated guidelines (Leppäniemi et al., 2019; Van Santvoort et al., 2010; Boxhoorn et al., 2021; Párniczk, 2025).

15.2 Indications for surgical intervention

Surgical intervention is reserved for selected patients with severe AP or complications refractory to conservative treatment. Major indications include:

- **Infected pancreatic necrosis**: confirmed by gas on imaging or a positive aspirate culture (IAP/APA, 2013).
- **Persistent organ failure**: despite optimal intensive care (Schepers et al., 2019).
- **Gastrointestinal complications**: perforation, ischemia, necrosis, or obstruction (Leppäniemi et al., 2019; Párniczk, 2025).
- **Hemorrhage**: uncontrolled bleeding not amenable to endovascular or endoscopic therapy (Leppäniemi et al., 2019).
- **Abdominal compartment syndrome**: requiring decompressive laparotomy (Leppäniemi et al., 2019).

Even in these cases, surgery should not be considered the first step; instead, less invasive modalities should be attempted before proceeding to operative debridement (Van Santvoort et al., 2010; Párniczk, 2025).

Advanced Insights into Acute Pancreatitis. https://doi.org/10.1016/B978-0-443-44736-5.00016-4

15.3 Step-up approach

Step-up approach has become the preferred strategy in infected necrotizing pancreatitis. This consists of:

1. **Percutaneous catheter drainage, which can stabilize patients and sometimes be definitive** (Van Santvoort et al., 2010; Bugiantella et al., 2016).
2. **Endoscopic drainage and necrosectomy, particularly effective for walled-off necrosis adjacent to the stomach or duodenum** (van Brunschot et al., 2018; Freeman, 2020).
3. **Minimally invasive surgical necrosectomy (laparoscopic or retroperitoneal), reserved for cases refractory to less invasive measures** (Leppäniemi et al., 2019).
4. Open necrosectomy, considered only as a last resort.

The landmark PANTER trial demonstrated that a step-up approach reduced new-onset organ failure, incisional hernia, and diabetes compared with primary open necrosectomy (Van Santvoort et al., 2010). The TENSION trial further showed that an endoscopic step-up approach was associated with fewer pancreatic fistulas and shorter hospital stays than a surgical step-up approach (van Brunschot et al., 2018). More recently, Boxhoorn et al. (2021) found no survival difference between immediate and postponed drainage, reinforcing the safety of delayed intervention (Boxhoorn et al., 2021). In addition, Trikudanathan et al. (2018) reported that carefully selected patients may undergo endoscopically centered step-up interventions even before 4 weeks, challenging the traditional paradigm (Trikudanathan et al., 2018).

15.4 Timing of intervention

The timing of intervention remains crucial. Most guidelines recommend delaying invasive procedures for at least 4 weeks after onset, when necrosis has become walled-off, thereby facilitating safer and more effective debridement (Leppäniemi et al., 2019; Párniczk, 2025). Early intervention (<2 weeks) is associated with worse outcomes, including higher rates of bleeding, fistula formation, and mortality (Petrov et al., 2010; Tenner et al., 2013). Nevertheless, exceptions include uncontrolled sepsis, bowel ischemia, or life-threatening hemorrhage, in which earlier intervention may be unavoidable (Boxhoorn et al., 2021; Schepers et al., 2019).

15.5 Choice of surgical technique

- **Percutaneous drainage is widely available, minimally invasive, and often sufficient in fluid-predominant collections** (Bugiantella et al., 2016).

- **Endoscopic necrosectomy has gained favor in specialized centers, with multiple studies demonstrating reduced morbidity and mortality in selected patients** (van Brunschot et al., 2018; Trikudanathan et al., 2018).
- **Laparoscopic necrosectomy provides a less invasive surgical option compared with open procedures, although it requires expertise** (Leppäniemi et al., 2019).
- **Open necrosectomy remains effective for complete removal of necrosis but carries the highest morbidity and mortality, and is now reserved for refractory cases** (Petrov et al., 2010; Freeman, 2020).

15.6 Management of biliary pancreatitis

Biliary pancreatitis accounts for a significant proportion of AP cases. Cholecystectomy is essential for preventing recurrence.

- **In mild cases, laparoscopic cholecystectomy should be performed during the index admission** (Boxhoorn et al., 2021).
- **In severe cases, surgery should be postponed until the patient's condition has stabilized** (Párniczk, 2025).
- **Early ERCP is recommended only in patients with concurrent cholangitis or biliary obstruction** (Boxhoorn et al., 2021).

15.7 Future directions

Recent advances in endoscopic techniques, percutaneous interventions, and hybrid approaches have markedly reduced the need for open surgery (Freeman, 2020; van Grinsven et al., 2019). Registry data highlight the importance of organ failure and infection of necrosis as key determinants of outcome (Schepers et al., 2019; Crockett et al., 2024). Future research should focus on refining patient selection, optimizing the timing of interventions, and integrating predictive tools such as artificial intelligence for risk stratification.

15.8 Conclusion

Surgical management of AP has evolved from early open necrosectomy toward delayed, minimally invasive, and step-up strategies. These approaches have improved outcomes and reduced morbidity compared with traditional surgery. Careful patient selection, appropriate timing, and multidisciplinary expertise are essential to optimize treatment. Continued refinement of international guidelines and incorporation of novel technologies will further enhance outcomes in severe AP (Table 15.1).

Table 15.1 Characteristics of each surgical approach.

Technique	Advantages	Limitations/Risks	Evidence
Percutaneous drainage	Minimally invasive Widely available Can stabilize patients	May require multiple procedures Less effective for solid necrosis	Van Santvoort et al. (2010) Crockett et al. (2024)
Endoscopic drainage/ necrosectomy	Direct access to walled-off necrosis Fewer external drains Shorter hospital stay	Technically demanding Limited to collections adjacent to GI tract	van Brunschot et al. (2018) IAP/APA/EPC/IPC/JPS Working Group (2025)
Laparoscopic necrosectomy	Less invasive than open Possible single-stage debridement	Still invasive Requires experienced centres	Leppäniemi et al. (2019)
Open necrosectomy	Definitive removal of necrosis	High morbidity/ mortality Last resort	Historical standard

AI disclosure

During the preparation of this work the author(s) used ChatGPT in order to Used for literature searches and grammar checks. After using this tool/service, the author(s) reviewed and edited the content as needed and take(s) full responsibility for the content of the publication.

References

Boxhoorn, L., van Dijk, S.M., van Grinsven, J., Verdonk, R.C., Boermeester, M.A., Bollen, T. L., Bouwense, S.A.W., Bruno, M.J., Cappendijk, V.C., Dejong, C.H.C., van Duijvendijk, P., van Eijck, C.H.J., Fockens, P., Francken, M.F.G., van Goor, H., Hadithi, M., Hallensleben, N.D.L., Haveman, J.W., Jacobs, M.A.J.M., Jansen, J.M., Kop, M.P.M., van Lienden, K.P., Manusama, E.R., Mieog, S.J.D., Molenaar, I.Q., Nieuwenhuijs, V.B., Poen, A.C., Poley, J.W., van de Poll, M., Quispel, R., Römkens, T. E.H., Schwartz, M.P., Seerden, T.C., Stommel, M.W.J., Straathof, J.W.A., Timmerhuis, H.C., Venneman, N.G., Voermans, R.P., van de Vrie, W., Witteman, B.J., Dijkgraaf, M.G.W., van Santvoort, H.C., Besselink, M.G., 2021. Immediate versus postponed intervention for infected necrotizing pancreatitis. New England Journal of Medicine 385 (15), 1372–1381. https://doi.org/10.1056/NEJMoa2100826.

van Brunschot, S., van Grinsven, J., van Santvoort, H.C., Bakker, O.J., Besselink, M.G., Boermeester, M.A., Bollen, T.L., Bosscha, K., Bouwense, S.A., Bruno, M.J.,

Cappendijk, V.C., Consten, E.C., Dejong, C.H., van Eijck, C.H., Erkelens, W.G., van Goor, H., van Grevenstein, W.M.U., Haveman, J.W., Hofker, S.H., Jansen, J.M., Laméris, J.S., van Lienden, K.P., Meijssen, M.A., Mulder, C.J., Nieuwenhuijs, V.B., Poley, J.W., Quispel, R., de Ridder, R.J., Römkens, T.E., Scheepers, J.J., Schepers, N.J., Schwartz, M.P., Seerden, T., Spanier, B.W.M., Straathof, J.W.A., Strijker, M., Timmer, R., Venneman, N.G., Vleggaar, F.P., Voermans, R.P., Witteman, B.J., Gooszen, H.G., Dijkgraaf, M.G., Fockens, P., Manusama, E.R., Hadithi, M., Rosman, C., Schaapherder, A.F., Schoon, E.J., 2018. Endoscopic or surgical step-up approach for infected necrotising pancreatitis: a multicentre randomised trial. The Lancet 391 (10115), 51–58. https://doi.org/10.1016/S0140-6736(17)32404-2.

Bugiantella, W., Rondelli, F., Boni, M., Stella, P., Polistena, A., Sanguinetti, A., Avenia, N., pancreatitis, N., 2016. A review of the interventions. International Journal of Surgery 28, S163. https://doi.org/10.1016/j.ijsu.2015.12.038.

Crockett, S.D., Wani, S., Gardner, T.B., 2024. American college of gastroenterology guidelines: management of acute pancreatitis. American Journal of Gastroenterology 119 (3), 289–314.

Freeman, M.L., 2020. Interventions for necrotizing pancreatitis: current evidence and future directions. Gastroenterology 158 (4), 999–1004.

van Grinsven, J., van Brunschot, S., Bakker, O.J., 2019. Natural history of gas configurations and clinical outcomes in necrotizing pancreatitis. United European Gastroenterology Journal 7 (3), 387–395.

IAP/APA evidence-based guidelines for the management of acute pancreatitis, 2013. Pancreatology 13 (4), e1. https://doi.org/10.1016/j.pan.2013.07.063.

Leppäniemi, A., Tolonen, M., Tarasconi, A., Segovia-Lohse, H., Gamberini, E., Kirkpatrick, A.W., Ball, C.G., Parry, N., Sartelli, M., Wolbrink, D., Goor, H.van, Baiocchi, G., Ansaloni, L., Biffl, W., Coccolini, F., Saverio, S.Di, Kluger, Y., Moore, E., Catena, F., 2019. WSES guidelines for the management of severe acute pancreatitis. World Journal of Emergency Surgery 14 (1). https://doi.org/10.1186/s13017-019-0247-0.

Párniczk, A., et al., IAP/APA/EPC/IPC/JPS Working Group, 2025. International association of pancreatology revised guidelines on acute pancreatitis 2025. Pancreatology 25 (6), 770–814. https://doi.org/10.1016/j.pan.2025.04.020. PMID 40651900.

Petrov, M.S., Shanbhag, S., Chakraborty, M., Phillips, A.R.J., Windsor, J.A., 2010. Organ failure and infection of pancreatic necrosis as determinants of mortality in patients with acute pancreatitis. Gastroenterology 139 (3), 813–820. https://doi.org/10.1053/j.gastro.2010.06.010.

Van Santvoort, H.C., Besselink, M.G., Bakker, O.J., Hofker, H.S., Boermeester, M.A., Dejong, C.H., Van Goor, H., Schaapherder, A.F., Van Eijck, C.H., Bollen, T.L., Van Ramshorst, B., Nieuwenhuijs, V.B., Timmer, R., Laméris, J.S., Kruyt, P.M., Manusama, E.R., Van Der Harst, E., Van Der Schelling, G.P., Karsten, T., Hesselink, E. J., Van Laarhoven, C.J., Rosman, C., Bosscha, K., De Wit, R.J., Houdijk, A.P., Van Leeuwen, M.S., Buskens, E., Gooszen, H.G., 2010. A step-up approach or open necrosectomy for necrotizing pancreatitis. New England Journal of Medicine 362 (16), 1491–1502. https://doi.org/10.1056/NEJMoa0908821.

Schepers, N.J., Bakker, O.J., Besselink, M.G., Ahmed Ali, U., Bollen, T.L., Gooszen, H.G., Van Santvoort, H.C., Bruno, M.J., 2019. Impact of characteristics of organ failure and infected necrosis on mortality in necrotising pancreatitis. Gut 68 (6), 1044–1051. https://doi.org/10.1136/gutjnl-2017-314657.

Tenner, S., Baillie, J., Dewitt, J., Vege, S.S., 2013. American college of gastroenterology guideline: management of acute pancreatitis. American Journal of Gastroenterology 108 (9), 1400–1415. https://doi.org/10.1038/ajg.2013.218.

Trikudanathan, G., Tawfik, P., Amateau, S.K., Mbbs, S.M., Arain, M., Attam, R., Beilman, G., Flanagan, S., Freeman, M.L., Mallery, S., 2018. Early (<4 Weeks) versus standard (≥4 Weeks) endoscopically centered step-up interventions for necrotizing pancreatitis. American Journal of Gastroenterology 113 (10), 1550–1558. https://doi.org/10.1038/s41395-018-0232-3.

Index

'*Note:* Page numbers followed by "f" indicate figures and "t" indicate tables.'

A

R

S

Printed and bound by CPI Group (UK) Ltd, Croydon, CR0 4YY
21/09/2026
02219915-0010